THE WEANER PIG
NUTRITION AND MANAGEMENT

The Weaner Pig
Nutrition and Management

Edited by

M.A. Varley
SCA Nutrition, Thirsk, UK

and

J. Wiseman
University of Nottingham, UK

CABI *Publishing*

CABI *Publishing* is a division of CAB *International*

CABI Publishing
CAB International
Wallingford
Oxon OX10 8DE
UK
Tel: +44 (0)1491 832111
Fax: +44 (0)1491 833508
Email: cabi@cabi.org
Web site: http://www.cabi-publishing.org

CABI Publishing
10 E 40th Street
Suite 3203
New York, NY 10016
USA
Tel: +1 212 481 7018
Fax: +1 212 686 7993
Email: cabi-nao@cabi.org

A catalogue record for this book is available from the British Library, London, UK.

Library of Congress Cataloging-in-Publication Data

The weaner pig : nutrition and management / edited by M.A. Varley and J. Wiseman.
p. cm.
Papers from the occasional Meeting of the British Society of Animal Science, held at Nottingham University, Sept. 2000
Includes bibliographical references (p.)
ISBN 0-85199-532-2 (hardback : alk paper)
1. Piglets--Nutrition--Requirements--Congresses. 2. Piglets--Feeding and feeds--Congresses. I. Varley, M. A. (Mike A.) II. Wiseman, J. (Julian) III. British Society of Animal Science. Meeting (2000 : Nottingham University

SF396.5.W43 2001
636.4′085--dc21 2001035454

ISBN 0 85199 532 2

Typeset in 10/12pt Adobe Garamond by Columns Design Ltd, Reading
Printed and bound in the UK by Cromwell Press, Trowbridge

Contents

Contributors

M. Bailey, Division of Molecular and Cellular Biology, Department of Clinical Veterinary Sciences, University of Bristol, Langford House, Langford, Bristol BS40 5DU, UK

P. Baynes, NuTec Ltd, Eastern Avenue, Lichfield, Staffordshire WS13 7SE, UK

J.D. Beal, Seale-Hayne Faculty, University of Plymouth, Newton Abbot, Devon TQ12 6NQ, UK

P.W. Bland, Department of Functional Anatomy, Hannover Medical School, 4120 Carl-Neuberg Strasse, 30625 Hannover, Germany

P.H. Brooks, Seale-Hayne Faculty, University of Plymouth, Newton Abbot, Devon TQ12 6NQ, UK

E.M.A.M. Bruininx, Research Institute for Animal Husbandry, PO Box 2176, 8203 AD, Lelystad, The Netherlands

A. Campbell, Seale-Hayne Faculty, University of Plymouth, Newton Abbot, Devon TQ12 6NQ, UK

M. Cole, SCA Nutrition Ltd, Maple Mill, Dalton, Thirsk, North Yorkshire YO7 3HE, UK

V. Demeckova, Seale-Hayne Faculty, University of Plymouth, Newton Abbot, Devon TQ12 6NQ, UK

S. Done, Veterinary Laboratories Agency, Woodham Lane, Addlestone, Surrey KT15 3BP, UK

M. Evans, Pig Production Training Ltd, 3 Trevose Close, Walton, Chesterfield S40 3PT, UK

D.M. Green, University of Edinburgh, King's Buildings, Edinburgh EH9 3JG, UK

D.J. Hampson, Division of Veterinary and Biomedical Sciences, Murdoch University, Murdoch WA 6150, Australia

K. Haverson, Division of Molecular and Cellular Biology, Department of Clinical

Veterinary Sciences, University of Bristol, Langford House, Langford, Bristol BS40 5DU, UK

S. Held, Centre for Behavioural Biology, Department of Veterinary Clinical Science, University of Bristol, Langford House, Langford, Bristol BS40 5DU, UK

D. Kelly, Rowett Research Institute, Bucksburn, Aberdeen AB21 9SB, UK

J.C. Kim, Division of Veterinary and Biomedical Sciences, Murdoch University, Murdoch WA 6150, Australia

T.P. King, Rowett Research Institute, Bucksburn, Aberdeen AB21 9SB, UK

J. Le Dividich, INRA, Unité Mixte de Recherches sur le Veau et le Porc, 35590 St-Gilles, France

D.E. McDonald, Division of Veterinary and Biomedical Sciences, Murdoch University, Murdoch, WA 6150, Australia

L. Melin, Department of Ruminant and Porcine Diseases, National Veterinary Institute, 751 89 Uppsala, Sweden

M. Mendl, Centre for Behavioural Biology, Department of Veterinary Clinical Science, University of Bristol, Langford House, Langford, Bristol BS40 5DU, UK

B.G. Miller, Division of Molecular and Cellular Biology, Department of Veterinary Clinical Sciences, University of Bristol, Langford House, Langford, Bristol BS40 5DU, UK

C.A. Moran, Seale-Hayne Faculty, University of Plymouth, Newton Abbot, Devon TQ12 6NQ, UK

G. Partridge, Finnfeeds International Ltd, PO Box 777, Marlborough, Wiltshire SN8 1XN, UK

C.M.C. van der Peet-Schwering, Research Institute for Animal Husbandry, PO Box 2176, 8203AD, Lelystad, The Netherlands

D.W. Pethick, Division of Veterinary and Biomedical Sciences, Murdoch University, Murdoch WA 6150, Australia

J. Pickard, Division of Agriculture and Horticulture, School of Biosciences, University of Nottingham, Sutton Bonington Campus, Loughborough, Leics LE12 5RD, UK

J.R. Pluske, Division of Veterinary and Biomedical Sciences, Murdoch University, Murdoch WA 6150, Australia

H.-J. Rothkötter, Department of Functional Anatomy, Hannover Medical School, 4120 Carl-Neuberg Strasse, 30625 Hannover, Germany

J.W. Schrama, Wageningen University and Research Centre, Department of Animal Science, PO Box 338, 6700 AH, Wageningen, The Netherlands

B. Sève, INRA, Unité Mixte de Recherches sur le Veau et le Porc, 35590 St-Gilles, France

H.A.M. Spoolder, Research Institute for Pig Husbandry, PO Box 83, 5240 AB Rosmalen, The Netherlands

M. Sprent, SCA Nutrition Ltd, Maple Mill, Dalton, Thirsk, North Yorkshire YO7 3HE, UK

C.R. Stokes, Division of Molecular and Cellular Biology, Department of Clinical

Veterinary Sciences, University of Bristol, Langford House, Langford, Bristol BS40 5DU, UK

J.W.G.M. Swinkels, Research Institute for Pig Husbandry, PO Box 83, 5240 AB Rosmalen, The Netherlands

M. Varley, SCA Nutrition Ltd, Dalton, Thirsk, North Yorkshire YO7 3HE, UK

M.A. Vega-Lopez, CINVESTAV-IPN, Experimental Pathology Department, Av. IPN 2508, Mexico 07360 DF, Mexico

H.M. Vermeer, Research Institute for Pig Husbandry, PO Box 83, 5240 AB Rosmalen, The Netherlands

P. Wallgren, Department of Ruminant and Porcine Diseases, National Veterinary Institute, 751 89 Uppsala, Sweden

C.M. Wathes, Silsoe Research Institue, Bio-Engineering Division, Wrest Park, Silsoe, Bedford MK45 4HS, UK

C.T. Whittemore, University of Edinburgh, King's Buildings, Edinburgh EH9 3JG, UK

J. Wiseman, Division of Agriculture and Horticulture, School of Biosciences, University of Nottingham, Sutton Bonington Campus, Loughborough, Leics LE12 5RD, UK

L. Zarkadas, Division of Agriculture and Horticulture, School of Biosciences, University of Nottingham, Sutton Bonington Campus, Loughborough, Leics LE12 5RD, UK

Preface

This book is the proceedings of a conference organized by the British Society of Animal Science (as part of their 'Occasional Meetings' series) held at the University of Nottingham, UK, in September 2000.

The post-weaned piglet presents a particular challenge both in modern production terms and in the context of the sciences related to production. The systems used in most of the pig-producing countries of the world include weaning at between 16 and 30 days of age using controlled environment housing and highly specialized diets. Whilst early weaning generates significant advantages for annual sow productivity, the system also demands a high level of management skills to make it work successfully.

The principal objective of the meeting was therefore to focus attention on the various disciplines involved in weaner science and production, and to review the research and development carried out recently in these areas. Accordingly the present work includes sections on: growth patterns, nutrition, feeding requirements, gut physiology, enteric health and the environmental requirements of the young post-weaned piglet. The papers were presented by acknowledged experts from around the globe and provide a solid foundation both for future research directions and also as guidelines for managers and consultants looking to improve their systems.

In addition to formal papers, the meeting also had submitted posters, abstracts for which can be found at: www.bsas.org.uk

Dr Mike Varley, SCA Nutrition Ltd, UK
Dr Julian Wiseman, The University of Nottingham, UK

Acknowledgements

Growth of the Young Weaned Pig

1

C.T. Whittemore and D.M. Green

University of Edinburgh, King's Buildings, Edinburgh EH9 3JG, UK

The Potential for Growth

Without human intervention, the pig will become nutritionally independent of its dam at 15–20 kg liveweight. Natural weaning will occur at 70 days or so of age. Earlier weaning than this creates a disruption to the growth of the weaned pig that is inversely proportional to pig age. The competence of the digestive system of the suckled pig to handle a non-milk diet begins to develop (aided by challenge/response) between 14 and 28 days of age. Under conditions of gradual diet change from liquid to solid feed, growth is likely to be fully supportable without sucked milk from around 56 days. The presentation of a mixture of milk and externally sourced solid feed to the gut of the young pig is relevant to natural development.

Abrupt weaning at 21 days of age is not conducive to the achievement of normal growth in pigs. Fifty years ago, conventional European practice was to wean at 56 days of age. Advances in nutritional knowledge and the manufacture of specialist housing for young pigs led to a rapid reduction in weaning age. After a number of failed flirtations with 7-day, 10-day and 14-day weaning, the 'industry standard' in the UK settled at 21-day weaning. A substantial proportion of successful practitioners nevertheless chose not to wean at ages below 28 days, and in other European countries 35-day weaning remains common. The UK industry standard has, since the 1980s, drifted upward from 21 days towards 28 days, with an apparent advantage in terms of numbers of pigs born per sow per year.

The description of growth following weaning requires at its core a prediction of protein mass and of its incrementation. Description of body composition further requires quantification of lipid mass and an understanding of any relationship that may exist between protein and lipid retention in the course of positive (and, in the case of the weaned pig, negative) growth.

Theoretical considerations

There is dissent over the nature of the curve that might best describe protein growth over time and weight. The conventional assumption of a sigmoidal form requires the rejection of the proposition still held by the de Lange school at Guelph (C.F.M. de Lange, personal communication, Guelph, 2000), which follows from the reviews of Kielanowski (1969) and Rerat (1972) and suggests that a single value be used to describe maximum daily potential protein retention rate (Pr_{max}) at all times during the active growing life of the pig. That the (single value) potential for growth is not achieved in early life, and particularly after weaning, may be ascribed to a failure on the part of the pig to be able to ingest sufficient feed. In addition to Pr_{max}, the original models of Whittemore and Fawcett (1974, 1976), and those that later sprang from them, required a second parameter: that of a minimum ratio between the daily rate of retention of lipid and protein in the gain (Lr:Pr). This latter was argued for on the grounds of a physiological imperative, and had the further (beneficial) effect of restraining the simulated daily rate of achieved protein retention (Pr) below Pr_{max} in young pigs, when appetite was limiting. The rule effectively enforces partitioning of energy from Pr to Lr. The setting of the minimum ratio of lipid to protein in the gain was necessary for the achievement of fit of the 'single value' model to recently weaned and young growing pigs, which would otherwise be predicted to perform at extravagant rates of protein growth. The Lr:Pr ratio was subsequently argued against (Whittemore, 1995), on the grounds that it was superfluous to modelling requirement to restrain early protein growth if a sigmoidal (Gompertz) function rather than a single value were used to describe Pr_{max}. The Gompertz function had an evident effect of limiting early protein growth potential. In both the original and amended (Whittemore,1995) eventualities, potential lipid retention (Lr_{max}) was unrestrained, and achieved lipid retention (Lr) was a function of energy supply. Emmans (1988) proposed that potential lipid retention (Lr_{max}) may also be described by the Gompertz function.

Functions other than the Gompertz in the sigmoidal series have been discussed by such as Huxley (1932), Hammond (1940), Brody (1945) and Schinckel (1999). The common characters of the sigmoidal descriptor are: (i) a period of increasing growth rate in early life; (ii) a period of decreasing growth rate in later life; (iii) a point of inflection (linearity) between the two; and (iv) an asymptote, at which point maturity is approached and growth ceases. The weaner pig finds itself in the middle of the phase of increasing growth rate, which in practice it rarely achieves.

The Gompertz (1825) function may be employed with the growing tissue (whole body, protein, or lipid, etc.) on the y axis, and the scale against which growth is to be expressed (time, or a body tissue) on the x axis. The function requires as parameters the asymptote of the y axis (an approximation of maturity for the tissue concerned) and a growth coefficient. Importantly, the y-coordinate of the point of inflection is fixed at 1/e of the asymptote. The general equation is:

$$y = A^* \exp\{-\exp[-B(x - x^{o})]\}$$

where A is the asymptote for y, x^{o} is the point of inflection measured on the x-axis and B is the growth coefficient.

It is of interest to express growth rate (dy/dx) as a function of both time and weight, either being potentially useful depending on circumstance.

Differentiation with respect to time of the general equation where y = weight and x = time leads to the derivative for gain as a function of time (x):

$$dy/dx = A^*B^* \exp\{-B^*(x - x^o) - \exp[-B^*(x - x^o)]\}$$

and for gain as a function of weight (y):

$$dy/dx = y^*B^* \ln(A/y)$$

where dy/dx is the gain, and A is the value for $y(W)$ when dy/dx (daily gain) has diminished to zero.

Whittemore (1998) suggested various values for B and A according to sex and genotype. These values range from 0.010 and 220 to 0.014 and 330, respectively, where A is expressed in terms of liveweight. Potential growth rates for pigs of 5 kg, 10 kg and 15 kg are predicted to be 189 g, 309 g and 402 g for the lower values and 293 g, 490 g and 649 g for the higher.

Alternatives to the Gompertz equation have been examined (and largely rejected) by Whittemore *et al.* (2001a). Amongst these were the functions of Bridges and Richards. The Bridges equation (Bridges *et al.*, 1986) differs from the Gompertz mainly in not having the inflection point fixed at 1/e. The Bridges equation is:

$$y = y_0 + A^*\{1 - \exp[-(m^*x^b)]\}$$

where y_0 is the start point for y (weight of growing tissue), A is the weight at maturity, m is the 'exponential growth decay constant', x is time and b is the 'kinetic order constant'.

The Richards equation (Richards, 1959; and described by France and Thornley, 1984) is more generalized and also has a variable inflection point, which gives it the flexibility to describe different pig types with points of inflection at differing proportions of their mature age (or weight). The Richards equation is expressed as:

$$y = (y_0{}^*A) / \{y_0{}^n + (A^n - y_0{}^n)^* \exp[-(k^*x)]\}^{1/n}$$

where, if y would be weight, y_0 is the start weight, A is the asymptote for y, and x measures time. Bridges and Richards are of interest as, although by no means as convenient as Gompertz, they can better describe data sets that do indeed have different inflection points. However, Knap (2000) found, rather convincingly, that although the point of inflection (as proportion of mature weight) did vary with data sets differing in their provenance, it did not deviate importantly from the 1/e determined with the Gompertz.

The description of growth through the relationship between a component (such as total protein, Pt) and the whole (such as live bodyweight, W) was suggested by Huxley (1932) and has been used with good effect by countless others since. The form used is:

$$z = a^*y^b$$

Whittemore *et al.* (1988) determined allometric relationships (commented upon by Schinckel, 1999) for z = protein (Pt), lipid, water and ash mass, and y = bodyweight. The pattern of daily protein retention rate, Pr, or dPt/dx, with increasing bodyweight was determined as follows. Daily liveweight gain, dy/dx, was first described with the Gompertz function:

$$dy/dx = B^{*}y^{*} \ln(A/y).$$

Subsequently, Pt was taken as a simple allometric function of weight ($Pt = a^{*}y^{b}$). It follows that:

$$Pr = [a^{*}b(y)^{b-1}]^{*}[B^{*}y^{*} \ln(A/y)].$$

As pointed out by Schinckel and de Lange (1996), the simple $z = ay^{b}$ allometric is potentially faulted in the assumption that the body component changes according to whole bodyweight in a uniform way.

Early growth as a special case

Being born with a body-fat content of perhaps 10–20 g kg^{-1}, the suckling pig partitions nutrients in favour of lipid deposition to reach 150–160 g kg^{-1} at the time of weaning, the ratio of lipid to protein in the body being around 1:1 at this time. The modern meat pig, slaughtered at less than 120 kg liveweight, is unlikely ever to be as fat again. Post-weaning feed intake inadequacies together with stress and disease challenge ensure a rapid loss of body lipid in support of maintenance (and protein synthesis). Whittemore *et al.* (1978) found no liveweight gain for 7 days in pigs weaned at 14 days of age. At 21 days of age, the body composition of these pigs comprised 150 g protein kg^{-1} and 76 g lipid kg^{-1}, compared with 146 g protein kg^{-1} and 148 g lipid kg^{-1} in unweaned 21-day-old pigs. In the case of both groups of weaned pigs (14- and 21-day weaned), the gains that occurred subsequent to post-weaning weight stasis and lipid loss were made in favour of protein (and not any recovery of body lipid). Lipid and protein gains after weaning were made in approximately equal proportion. Thus, by 50 days of age the composition of the body was little changed, with some 60–70 g lipid kg^{-1} and 150–180 g protein kg$^{-1.}$ These authors noted that lipid losses had taken place commensurate with weight stasis (not weight loss), and concluded that water enhancement had compensated for the removal of lipid. This proposition was put to a more severe test by Whittemore *et al.* (1981), who found that zero daily weight change was associated with 56 g of lipid loss and 53 g of water gain. Only when the daily weight gain exceeded 193 g did lipid gains begin to become positive.

Water gain (g day^{-1}) = 0.56 empty bodyweight gain + 53
Lipid gain (g day^{-1}) = 0.29 empty bodyweight gain − 56
Protein gain (g day^{-1}) = 0.15 empty bodyweight gain − 4

Between zero and 200 g daily liveweight gain, it would appear that the pigs catabolized body lipid in favour of the anabolism of body protein. Not until the rather appreciable rate of gain of some 60% of what would normally be expected at this time did the pigs return to lipid retention. Whittemore *et al.* (1978) also noted the relative intransigence of the body proportion made up of protein and the stability of protein content in the face of perturbations in feed intake.

Whittemore (1998) pointed out that, given appropriate conditions, weaned pigs will grow at rates substantially above the commercial norm (Table 1.1). Healthy pigs of little more than 5 kg have the potential, given unrestrained feed intake, to grow at 500 g daily, and *B* coefficients for the Gompertz function as high as 0.020 have been measured at Edinburgh. It would seem, therefore, that the negative consequences of post-weaning growth suspension are in part, or in whole, avoidable.

Given the propensity of contemporary pig-keeping systems to predispose weaned pigs to a period of slow, zero or negative growth, it is natural that great interest is shown in the possibility of compensatory growth. By such means may nature gratuitously make amends for the inadequacies of husbandry. That animals will grow faster when feed is plentiful, and make provision for times when feed is scarce, is undeniable. It is also undeniable that lipid losses in support of tissue accretions is physiologically normal (as, for example, in lactation). The seminal question in the case of the weaned pig, however, is whether or not supra-normal gains following a period of deprivation can make good the earlier growth losses and whether, in addition, these gains can be achieved at supra-normal efficiency. This, of course, makes 'normal' a begged definition.

Tullis and Whittemore (1986) conducted a carefully constructed trial, some of the findings of which are given in Table 1.2. Pigs on restricted feeding from 25 to 55 days expressed reduced liveweight gains and little or no lipid gains, but achieved no greater liveweight gains on realimentation, nor gains of differing composition, than unrestricted pigs. Compensatory gains were not in evidence. Kyriazakis and his group at Edinburgh have put forward the concept of the young pig having a preferred ratio of lipid to protein in the body. Post-weaning feed restriction will move the pig away from that preferred ratio through the catabolism of lipid, and there will be a natural predisposition to readjust the balance. It should not be assumed, however, that the lipid content of the body at weaning is necessarily expressive of a preferred ratio; the storage of lipid in excess of this ratio (in case of hard times to come) would be a normal expectation. Were compensatory growth to be found, its benefaction would lie in the recovery of protein mass by an enhancement in the rate of protein deposition. The difficulty with the test is in proving that the 'control' group was maximizing in the first place. Animals may readily show

Table 1.1. Post-weaning growth performance at University of Edinburgh.

Liveweight at start (kg)	Liveweight at finish (kg)	Days	Daily feed intake (g)	Daily liveweight gain (g)
6	12	13	500	450
6	24	31	800	581
8	16	14	650	590
12	24	16	900	760

Table 1.2. Growth of weaned pigs (g day^{-1}) following feed intake restriction from 25 to 55 days of age and fed to appetite from 55 to 70 days of age. (The percentage of lipid in the gain is given in parentheses.)

Days of age	Fed to appetite 25–70 days	Restricted 25–55 days
25–40	321 (9.6)	192 (4.8)
40–55	532 (9.0)	162 (0.6)
55–70	601 (14.5)	508 (16.5)

enhancement of appetite and growth above that achieved under previous restriction, and no 'compensatory' benefit could be claimed. Further, over the total period under review (restriction followed by compensation) there is no likelihood of any recovery of the efficiency lost in the former period. As a management tool of convenience, compensatory growth must therefore be rejected, however attractive its acceptance might be to those espousing less than adequate husbandry techniques. This is not to deny the propensity of the pig to make the most of growth opportunities. In a classical experiment Kyriazakis *et al.* (1991) fed young pigs to achieve luxury levels of lipid deposition, and then presented the pigs with a diet of exceptionally high protein content. The pigs used body stores of lipid to augment ingested energy levels (limited by the constraints of gut capacity) and achieved remarkable rates of protein retention and liveweight gain. Whether the 925 g of daily liveweight gain attained at a live bodyweight of 13 kg should be considered as indicative of a potential normally available or of compensatory gains remains conjecture.

It may be concluded that, in the case of the young weaned pig, the driver for growth is feed intake and that this inevitably constrains growth to a level that is below potential.

Feed Intake

Although gut capacity is dependent on body size, Parks (1982) states the inevitable truth that in the actively growing animal it is the (increasing) bodyweight that is dependent on feed intake:

$$W = (A - W_0)\{1 - \exp[-(AB)F/A]\} + W_0$$

where W is liveweight (accumulated gain), A is the liveweight at maturity, W_0 is the initial liveweight, B is the efficiency coefficient and F is the accumulated feed intake. Whittemore *et al.* (1978) showed that a 1 g increment in the intake of digestible crude protein was associated with 2.5 g bodyweight gain, 0.49 g protein gain and 0.28 g lipid gain. An increment of 1 MJ in digestible energy (DE) intake was associated with 22 g bodyweight gain, 3.3 g protein gain and 8.2 g lipid gain. These functions emphasize that it is feed intake that regulates growth in the weaner pig, and growth is invariably curtailed below the potential through the adverse modulation of a constrained appetite.

For a few days immediately after weaning, feed intake may be strictly rationed. This is usually in the interests of actual or presumed enteric disease limitation. Where disease susceptibility is low, or the threat presumed to be passed, weaned pigs may optimize their growth by maximizing their voluntary feed intake. The *ad libitum* appetite may be indicative of the potential appetite prevailing for a given pig type and its nutrient requirement (conventionally the sum of the requirements for maintenance, protein retention, lipid retention and cold thermogenesis). However, appetite in weaner pigs is far more likely to be indicative of the circumstances of feed type, management quality, herd health and housing environment, including: (i) living space, feeder space, and peer competition; (ii) health and individual assertiveness of the animal; (iii) environmental temperature; (iv) pig size; and (v) gut capacity (volume, but usually expressed as feed weight).

It is not evident to what extent the type of pig, and selection pressure upon the type, might affect voluntary feed intake of weaner pigs. It is clear, however, that in the case of growing pigs certain breeds (such as the Chinese types) may have a superior capacity for fibrous feeds, whilst halothane-positive pigs eat some 15–20% less than negative types (Henry, 1985; Kalm, 1986; Webb, 1989). Schinckel and de Lange (1996) observed a 30% difference between genotypes of pigs fed under similar conditions. Unimproved genotypes have appetites up to one-third greater than those selected for leanness (Meat and Livestock Commission, 1982; Cole and Chad, 1989; Webb, 1989).

Feed intake capacity

Despite all of the above together with the assumption that little or no feed is eaten in the first day, NRC (1998) presented the following alternative empirical expressions to describe feed intake in young pigs of 5–15 kg:

$$\text{Energy intake (MJ DE day}^{-1}) = -6.40 + 1.93W - 0.0407W^2$$

$$\text{Energy intake (MJ DE day}^{-1}) = -0.556 + 1.05W - 0.00413W^2,$$

using the latter in their model. NRC (1987) risked the suggestion of a single equation for the liveweight range 4.5–117 kg:

$$\text{Energy intake (MJ DE day}^{-1}) = 55\ (1 - e^{-0.0176W}).$$

The form $y = y_0 + ax^b$, where y = feed intake (kg) per day, has been found useful to deliver a curvilinear response fitting much of the data for older growing pigs. Cole *et al.* (1967) proposed:

$$\text{Energy intake (MJ DE day}^{-1}) = 2.4W^{0.68}$$

and Whittemore (1983) suggested:

$$\text{Feed intake (kg day}^{-1}) = 0.12W^{0.75}$$

as descriptive of practical feed intakes from *ad libitum* dry-feed hoppers. These latter two may also be reasonable for some circumstances of weaned pigs of less than 20 kg.

Fowler and Gill (1989) pointed out that to maintain a pre-weaning growth rate of some 280 g daily, a 6 kg weaned pig would need to eat 475 g of a diet of high energy concentration (16.5 MJ DE kg^{-1}). Weaned pigs may eat less than their maintenance requirement for some 3 days or more after weaning, and a feed intake commensurate with pre-weaning growth rates is only likely to be achieved between the second and third week after weaning in the case of pigs weaned at 21 days of age. Fowler and Gill (1989) gave the guidance as presented in Table 1.3. Whittemore (1998) suggested a rather similar outlook (Table 1.4).

Whittemore *et al.* (1995) pointed out that in considering the prediction of feed intake, bodyweight is reflective only of maintenance, a minor component of feed usage in the growing pig. Pigs of similar weight may ingest nutrients and grow at widely dissimilar rates of growth and body composition. Kyriazakis and Emmans (1999) put forward the logic also favoured by Black *et al.* (1986) that feed intake will equal that which is desired to fulfil the metabolic requirements of the day in question. These requirements are a function of pig weight (maintenance) *and* tissue weight gains:

$$\text{Daily feed intake} = (1/\text{feed energy content}) * (E_m + E_{Pr}/k_{Pr} + E_{Lr}/k_{Lr})$$

where E_m is the energy for maintenance, and E_{Pr} and E_{Lr} are the daily rates of energy retention in protein and lipid, together with their respective energy cost factors (k). In cold environments an energy cost for cold thermogenesis may also be added: 0.016 MJ $W^{0.75}$ for each °C of cold ($Tc - Te$) (ARC, 1981).

Observed feed intakes in weaned pigs fail to differentiate between limits imposed by satisfaction of nutritional desire and those imposed by the capacity of the gut and other environmental constraints. The concept of the young weaned pig eating to its energy requirement is therefore likely to be redundant.

In the early stages of growth, young pigs show characteristics that are indicative of gut capacity limiting intake below that desired. These include rapid increase in feed intake commensurate with rapid increase in bodyweight and immediate positive growth response to increasing diet nutrient concentration (Cole and Chadd, 1989). The capacity of the gut is a function of its size or volume, the extent of habituation to low nutrient concentration diet, and the rate of throughput of digesta (diet digestibility). There is likely to be a genetic component to gut capacity (Fowler and Gill, 1989), but it remains largely unquantified. Black *et al.* (1986) reviewed data showing that the ability of pigs to increase their voluntary intake in the face of decreasing nutrient concentration is strongly related to pig size. Thus there is little or no accommodation possible for pigs of less than 50 kg whose capacity for growth appears in excess of their capacity for feed at any nutrient concentration. Tybirk (1989) presented curves suggesting that, at 20 kg liveweight, pigs will be able to eat to 70% of their 'energy capacity' when fed a diet of 11 MJ ME kg^{-1}, 85% when fed a diet of 12 MJ ME kg^{-1}, and 95% when fed a diet of 13 MJ ME kg^{-1}. These values need to be appropriately proportioned for pigs of lesser weight. Fowler and Gill (1989) presumed a DE concentration of 15.5 MJ for 6 kg weaners, and it would appear likely that concentrations below this cannot be accommodated by young weaned pigs of less than 15 kg. Such pigs will therefore invariably eat below their requirement unless offered diets of high nutrient concentration and digestibility. NRC (1987) pointed to minimum DE concentration of around 14 MJ, and a reduction of intake of some 1.5 MJ daily for each dietary DE concentration decrease of 1 MJ kg^{-1}. Black *et al.* (1986) set a maximum (gut capacity) feed intake limit of:

$$\text{Feed intake (kg day}^{-1}) = 0.111\,W^{0.803}$$

where the feed is of 900 g dry matter (DM) kg^{-1}.

Whittemore (1993) explored the influence of diet digestibility, through the medium of the equation:

$$\text{Feed intake (kg day}^{-1}) = 0.013\,W/(1 - \text{digestibility coefficient}).$$

The term $0.013W$ emanates from an estimate of faecal organic matter dry matter output by pigs. This equation suggests, for weaned pigs of 5, 10 and 15 kg liveweight, physical limits of 325 g, 650 g and 980 g, respectively, if a digestibility of 0.80 is used. Ferguson *et al.* (1994) proposed that feed intake be constrained by:

$$\text{Feed intake (kg day}^{-1}) = 0.090(Pt/\text{BULKDN})$$

where Pt is the protein mass of the pig and BULKDN is the estimated indigestible organic matter content of the feed and therefore rather similar to (1 − digestibility coefficient); and as Pt is usually around $0.16W$, $0.09Pt$ is rather similar to $0.013W$.

Tsaras *et al.* (1998) proposed that water-holding capacity (WHC) can adequately describe the 'bulkiness' of fibrous feeds. Thus an experiment using differing levels of inclusion of feed ingredients of differing fibrosity and WHC (measured by centrifugation) yielded:

$$\text{Feed intake (g DM kg}^{-1}\ W\text{ daily}) = 207\ (1/\text{WHC})$$

where WHC was measured in g water per g feed and ranged from 5.1 to 8.5 (the control value was 3.9). Expressed in this way, the equation is presumptive upon a

Table 1.3. Feed intake of weaned piglets.

	Pig weight (kg)	Feed intake (g day^{-1})	Energy intake (MJ DE day^{-1})	Daily liveweight gain (g)
First day	6.50	26	0.385	0.00
First week	6.00	210	2.72	90
Second week	7.10	410	5.29	220
Third week	9.34	630	8.15	410

Table 1.4. Feed intake of weaned piglets.

Pig liveweight (kg)	Feed intake (g day^{-1})		
	Nutrient need	Gut capacity	Good commercial practice
5	375	350	< 100
10	750	700	400
15	1000	800	800

'base-line' feed intake of around 50 g feed kg^{-1} liveweight. The experiment of Tsaras *et al.* (1998) would suggest that a unit (g g^{-1}) increase in the WHC value of a feed is associated with a decrease in feed intake of some 6 g feed DM kg^{-1} liveweight. An earlier experiment (Kyriazakis and Emmans, 1995) also indicated that one unit (g g^{-1}) increase in WHC was associated with a decrease in feed intake of around 6 g feed DM kg^{-1} liveweight.

It has been proposed that gut constraints may become plastic when animals are in frank nutrient deficit through being given bulky feeds (an effective definition of the weaned pig transferring from milk to a solid cereal-based diet). This means that the influence of feed bulk upon feed intake can be overridden (Tolkamp and Ketelaars, 1992). If this proposition were to hold for young weaned pigs, then any model of feed intake would need to accommodate the concept of eating to optimize the efficiency of the metabolic processes as expressed by the ratio of net energy consumption to oxygen utilization rate (Tolkamp and Ketelaars, 1992). However, a recent series of carefully constructed experiments (Whittemore *et al.*, 2001a,b) has shown that the balance of evidence is in favour of the assumption that, even when nutritionally embarrassed by being presented with a diet of lower nutrient concentration than preferred, the young pig cannot readily override the constraint of gut capacity. The pigs appeared limited at around 50 g daily feed intake kg^{-1} W (Whittemore *et al.*, 2000). It would appear safe to assume that on all but the highest DE concentration diets (those in excess of 15 MJ ME kg^{-1}), the feed intake of the recently weaned pig is constrained by the limitations of its gut capacity.

Influence of environmental temperature and stocking density

Pigs at first eat more, and then less, in a quadratic response of feed intake to environmental temperature (Close, 1989; Whittemore, 1998). Effective temperatures (Te) below that required for the metabolic comfort of the pig will increase energy demand for cold thermogenesis. To calculate effective temperature, Whittemore (1983, 1998) followed the ARC lead by developing the findings of the Cambridge School (Mount and colleagues) to give:

$$Te = Ta * Ve * Vl$$

where Ta is the ambient temperature and Ve and Vl are described with a series of coefficients ranging from 0.7 to 1.4, depending upon rate of air movement, draught and lying conditions. The energy cost of cold thermogenesis (MJ) when the effective temperature (Te) is below the animal comfort temperature (Tc) is probably between 0.011 and 0.016 per degree difference between effective (Te) and comfort (Tc) temperatures per kg $W^{0.75}$ (Verstegen *et al.*, 1973). Whittemore and Fawcett (1976) give:

$$\text{Energy cost of cold thermogenesis} = 0.016W^{0.75}(Tc - Te).$$

Appetite will therefore increase as the ambient temperature is reduced.

Effective temperatures (*Te*) *above* that required for the metabolic comfort of the pig will reduce appetite as a response to an increasing embarrassment of body heat that cannot be dissipated into the environment. This has the consequence of an increasing imperative to reduce the rate of heat formation by limiting nutrient input, and/or by ingesting energy in the form of substrates that generate a lower heat output (such as highly digestible starches and fats).

Whittemore (1998) suggested a feed reduction of 1 g kg^{-1} bodyweight per degree of heat above comfort level, or:

$$\text{Reduction in energy intake (MJ DE day}^{-1}) = 0.014W(Te - Tc)$$

where *Te* is the effective ambient temperature and *Tc* the comfort temperature of the pigs in question, estimated as $Tc = 27 - 0.6H$, with *H* expressing the total heat output generated from the body of the pig. It would appear that each degree change in temperature results in a 2–3% change in feed intake.

Decreasing the rate of air movement and increasing the relative humidity will further depress the appetite of young weaned pigs. Close (1989) suggested that a change in air speed of 0.2 m s^{-1} is equivalent to a change in *Te* of 1°C. The effects of relative humidity (militating against evaporative heat loss from the body of the pig) are greater at higher temperatures. The review of Close (1989) pointed to a proportional increase in relative humidity of 0.15 being equivalent to an increase of 1°C in *Te*.

It is recognized that pigs will grow more slowly as the stocking density increases (NRC, 1998). A substantial element of this response is a negative effect on feed intake. Whittemore (1998) indicated that where the area occupied by the pig (m^2) $= kW^{0.67}$, a change in *k* value of 0.005 below optimum can be associated with a 4% change in feed intake. Acceptable space allowances in intensive housing conditions (with respect to feed intake considerations) are associated with values for *k* of approximately 0.050.

In practice, the ambient temperature of rooms housing weaned pigs of 5–7 kg may be as high as 30°C, the air speed low, the relative humidity high and stocking density excessive. A temperature of 30°C is appropriate for a pig eating little or no feed, but is an embarrassment to a pig attempting to eat sufficient to grow at 400 g daily. Reduction in room temperature is therefore a prerequisite for satisfactory feed intake. That there is reluctance to do this is often associated with persistent failure on the part of the pigs to pick up feed intake after weaning, as a result of stress and disease challenge. The virtuous spiral of decreasing temperature and increasing feed intake is therefore rarely properly exploited. Ideally, room temperature should be around 22°C by the time the pigs are 15 kg liveweight (a rate of reduction of about 0.5°C day^{-1} for healthy pigs).

Mitigation of the Post-weaning Growth Check

Even in the absence of disease challenge, the weaned pig nevertheless fails voluntarily to eat to the level of its nutrient requirement. Given the foregoing discussion, the following points may mitigate the post-weaning disruption to normal growth and avoid the diseconomies of lost protein growth and of lipid catabolism: (i) An increase in pig size at the time of their weaning; (ii) weaning pigs according to their size and ability to cope (including the possibility of weaning different pigs in a single litter at different times); (iii) accustoming pigs to the consumption of viable quantities of solid feed before they are weaned commensurate with the possibility of continuation of normal growth after weaning; (iv) use of solid diets that are highly digestible and which can be accommodated in sufficient quantity within the capacity of the gut of the young pig; (v) avoidance of environmental (especially temperature) and competitive stress.

Conclusions

The growth of the young weaned pig can be satisfactorily described within the context of overall growth to maturity by means of the Gompertz function:

$$\text{Daily gain} = \text{liveweight} * B * \ln(\text{weight at maturity/liveweight})$$

where the growth coefficient, B, conventionally ranges between 0.010 and 0.015, depending on sex and genotype. Experiments achieving feed intakes that approach the satisfaction of nutrient requirement have demonstrated B values of 0.020, and potential daily gains approaching 1000 g at 15 kg liveweight. Growth rate is compromised by post-weaning stress and disease but, even when such challenges are absent, achieved growth rates in young (21-day-old) weaned pigs of 100 g, 200 g and 400 g in the first, second and third week after weaning represent substantial underperformance.

The driver for post-weaning growth is feed intake, and this is inevitably curtailed by the limitations of gut capacity on diets of < 15 MJ ME kg^{-1}. The description of growth in the young weaned pig must therefore be the description of feed intake. It is not evident that gut capacity is the operational modulator of feed intake; the pig invariably volunteers to consume less than it could for periods of up to 3 weeks after weaning. This propensity may be attributed to a frank inability of young weaned pigs to cope effectively with the process of weaning at the relatively tender degree of maturity and body size associated with weaning at 28 days of age or less. Normal (uninterrupted) growth patterns are seen only under the highest standards of husbandry and nutrition, and their likelihood is positively associated with bodyweight at the time of weaning.

References

ARC (1981) *The Nutrient Requirements of Pigs.* Agricultural Research Council, Commonwealth Agricultural Bureaux, Farnham Royal, UK.

Black, J.L., Campbell, R.G., Williams, I.H., James, K.J. and Davies, G.T. (1986) Simulation of energy and amino acid utilisation in the pig. *Research and Development in Agriculture* 3, 121–146.

Bridges, T.C., Turner, U.W., Smith, E.M., Stahly, T.S. and Loewer, O.J. (1986) A mathematical procedure for estimating animal growth and body composition. *Transactions of the American Society of Agricultural Engineers* 29, 1342–1347.

Brody, S. (1945) *Bioenergetics and Growth.* Reinhold, New York.

Close, W.H. (1989) The influence of the thermal environment on the voluntary food intake of pigs. In: Forbes, J.M., Varley, M.A. and Lawrence, T.L.J. (eds) *The Voluntary Feed Intake of Pigs.* Occasional Publication No. 13, British Society of Animal Production, Edinburgh, pp. 87–96.

Cole, D.J.A. and Chadd, S.A. (1989) Voluntary food intake of growing pigs. In: Forbes, J.M., Varley, M.A. and Lawrence, T.L.J. (eds) *The Voluntary Feed Intake of Pigs.* Occasional Publication No. 13, British Society of Animal Production, Edinburgh, pp. 61–70.

Cole, D.J.A., Duckworth, J.E. and Holmes, W. (1967) Factors affecting voluntary feed intake in pigs. *Animal Production* 9, 141–148.

Emmans, G.C (1988) Genetic components of potential and actual growth. In: Land, R.B., Bulfield, G. and Hill, W.G. (eds) *Animal Breeding Opportunities.* Occasional Publication No. 12, British Society of Animal Production, Edinburgh, pp. 153–181.

Ferguson, N.S., Gous, R.M. and Emmans, G.E. (1994) Preferred components for the construction of a new simulation model of growth, food intake and nutrient requirements of growing pigs. *South African Journal of Animal Science* 24, 10–17.

Fowler, V.R. and Gill, B.P. (1989) Voluntary food intake in the young pig. In: Forbes, J.M., Varley, M.A. and Lawrence, T.L.J. (eds) *The Voluntary Feed Intake of Pigs.* Occasional Publication No. 13, British Society of Animal Production, Edinburgh, pp. 51–60.

France, J. and Thornley, J.H.M. (1984) *Mathematical Models in Agriculture.* Butterworths, London.

Gompertz, B. (1825) On the nature and the function expressive of the law of human mortality and a new method of determining the value of life contingencies. *Philosophical Transactions of the Royal Society* 115, 513–585.

Hammond, J. (1940) *Farm Animals: Their Breeding, Growth and Inheritance.* Edward Arnold, London.

Henry, Y. (1985) Dietary factors involved in feed intake regulation in growing pigs: a review. *Livestock Production Science* 12, 339–354.

Huxley, J. (1932) *Problems of Relative Growth.* Methuen, London.

Kalm, E. (1986) Evaluation and utilisation of breed resources: as sire lines in cross-breeding. In: *Proceedings of the 3rd World Congress on Genetics Applied to Livestock Production,* Vol. 10. University of Nebraska, Lincoln, pp. 35–44.

Kielanowski, J. (1969) Energy and protein metabolism in growing pigs. *Revista Cubana de Ciencia Agricola* 3, 207–216.

Knap, P.W. (2000) Time trends of Gompertz growth parameters in 'meat-type' pigs. *Animal Science* 70, 39–49.

Kyriazakis, I. and Emmans, G.C. (1995) The voluntary feed intake of pigs given feeds based on wheat bran, dried citrus pulp and grass meal, in relation to measurements of feed bulk. *British Journal of Nutrition* 73, 191–207.

Kyriazakis, I. and Emmans, G.C. (1999) Voluntary feed intake and diet selection. In: Kyriazakis, I. (ed.) *A Quantitative Biology of the Pig.* CAB International, Wallingford, UK, pp. 9–38.

Kyriazakis, I., Stamataris, C., Emmans, G.C. and Whittemore, C.T. (1991) The effects of food protein content on the performance of pigs previously given foods with low or moderate protein contents. *Animal Production* 52, 165–173.

Meat and Livestock Commission (1982) *Commercial Product Evaluation, 8th Test Report.* Pig Improvement Services, MLC, Milton Keynes.

NRC (1987) *Predicting Feed Intake of Food-producing Animals.* National Academy Press, Washington, DC, 85pp.

NRC (1998) *Nutrient Requirements of Swine,* 10th edn. National Academy of Sciences, National Academy Press, Washington, DC, 189pp.

Parks, J.R. (1982) *A Theory of Feeding and Growth in Animals.* Springer Verlag, New York.

Rerat, A. (1972) Protein nutrition and metabolism in the growing pig. *Nutrition Abstracts and Reviews* 42, 13–39.

Richards, F.J. (1959) A flexible growth function for empirical use. *Journal of Experimental Botany* 10, 290–300.

Schinckel, A.P. (1999) Describing the pig. In: Kyriazakis, I. (ed.) *A Quantitative Biology of the Pig.* CAB International, Wallingford, UK, pp. 9–38.

Schinckel, A.P. and de Lange, C.F.M. (1996) Characterization of growth parameters needed as inputs for pig growth models. *Journal of Animal Science* 74, 2021–2036.

Tolkamp, B.J. and Ketelaars, J.J.M.H. (1992) Toward a new theory of feed intake regulation in ruminants. 2. Cost and benefits of feed consumption: an optimisation approach. *Livestock Production Science* 30, 297–317.

Tsaras, L.N., Kyriazakis, I. and Emmans, G.C. (1998) The prediction of the voluntary food intake of pigs on poor quality foods. *Animal Science* 66, 713–723.

Tullis, J.B. and Whittemore, C.T. (1986) Body composition and feed intake of young pigs post weaning. *Journal of the Science of Food and Agriculture* 37, 1178–1184.

Tybirk, P. (1989) A model of food intake regulation in the growing pig. In: Forbes, J.M., Varley, M.A. and Lawrence, T.L.J. (eds) *The Voluntary Feed Intake of Pigs.* Occasional Publication No. 13, British Society of Animal Production, Edinburgh, pp. 105–110.

Verstegen, M.W.A., Close, W.H., Start, I.B. and Mount, L.E. (1973) The effects of environmental temperature and plane of nutrition on heat loss, energy retention and deposition of protein and fat in groups of growing pigs. *British Journal of Nutrition* 30, 21–35.

Webb, A.J. (1989) Genetics of food intake in the pig. In: Forbes, J.M., Varley, M.A. and Lawrence, T.L.J. (eds) *The Voluntary Feed Intake of Pigs.* Occasional Publication No. 13, British Society of Animal Production, Edinburgh, pp. 41–50.

Whittemore, C.T. (1983) Development of recommended energy and protein allowances for growing pigs. *Agricultural Systems* 11, 156–186.

Whittemore, C.T. (1993) *The Science and Practice of Pig Production.* Longman, London, 661pp.

Whittemore, C.T. (1995) Modelling the requirement of the young growing pig for dietary protein. *Agricultural Systems* 47, 415–425.

Whittemore, C.T. (1998) *The Science and Practice of Pig Production,* 2nd edn. Blackwell Science, Oxford, 624pp.

Whittemore, C.T. and Fawcett, R.H. (1974) Model responses of the growing pig to dietary intake of energy and protein. *Animal Production* 19, 221–231.

Whittemore, C.T. and Fawcett, R.H. (1976) Theoretical aspects of a flexible model to simulate protein and lipid growth in pigs. *Animal Production* 22, 87–96.

Whittemore, C.T., Aumaître, A. and Williams, I.H. (1978) Growth of body components in young weaned pigs. *Journal of Agricultural Science, Cambridge* 91, 681–692.

Whittemore, C.T., Taylor, H.M., Henderson, R., Wood, J.D. and Brock, D.C. (1981) Chemical and dissected composition changes in weaned piglets. *Animal Production* 32, 203–210.

Whittemore, C.T., Tullis, J.B. and Emmans, G.C. (1988) Protein growth in pigs. *Animal Production* 46, 437–445.

Whittemore, C.T., Kerr, J.C. and Cameron, N.D. (1995) An approach to prediction of feed intake in growing pigs using simple body measurements. *Agricultural Systems* 47, 235–244.

Whittemore, E.C., Kyriazakis, I., Emmans, G.C. and Tolkamp, B.J. (2001a) Tests of two theories of food intake using growing pigs. 1. The effect of ambient temperature on the intake of foods of differing bulk content. *Animal Science* 72, 351–360.

Whittemore, E.C., Emmans, G.C., Tolkamp, B.J. and Kyriazakis, I. (2001b) Tests of two theories of food intake using growing pigs. 2. The effect of a period of reduced growth rate on the subsequent intake of foods of differing bulk content. *Animal Science* 72, 361–373.

Energy Requirements of the Young Pig

2

J. Le Dividich and B. Sève

INRA, Unité Mixte de Recherches sur le Veau et le Porc, 35590 St-Gilles, France

Introduction

The growth performance of the young pig from birth to 8–9 weeks of age is critical in determining subsequent weight-for-age relationships and, as such, ultimate weight at slaughter. The young pig has a high potential for growth (Hodge, 1974; Harrell *et al.*, 1993). Numerous factors influence the extent to which this potential is expressed, among which energy (food) intake is the most important. The second half of lactation and the immediate post-weaning period are commonly identified as major periods during which the deficit in energy intake is the most pronounced. However, a low milk intake during the neonatal period also influences the ability of the piglet to survive and to thrive.

A complete discussion of the energy requirements of the young pig is beyond the scope of a single chapter. For a more comprehensive coverage, reviews are mentioned in various sections. This chapter will consider four main sections: (i) an overview of the assessment of the energy requirements while discussing the specificity related to the young pig; (ii) the growth pattern of the suckling and weaned pig, with emphasis being given to periods of growth check; (iii) energy supply and factors through which it is influenced; and (iv) means to improve the energy intake in order to approach the growth potential. Unless specified, piglet age at weaning is between 18 and 24 days, as practised in major pig-producing areas.

Assessment of Energy Requirements

Significance of energy requirements in the young pig

In growing/finishing pigs, energy requirements are designed to promote maximum protein deposition while providing adequate carcass composition at slaughter.

Adequate carcass composition is achieved through appropriate dietary manipulation (protein to energy ratio, restricted feeding) during the growing/finishing period. However, it is doubtful that body composition of the young pig subsequently affects the final carcass composition (Sève and Bonneau, 1986; Caugant and Guéblez, 1993; Kavanagh *et al.*, 1997). It is likely that the extra fat deposited in the young pig makes a small contribution to the total stored fat and so the major objective during the suckling and weaning phases should be to maximize piglet growth. Heavy piglets at weaning have: (i) more body fat (Sloat *et al.*, 1985) and hence a better ability to withstand the period of underfeeding following weaning; and (ii) a more developed digestive tract (de Passillé *et al.*, 1989; Cranwell *et al.*, 1997) and therefore a better ability to cope with transition to the post-weaning diet. More important are the findings that a high growth rate during both suckling and weaning phases usually persists until slaughter (Mahan and Lepine, 1991; Mahan, 1993; Kavanagh *et al.*, 1997; Mahan *et al.*, 1998). Therefore, because protein deposition increases linearly with energy intake in the suckling pig, maximization of protein deposition also implies maximization of energy intake.

Components of energy requirements

Changes in body composition

During the immediate postnatal period there is a very rapid increase in both protein and fat accretion (Table 2.1). Between birth and weaning at 21 days of age, body protein and fat contents increase linearly at a mean rate of 34–38 and 30–35 g day^{-1}, respectively. Due to the low food intake associated with the nursing–weaning transition period, protein accretion slows temporarily, while fat accretion is usually negative (Whittemore *et al.*, 1978). During this period, the rate of protein accretion is related to both the food intake and its protein content (Sève *et al.*, 1986). The rate of body fat mobilization depends on the level of food intake and its protein content, and on the environmental temperature. It is more pronounced at low food intake and low environmental temperature and is increased with dietary protein supply (for a review, see Le Dividich and Sève, 2000). Once regular food intake is established, accretion of both protein and fat accelerates; however, fat deposition does not recommence until a bodyweight (BW) gain of about 200 g day^{-1} is reached (Whittemore *et al.*, 1981).

Energy cost of growth

The bases for estimations of the energy requirements of the pig have been reviewed in recent years (ARC, 1981; National Research Council, 1998). Briefly, estimations based on the factorial approach include determination of the energy cost of each component of the requirements, i.e. maintenance (ME_m) and protein together with fat deposition. Compared with the growing/finishing pig, there is little information in the literature on the estimations of energy cost of maintenance and growth in the suckling piglet, due to the difficulties in determining milk energy intake precisely.

These estimations are also difficult to establish immediately after weaning, because of the usual low food intake and the ensuing negative energy balance.

MAINTENANCE ENERGY REQUIREMENTS. A summary of values for metabolizable energy (ME) requirements for maintenance (ME_m) and efficiencies of ME intake for total energy retention (k_g) and for energy retained as protein (k_p) and fat (k_f) is given in Table 2.2. ME_m includes the energy cost of basal metabolism, physical activity and thermoregulation. In the suckling pig ME_m amounts to 470 kJ ME kg^{-1} $BW^{0.75}$ per day (Jentsch *et al.*, 1992). A lower value (340 kJ ME kg^{-1} $BW^{0.75}$ per day) was reported by Marion and Le Dividich (1999) but this value may be largely underestimated, because piglets were restrained in small cages with ensuing minimal energy expenditure for feeding and physical activity. In the suckling pig, the energy cost of standing is 0.18 kJ $BW^{0.75}$ per minute (Herpin and Le Dividich, 1995), but in practice the total energy cost of suckling and standing activity is very variable and difficult to assess, especially in outdoor rearing conditions. In the weaned pig, the average value for ME_m is the same as in the suckling pig (470 kJ ME kg^{-1} $BW^{0.75}$ per day), with the energy for activity accounting for 15–30% of the ME_m (Halter *et al.*, 1980).

Table 2.1. Relationship between bodyweight, body protein and body fat in the young pig.

Age (days)	Bodyweight (kg)	Body protein (kg)	Body fat (kg)	Protein/fat
Birth	1.23	0.130	0.020	6.5
1	1.45	0.175	0.030	5.8
7	2.80	0.410	0.190	2.2
14	4.50	0.650	0.510	1.3
21 (weaning)	6.30	0.830	0.640	1.3
28	7.00	0.890	0.540	1.6
35	8.80	1.130	0.730	1.5

Data calculated from Close and Stanier (1984a); Sloat *et al.* (1985); Le Dividich *et al.* (1997); Marion and Le Dividich (unpublished).

Table 2.2. Summary of the estimated values for maintenance energy requirements (ME_m) and for efficiency of metabolizable energy intake for growth k_g, protein k_p and fat k_f deposition.

	ME_m (kJ ME kg^{-1} $BW^{0.75}$)	k_g	k_p	k_f
Suckling pig	470	0.70–0.73	0.56	1.0
Weaned pig				
Mean	470	0.72	0.60	0.80
Range	420–550	0.60–0.79	0.50–0.66	0.73–0.84

Data on the suckling pig are from Jentsch *et al.* (1992) and Marion and Le Dividich (1999). Data on weaned pigs are from de Goey and Ewan (1975); Halter *et al.* (1980); Le Dividich *et al.* (1980); Noblet and Le Dividich (1982); Stibic (1982); Close and Stanier (1984b) and Gädeken *et al.* (1985).

Below the lower critical temperature (LCT), heat production (HP) increases to meet the energy requirement for thermoregulation. The extra ME requirements for thermoregulation decrease from 48 kJ kg^{-1} $BW^{0.75}$ per °C in the newborn to 16–18 kJ kg^{-1} $BW^{0.75}$ per °C at the time of weaning (Le Dividich *et al.*, 1998a). In indoor rearing conditions, the ambient temperature is expected to have little effect on the ME_m of the suckling pig. However, in piglets reared in outdoor conditions, and in the weaned pig, the environmental temperature may have a major impact on the ME_m. For example, at 5°C below LCT, the extra ME requirement amounts to 330 kJ day^{-1} for a 7 kg pig. On the basis that 1 g sow milk supplies 4.75 kJ ME and 1 g weaning diet supplies 14.5 kJ ME, this represents an additional 70 g milk and 23 g weaning diet per day.

Energy cost of protein and fat deposition. The efficiency of sow milk ME for growth (k_g) is between 0.70 and 0.73 (Jentsch *et al.*, 1992; Marion and Le Dividich, 1999) (Table 2.2) and similar to that reported for cow milk and ewe milk. In the suckling pig, fat deposition results primarily from direct incorporation of absorbed milk fatty acids into the adipose tissue (Gerfault *et al.*, 2000), with an ensuing k_f of 1.0. This value is confirmed by the finding that calculated k_f is not different from 1.0 (Marion and Le Dividich, 1999). According to the same authors, the efficiency of ME for protein deposition (k_p) is 0.56. In the weaned pig, the average efficiencies of ME intake for growth and for energy retained as protein and fat are 0.72, 0.66 and 0.77, respectively, i.e. values close to those reported in the growing pig (ARC, 1981). Furthermore, there is a clear tendency for k_g to increase with the decrease in environmental temperature, suggesting that part of the HP associated with the productive processes is used for thermoregulation in the cold.

Energy system

The relevance of the energy systems has been discussed (ARC, 1981; Noblet, 1997; National Research Council, 1998). The net energy (NE) system takes into account the metabolic utilization of energy and hence is likely to be more desirable than the digestible energy (DE) or ME systems. However, there is very little information on the NE values of diets for the young pig (de Goey and Ewan, 1975). Construction of an energy system for the young pig presents difficulties because diet digestibility, a key step in any energy system, changes markedly after weaning (Sève, 1985; Owsley *et al.*, 1986), while being largely dependent on the health status of the animal, which is very variable after weaning. Further, even though sow milk is highly digestible (0.99–1.00), its DE or ME value varies widely in relation to its fat content. Also, for the growing pig a main interest of an energy system lies in formulating diets at least cost. In contrast, in the weaned pig the choice of dietary ingredients must primarily fit the digestive capacity of the animal, maintain the gut health and promote food intake. Therefore, it is doubtful that construction of an energy system is a priority for the young pig, but once adaptation to the digesting of solid food is completed, the piglet can be considered as a growing pig with regard to the energy system. In this chapter, the energy values of diets are expressed in gross energy (GE), DE or ME, while keeping in mind that DE and ME values provided by tables of raw material composition are liable to large errors.

Potential Growth Performance of the Young Pig

From literature data, the recorded growth rate (GR) patterns of pigs during both suckling and weaning phases (from birth to 18–22 kg) are given in Fig. 2.1 together with the growth potential of piglets artificially reared from birth. The GR pattern during the suckling phase was recorded on piglets from modern genotypes, weighed daily (265 piglets from 23 litters with an average individual weaning weight of 6.6 kg at the age of 20 days).

Both patterns include an early period of variable duration during which growth rate is very variable and gradually increases to a plateau of 250–270 g day^{-1} and 600–650 g day^{-1} during the suckling and the weaning phase, respectively. Mean growth rates during the suckling phase and weaning phases are 240 and 460 g day^{-1}, respectively. Data obtained on a conventional genotype provided similar GR patterns during the suckling period but with a lower plateau value for GR (180–200 g day^{-1}) (King *et al.*, 1999). At birth, the period of moderate GR corresponds to the establishment of lactation in the sow and the formation of the

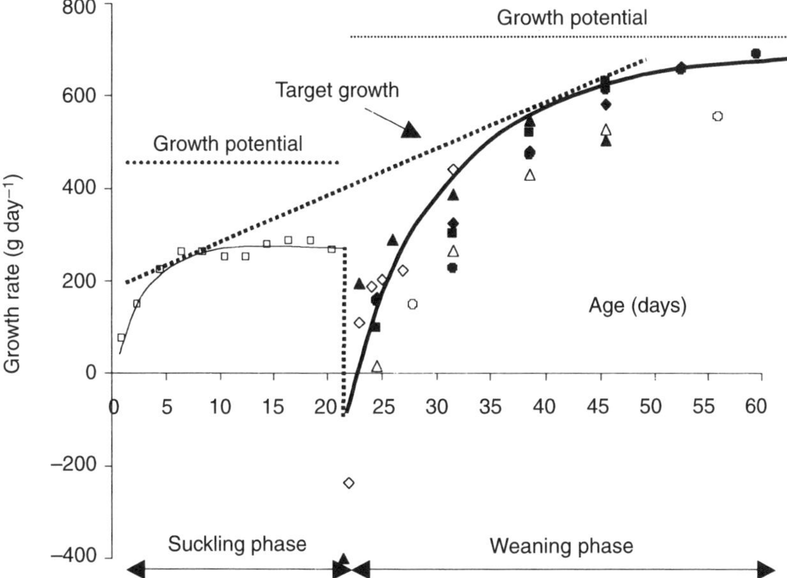

Fig. 2.1. Pattern of growth rate of piglets during the suckling and weaning phases. Data of the suckling phase (□) (265 piglets from 23 litters weighed daily) are from Teurnier (personal communication). Weaning phase data are from: (◆) Mahan *et al.* (1998); (◇) Le Dividich and Orgeur (unpublished data); (▲) Ogunbameru *et al.* (1992); (△) Leibbrandt *et al.* (1975a); (●) and (○) Le Dividich (1981); and (■) Russell *et al.* (1996). The least-squares fitted equations were as follows:

Suckling phase: $Y = 273\ (\pm 3.6) - 289\ (\pm 28)\ e^{-0.38\ (\pm 0.05)\ x_1}$

Weaning phase: $Y = 650\ (\pm 53) - 784\ (\pm 57)\ e^{-0.103\ (\pm 0.02)\ x_2}$

where Y = growth rate (g day^{-1}); x_1 = day post-natal and x_2 = day post-weaning.

nursing order in the litter. After weaning, it relates to the formation of a new social order and the establishment of a regular food intake. Growth rates attained at the plateau can also be very variable, as they are dependent on sow milk production, litter size, parity and birthweight during the suckling phase, and on piglet bodyweight, food intake, environmental factors and health conditions after weaning.

By comparison, results of artificial rearing studies (Hodge, 1974; Harrell *et al.*, 1993) reported that values for liveweight gain as high as 400–550 g day^{-1} and 700–800 g day^{-1} are possible during the suckling and the weaning phases, respectively. This suggests that the biological growth potential of both suckling and weaned pigs through the supply of sufficient nutrients is much greater than that achieved in practice. The potential is difficult to attain, due to the inability to optimize all factors that influence performance, but it is clear that attention must be paid to the periods where the deficit in GR is the most pronounced. As suggested in Fig. 2.1, overcoming periods of GR deficit would provide a realistic target for GR of 280–300 and 550–580 g day^{-1}, in the suckling and weaning phase, respectively. Corresponding target bodyweight at weaning (20–22 days) and at the end of the weaning phase (60–63 days) should be 7.2–7.6 and 31.5–33.0 kg, respectively.

Energy Supply to the Young Pig

The suckling pig

Once suckling is established in the newborn, energy intake and hence growth depend on the availability and composition of colostrum and milk and on their efficiency of conversion into gain. This is strongly supported by the high relationship (r^2 = 0.87–0.90) between pre-weaning growth and sow milk nutrient output (Noblet and Etienne, 1989). The average composition of colostrum and milk is given in Table 2.3. Fat is the main source of energy accounting for 30–40% and 55–60% of the total energy of colostrum and milk, respectively. It is a major constituent, readily influenced by the sow diet. For example, feeding more fat to the sow during late gestation or during lactation is associated with a higher content of fat in colostrum and milk (Shurson *et al.*, 1986). In contrast, milk protein and lactose are marginally dependent on the composition of the sow diet, while amino acid composition is remarkably constant (King *et al.*, 1999).

Colostrum and milk are remarkably well utilized by the piglet, with both having a ratio of ME/GE between 0.95 and 0.97 (Jentsch *et al.*, 1992; Le Dividich *et al.*, 1994; Marion and Le Dividich, 1999) and a ratio of N retained/N intake between 0.88 and 0.91 (Noblet and Etienne, 1987; Marion and Le Dividich, 1999). If endogenous losses and digestibility of protein are taken into account, the biological value of protein of sow milk must approach 1.0. Efficiency of milk ME for gain increases during the course of lactation, averaging 17 kJ ME g^{-1} bodyweight gain (BWG) in the first week (Marion and Le Dividich, 1999) and 19 kJ ME g^{-1} BWG in the following 2 weeks (Noblet and Etienne, 1986; Beyer and Jentsch, 1994).

Pattern of energy intake

The pattern of ME intake (kJ ME kg^{-1} $BW^{0.75}$ per day) in suckling pigs is presented in Fig. 2.2. ME intake is maximum (> 5 times ME_m) at 7–10 days of age and decreases afterwards. At the time of weaning, it amounts to about 2.5–2.7 times ME_m. Numerous factors influence piglet milk consumption and detailed analysis of these factors (including the genetic potential, litter characteristics and provision of adequate nutrients required for the synthesis of milk components) have been reviewed by Verstegen *et al.* (1998). They will not be specifically discussed here but it is relevant to note that piglet milk consumption is highly variable, both within and among litters, and is dependent on environmental conditions.

The within-litter variations in piglet milk consumption reflect differences in the functionality of the mammary glands and piglet ability to compete at the udder. The anterior and middle glands have greater wet and dry weights and greater protein and DNA contents compared with the posterior glands (Kim *et al.*, 2000). Milk yield decreases from the most anterior to the most posterior teats (Fig. 2.3) (Pluske and Williams, 1996a). It follows that piglets adopting the anterior teats are heavier at weaning than those adopting the posterior teats (Hoy and Puppe, 1992; Pluske and Williams, 1996a). In addition, piglet bodyweight itself may contribute to differences in milk intake. At 17–24 days of lactation, heavy piglets consume 28–30% more milk than do small piglets (Pluske and Williams, 1996a). It is suggested that, compared with small piglets, those of high birthweight are doubly advantaged by their ability to select the most productive anterior glands and to extract more milk from the teats (King *et al.*, 1997). This emphasizes the importance of birthweight in terms of milk consumption and postnatal growth. Overall, the weaning weight increases by 0.35 to 1.07 kg for every 100 g increase in birthweight (Mahan and Lepine, 1991; Caugant and Gueblez, 1993; Rousseau *et al.*, 1994; Dunshea *et al.*, 1997).

Table 2.3. Average composition (g kg^{-1}) of sow colostrum and milk (according to King *et al.*, 1999).

	Colostrum	Milk
Dry matter	215	187
Crude protein (N × 6.38)	105	54
Fat	54	71
Lactose	38	54
Gross energy (MJ kg^{-1})	5.4	5.0
Essential amino acid (g per 16 g N)		
Lysine	7.32	7.33
Methionine + cystine	3.24	3.30
Tryptophan	1.86	1.24
Threonine	5.46	4.52
Leucine	9.81	8.43
Isoleucine	3.87	3.98
Valine	5.95	5.25
Histidine	3.17	3.23
Phenylalanine	4.56	4.09
Tyrosine	5.63	4.26

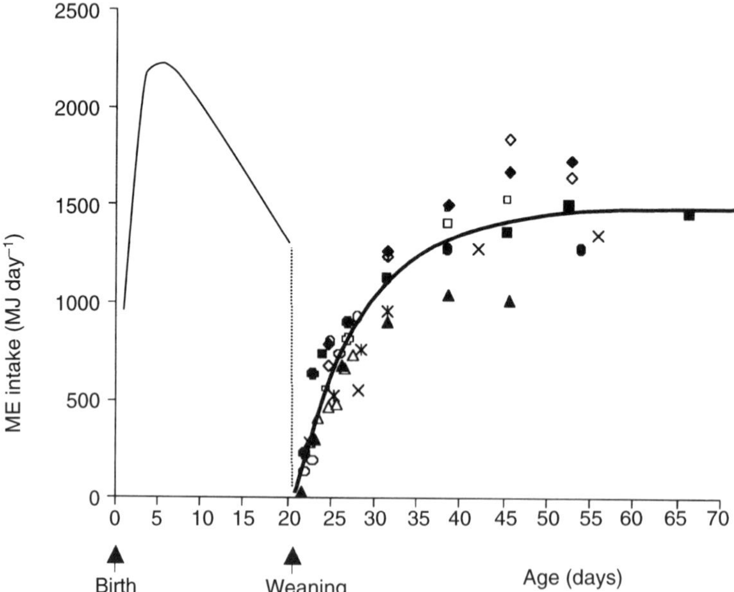

Fig. 2.2. Pattern of metabolizable energy intake in young pigs. Suckling phase data are calculated from Teurnier (unpublished data) assuming milk ME conversion rate of 17 kJ milk ME g^{-1} BWG during the first post-natal week (Marion and Le Dividich, 1999) and 19 kJ milk ME g^{-1} BWG thereafter (Noblet and Etienne, 1986; Beyer and Jentsch, 1994). Weaning phase data are from: (◆) Mahan *et al.* (1998); (◇) Leibbrandt *et al.* (1975a); (▲) Ogunbameru *et al.* (1992); (△) Bark *et al.* (1986); (●) Orgeur *et al.* (2000); (□) Le Dividich *et al.* (1980); (*) Leibbrandt *et al.* (1975); (x) Le Dividich (1981); (○) McCracken (1989); (■) Rinaldo (1989). The least-squares fitted equation of ME intake during the post-weaning phase is as follows: $Y = 1509\ (\pm 65) - 1479\ (\pm 80)\ e^{-0.127\ (\pm 0.02)\ x}$ where Y = ME intake (kJ ME kg^{-1} BW$^{0.75}$ day^{-1}) and x = day post-weaning.

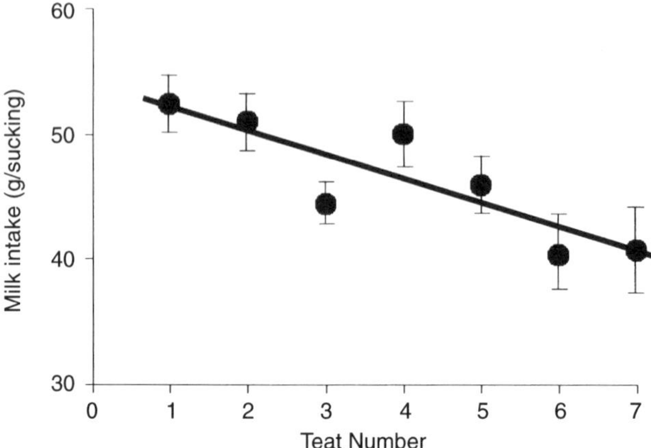

Fig. 2.3. Relationship between average milk intake of piglets and teat order, with teats numbered 1–7 from the anterior to the posterior end of the udder. (From Pluske and Williams, 1996a.)

The between-litter differences in milk intake and therefore in growth are also large. In commercial herds a twofold difference in piglet mean weaning weight is very common. Numerous factors account for these variations but it is relevant to note that these variations are detected soon after birth, as exemplified by the results of Thompson and Fraser (1988), which showed piglet growth ranging from 5 to 227 g day^{-1} during the first 3 days of life. The distribution of litter growth during the first postnatal week, recorded in a commercial farm, is shown in Fig. 2.4. The average GR of litters was 210 g per piglet. It was noticeable that 38% of litters achieved GR > 200 g per piglet, 49% had GR ranging between 150 and 200 g per piglet, and 13% had GR < 150 g per piglet. According to Thompson and Fraser (1986), approximately 45% of the variation in the litter mean weight at day 14 is explained by the gain during days 0–2, with differences established by day 14 being perpetuated and sometimes enhanced during the remainder of lactation. Furthermore, a low intake of colostrum then milk immediately after birth may have important consequences on the ability of piglets to survive and to thrive. Insufficient intake of colostrum and hence of energy and immunoglobulins is a major cause of neonatal mortality (Herpin and Le Dividich, 1995). Low milk intake during the first postnatal week also has marked effects on growth, protein metabolism, and muscle and gut development of piglets. Whole-body protein synthesis is reduced by 47% in piglets fed sow milk at the rate of 100 g kg^{-1} BW per day during the first week after weaning, compared with those fed 300 g kg^{-1} BW per day (Marion *et al.*, 1999). The total number of muscle fibres in longissimus dorsi and rhomboïdeus muscles is not affected by the level of milk intake, but the cross-sectional area of secondary fibres is dramatically decreased in both muscles of piglets fed the low level of milk. These piglets also have a higher proportion of fibres containing fetal myosin heavy chains in their muscles, suggesting that a low amount of milk intake soon after birth delays muscle maturation (Lefaucheur *et al.*, 1999). Segmental activity of lactase and height of villi of the small intestine are also reduced in piglets fed a low level of milk, suggesting that the digestive capacity is lowered in piglets consuming less milk (Le Huérou-Luron *et al.*, 1999), though to what extent these effects persist thereafter remains to be determined.

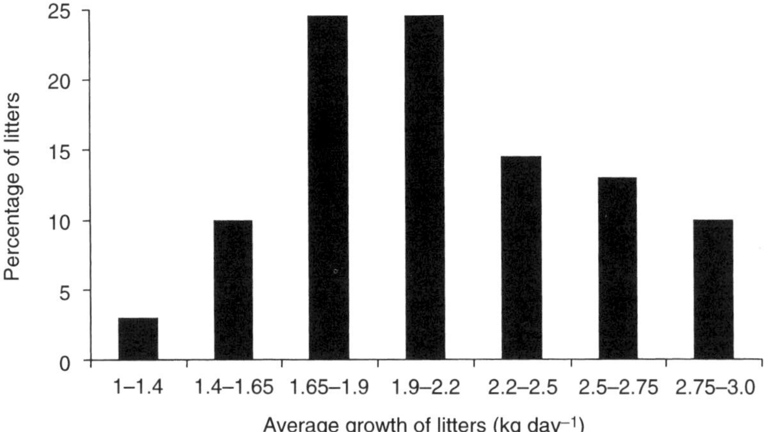

Fig. 2.4. Distribution of mean growth of litters during the first week of lactation. Date recorded in 69 litters (litter size adjusted to 11 piglets) from a commercial farm. (From M. Drillet, 1999, personal communication.)

Altogether, these observations emphasize how important is the period of the onset of lactation (a previously neglected area) on the subsequent performance of piglets. It is suggested that more attention should be paid to this period.

Environmental temperature may have detrimental effects on milk intake in piglets both directly, when subjected to cold stress at birth, and indirectly, through the reduced milk yield of the sow at high environmental temperatures. Although supplemental heat is routinely provided in the piglet area, newborn pigs (especially those suffering from hypoxia during parturition) (Herpin *et al.*, 1996) can be subjected to cold stress. However, unlike the older animal, the newborn pig is unable to increase its energy intake in a cold environment. At an ambient temperature of 18–20°C, such piglets are reported to consume 27% less colostrum than their littermates kept at 30–32°C (Le Dividich and Noblet, 1981). The resulting impairment in thermostability reduces the vigour of the piglets, leading to less aggressive nursing behaviour and thus compromising their survival and subsequent growth. On the other hand, high environmental temperatures have direct effects on sow milk yield. During the whole lactation, sows maintained at 27–30°C produced 10–22% less milk than at 18–20°C, with the individual weight of piglets at weaning being decreased to a similar extent (Stansbury *et al.*, 1987; Schoenherr *et al.*, 1989; Prunier *et al.*, 1997). However, management practices aiming at reducing the severity of heat stress, such as snout coolers or drip water, have proved to be efficient (Stansbury *et al.*, 1987; McGlone *et al.*, 1988)

Sow milk: the main limiting factor on piglet growth?

Despite the high biological value of sow milk, growth of piglets reared by the sow is much lower than that achieved by artificially reared ones. As reviewed by Pluske *et al.* (1995), availability and composition of milk are the major reasons for the relatively slower growth rates of sow-reared piglets.

Evidence that the milk produced by the sow may be limiting is shown in Fig. 2.2. In agreement with Harrell *et al.* (1993), milk production becomes limiting at 7–10 days of lactation, with the difference between need and supply increasing as lactation proceeds. However, it is also obvious that milk production can be limiting in early lactation. Piglets have the capacity to ingest more milk than usually observed. Thus, in bottle-fed piglets, colostrum consumption during the first 27 h can be as high as 460 g kg^{-1} BW, i.e. 30–40% more than values usually reported (Le Dividich *et al.*, 1997). Similarly, piglets fed artificially with a milk substitute consume up to 60% more ME than control sow-reared piglets (Harrell *et al.*, 1993). Modern sows produce 10–12 kg milk day^{-1} (Etienne *et al.*, 1998), i.e. approximately double the yield of 20–30 years ago, but a sow would have to produce in excess of 18 kg milk day^{-1} to meet the energy requirements of a ten-pig litter at 21 days (Boyd *et al.*, 1995). Super-alimentation of the sow does not improve milk production (Pluske *et al.*, 1998); however, Auldist *et al.* (1995) and Sauber *et al.* (1995) reported that, under conditions of high nursing demand, modern sows have the capacity to produce almost double the milk currently achieved, suggesting that full expression of this capacity is probably influenced by the demand of the piglets. The possibility also exists that sow milk contains factors that influence this demand. For example, high levels of leptin are found in sow milk (Estienne *et al.*, 2000) and leptin influences food intake, though to what extent ingested leptin influences milk intake in the suckling pig remains to be assessed.

Secondly, the growth potential of the suckling pig may be limited by the composition of milk and mainly by its protein content. Protein deposition rates of 20 and 15 g kg^{-1} $BW^{0.75}$ per day have been reported in 2 kg and 4 kg piglets, respectively (Williams, 1976; Marion and Le Dividich, 1999). The potential for protein deposition is probably higher than these values and beyond that allowed by milk intake. Salmon Legagneur and Aumaître (1962) and Noblet and Etienne (1987) suggested that the protein to energy ratio in sow milk is lower than that which provides maximum growth in the young pig. In artificially reared pigs, the dietary protein requirements for maximum protein deposition is between 12.4 and 15.6 g protein MJ^{-1} DE (Williams, 1976; McCracken *et al.*, 1980), whereas in sow milk the protein to energy ratio is between 8 and 10, depending on the fat content of milk. Similarly, the dietary lysine requirement for piglets during their first 3 weeks of life is about 0.90–0.95 g lysine MJ^{-1} GE (Nam and Aherne, 1994; Auldist *et al.*, 1997), whereas sow milk contains only 0.67–0.85 g lysine MJ^{-1} GE. In the young pig, the rate of protein deposition increases linearly with the increase in energy intake, while the composition of daily weight gain remains constant (Noblet and Etienne, 1987).

It is suggested that sow milk is designed to promote fat accretion in the young pig. Fats are of vital importance for normal growth and development of the suckling pig. Body fat stores are very low at birth and milk fat is the sole source of fat of the young pig because lipogenesis is limited during the suckling phase by the availability of substrates (Gerfault *et al.*, 2000). Therefore body fat accretion is closely dependent on ingested milk fat. Fat is the main source of energy of the suckling pig, as indicated by the observation that the respiratory quotient (RQ) ranges from 0.73 and 0.84, depending on the amount of milk consumed (Noblet and Etienne, 1987; Marion and Le Dividich, 1999).

Finally, stored fat is mostly subcutaneous fat that provides thermal insulation to the young pig and energy through its mobilization during the weaning period. For example, in piglets of low weaning weight, body fat stores can be mobilized at a rate of 32% during the first week after weaning (Sloat *et al.*, 1985). Manipulation of sow milk to match the dietary protein and amino acids requirements of the piglets more closely would increase protein deposition by 4.3 g day^{-1} for each 0.1 g MJ^{-1} increase in the lysine:GE ratio (Auldist *et al.*, 1997). However, as mentioned above, the protein concentration in sow milk is only marginally dependent on the composition of the sow's diet and it is unlikely that the protein to energy ratio in sow milk can be manipulated by conventional means.

The weaned pig

At weaning at about 3 weeks of age, most piglets have consumed very little solid food (Pluske *et al.*, 1995); hence they are unfamiliar with the weaning diet. It follows that the nursing–weaning transition is associated with a critical period of underfeeding during which piglets adapt to digest the solid food. Both the extent and duration of underfeeding are very variable (Fig. 2.2). Behavioural studies reported by Brooks (1999a) indicate that, although the majority of pigs start to eat solid food within 5 h of weaning, some of them take up to 54 h before eating their first meal. Thereafter, ME intake gradually increases at a rate of 100–120 kJ ME kg^{-1} $BW^{0.75}$ per day and reaches a plateau at 14–21 days after weaning. It can be seen from Fig. 2.2 that ME_m requirement is not met until days 3–4 after weaning and that the level of ME intake attained at the end of the first week after weaning amounts to about 800–900 kJ ME kg^{-1} $BW^{0.75}$ per day, which accounts for 60–70% of the pre-weaning milk ME intake. Voluntary ME intake attained at plateau averages 1.5 MJ ME kg^{-1} $BW^{0.75}$ per day, i.e. about three times ME_m. Again, the level attained may be variable, depending on both nutritional and environmental factors.

Nutritional factors

DIETARY ENERGY. Growing/finishing pigs are relatively well adapted to adjust their dry food intake to meet their energy requirements, so that dietary energy concentration may vary within a fairly large range of values. In contrast, at weaning, the ability of the young pig to use diets of variable energy concentration is more closely related to its tolerance to the fibre used for the dilution of energy, or to the fat used for the concentration of energy. The addition of fibre results in a decrease in total energy intake (McConnell *et al.*, 1982). The same trend is noticed especially during the first 2 weeks following weaning, when fat was added to the diet at levels higher than 50 g kg^{-1} (Leibbrandt *et al.*, 1975a; McConnell *et al.*, 1982; Zhang *et al.*, 1986; Tokach *et al.*, 1995).

Fat and fibre are both known to decrease the rate of passage of the digesta in the upper part of the digestive tract, particularly at the stomach level, and this may explain the depression of food intake. The tolerance to high-fibre diets may depend on the type of fibre; for example, barley, a high-fibre cereal grain, is sometimes preferred to wheat or maize in the weaning diet (phase I) to prevent diarrhoea, though the reason for this beneficial effect is not well understood. For fat, it is accepted that, despite a longer digestion process, the apparent digestibility is particularly low during that period (Sève and Aumaître, 1983; Cera *et al.*, 1988). The result is an accentuation of the depressive effect of feeding dry diets at weaning on the digestibility of energy (Nam and Aherne, 1994). Metabolic factors may also be involved in the low appetite of newly weaned pigs for high-fat diets. Piglets from sows fed high-fat diets are reported to be heavier and fatter (Stahly *et al.*, 1980; Jones *et al.*, 1999). They are able to mobilize more fat at weaning and may transiently display lower appetite than lighter piglets. In that physiological state, insulin secretion and tissue sensitivity to insulin are both reduced, concomitantly affecting the metabolic utilization of energy for growth and appetite.

According to McConnell *et al.* (1982) the depressive effect of fat on energy intake is already observed at 62 g fat kg^{-1} (14.7 MJ ME kg^{-1}) during weeks 1 and 2 following weaning, while it is observed at 113 g fat kg^{-1} (15.6 MJ ME kg^{-1}) during weeks 3–5. It follows that, at least with conventional ingredients, the optimal dietary energy concentration is higher during the post-adaptive (14–15 MJ ME kg^{-1}) than during the adaptive (13.5–14 MJ ME kg^{-1}) stage of weaning (Fig. 2.5). In fact, once regular food intake is established, the piglet responds to variations in energy concentration in the same way as the growing/finishing pig.

PROTEIN TO ENERGY RATIO. In the earliest piglet experiments, comparisons of energy concentrations were biased, because the addition of fat was made without adjustment of the dietary protein to energy ratio (Allee and Hines, 1972). The concept of an optimal protein to energy ratio extends to amino acids. The requirement for each amino acid is discussed in Chapter 3 and, regarding the supply of energy to piglets, three points deserve attention here in relation to total ideal protein or to the limiting amino acid supply.

The first point regards the adaptive period of weaning involving an important loss of fat, mainly during the first week after weaning. During this period, dietary protein is preferentially used for protein deposition (Table. 2.4) rather than for heat production. Hence, the increase in protein supply is associated with an enhanced fat mobilization. Because protein deposition and growth rate are low at this period, this observation would be of limited interest, but it appears that the early additional protein supply is necessary to achieve pig growth potential (Sève, 1984; Sève and Ballèvre, 1991).

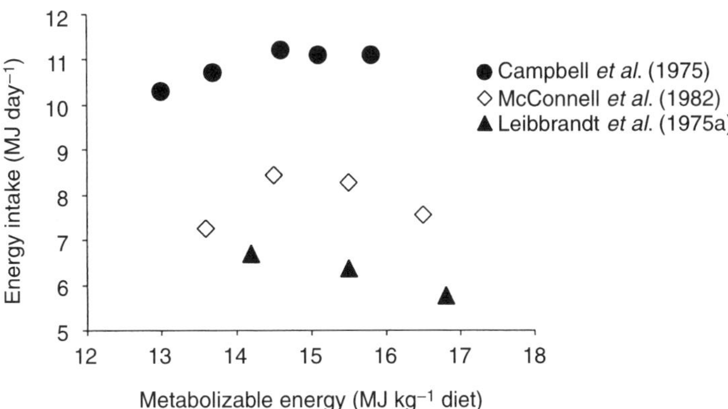

Fig. 2.5. Effect of dietary energy concentration on voluntary energy intake in weaned pigs.

Table 2.4. Effect of dietary protein content on protein and fat deposition in newly weaned pigs.

	Dietary protein content (g kg^{-1})	
	155	296
ME intake	454	471
Energy retained		
Total	−255	−262
As protein	17	89
As fat	−272	−351

Data are expressed in kJ ME kg^{-1} BW$^{0.75}$ per day.
Piglets were weaned at 10 days of age; they were killed 7 days after weaning (Sève *et al.*, 1986).

The second point concerns the post-adaptive stage of weaning. In growing pigs, according to Campbell (1988), the response of nitrogen retention to dietary protein may be considered as linear up to a plateau value that is reached with the optimal protein to energy ratio, with the slope of the response being similar regardless of the dietary energy density. This did not appear to be the case in piglets weighing between 9 and 25 kg (van Lunen and Cole, 1998). The slope of the response to lysine was lower with the higher energy concentration of the diet, in such a manner that, below the requirements, the response of nitrogen retention to lysine intake was increased. This suggests that the increase in dietary energy concentration improves the efficiency of lysine utilization below the requirements. Others (Zhang *et al.,* 1986; Nam and Ahernes, 1994) have reported similar results at the same stage, probably related to the compensatory growth following the adaptive period. These observations are in favour of the use of high-energy low-protein diets during the post-adaptive period of weaning. In such pigs, the early enhancement of fat deposition would not result in depreciation of carcass quality at commercial weight, as observed with more conventional pigs (Sève *et al.*, 1985).

The third point relates to the amino acid balance of protein, which is an important factor affecting energy intake. The imbalance mechanism of action involves the accumulation in the plasma of essential amino acids provided in excess of the requirement. These amino acids may be taken up in the brain, where they have toxic effects that will be prevented through a reduction in voluntary feed intake. Evidence for such effects in piglets was mainly obtained when tryptophan was limiting, with relative excess of other neutral amino acids or threonine (Meunier-Salaün *et al.*, 1991; Sève *et al.*, 1991). Recent work in the post-adaptive period of weaning strongly suggests that the risk of imbalance is more important in low-protein diets, due to the low induction of amino acid catabolism enzymes in the liver (Sève *et al.*, 1999).

Environmental factors

Among the environmental factors affecting voluntary food intake (VFI) and dietary energy available for growth in the weaned piglet, ambient temperature (T_a) is the most important: it has a direct effect on heat loss and VFI and therefore on performance. Maintaining the animal at its LCT maximizes the energy available for production while avoiding excessive loss of body fat at weaning.

The combination of low feed intake and decreased thermal body insulation results in a transient increase in LCT from 22–23°C at weaning to 26–28°C in the first week after weaning, decreasing to 23–24°C in the second week after weaning (for a review, see Le Dividich *et al.*, 1998a). Whether T_a affects VFI during this period is not known. However, once regular feed intake is established, pigs adjust their VFI to compensate for changes in T_a. From data of Nienaber *et al.* (1982) and Rinaldo and Le Dividich (1991) (Fig. 2.6), it can be calculated that, between 25 and 20°C, VFI increases at a mean rate of 19.3 g day^{-1} per °C whereas growth rate remains essentially constant, suggesting that the extra energy consumed compensates for the extra thermoregulatory demand. However, change

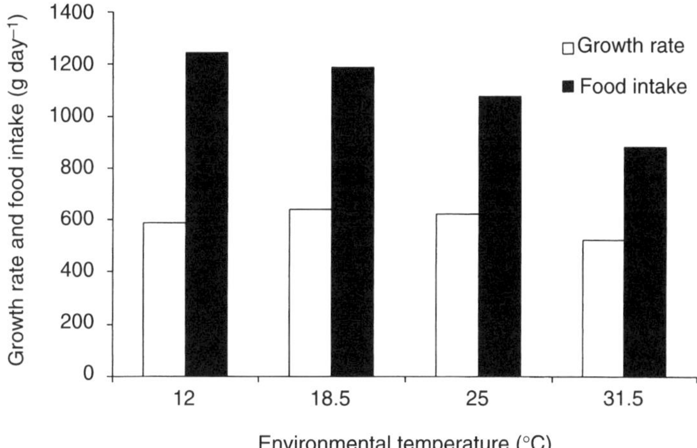

Fig. 2.6. Effect of environmental temperature on performance of the weaned pig (bodyweight range 10–30 kg). (From Rinaldo and Le Dividich, 1991.)

in VFI is marginal between 20 and 15°C, suggesting that maximum VFI is reached between these temperatures. Above 25°C, VFI decreases markedly at the rate of 26 g day^{-1} per °C and results in a reduction of growth performance. Interestingly, increasing dietary energy concentration enhances energy intake at high T_a which helps, to some extent, to counterbalance the reduction in VFI occurring at high T_a (Le Dividich and Noblet, 1986). In contrast, excessive levels of protein or amino acid may create problems of heat stress through the thermal effect of feeding as a result of the catabolism of excess dietary protein or amino acids. This results in a reduction in energy intake, thereby accentuating the energy deficit (Le Dividich *et al.*, 1998a).

Other environmental variables such as hygiene, group size and space allocation influence VFI. For example, at a constant optimum floor allowance (0.25–0.30 m^2 per piglet raised on a totally slatted floor), there was a small decrease in performance as the number of piglets increased (Kornegay and Notter, 1984). However, when space allowance was below the optimum, GR was decreased as a result of a decreased food intake. It may be calculated that each 0.1 m^2 reduction in space allocation results in a 45 g day^{-1} reduction in food intake and this (assuming piglets are fed a conventional diet containing 14 kJ ME) is equivalent to reduction in ME intake of 630 kJ day^{-1}.

Energy Supply to Approach the Potential for Growth

From the above, the question arises: to what extent can provision of supplemental nutrients to the young pig help to achieve potential for growth? This implies provision of supplemental energy during periods of energy deficit and hence of growth deficit.

The suckling phase

The neonatal phase

As mentioned below, supplementation of piglets during lactation with a liquid feed offers the potential to improve growth. However, consumption of supplemental artificial milk or cow milk is very low during the neonatal period, with an ensuing marginal effect on growth (King *et al.*, 1998).

The remaining suckling phase

From 10 to 14 postnatal days, piglets are usually offered creep food in solid form. Creep food consumption varies enormously between and within litters but is usually low. As reviewed by Pluske *et al.* (1995), pre-weaning creep food accounts for about 1.4–5.4% of daily ME intake in piglets weaned at 21 days. Attempts made to stimulate creep food intake have had variable results but increasing dietary energy concentration (Leibbrandt *et al.*, 1975a), diet complexity (Fraser *et al.*, 1994) or palatability (Yang *et al.*, 1997) does not usually result in a substantial increase in creep food intake. In addition, piglet performance in the immediate period following weaning is poorly related to pre-weaning creep food intake (Fraser *et al.*, 1994). On this basis, the benefit of this practice would be questionable for pigs weaned at about 21 days of age.

In contrast, providing litters with liquid diets (milk substitute or cow milk) has been proved to increase growth. Litters having access to supplemental liquid milk substitute grow 10–31% faster than non-supplemented litters during the whole suckling phase (Fig. 2.7), with the improvement increasing as lactation proceeds. According to Australian studies (King *et al.*, 1998; Dunshea *et al.*, 1999), piglets fed supplementary milk were 0.6 and 1.5 kg heavier at weaning at 20 and 28 days of age, respectively, than the non-supplemented ones. Daily supplementary milk consumption amounted to 1.5–1.6 l per litter between birth and 14 days of age and up to 9.1 l per litter between 10–14 days and weaning at 28 days of age. This practice affects neither sow milk production (King *et al.*, 1998; Dunshea *et al.*, 1999) nor piglet body composition at weaning (King *et al.*, 1998) but it improves the body condition of sows weaned at 35 days (Lindberg *et al.*, 1997). The provision of supplemental milk may be particularly useful under conditions such as heat stress, when sow food intake and milk production and hence growth rate of piglets are reduced. However, when provided with supplementary milk, the piglets are reported to consume sufficient energy for their GR and weaning weight to be similar to those of piglets reared under cooler conditions (Azain *et al.*, 1996). The practice needs to be evaluated against the cost of the extra labour and the relative cost of the milk substitute.

The weaning diet transition

The abrupt change from milk to starter solid diets in pigs weaned at about 3 weeks is associated with an underfeeding period immediately after weaning. The resulting growth check can have a major impact on the subsequent performance. This is illus-

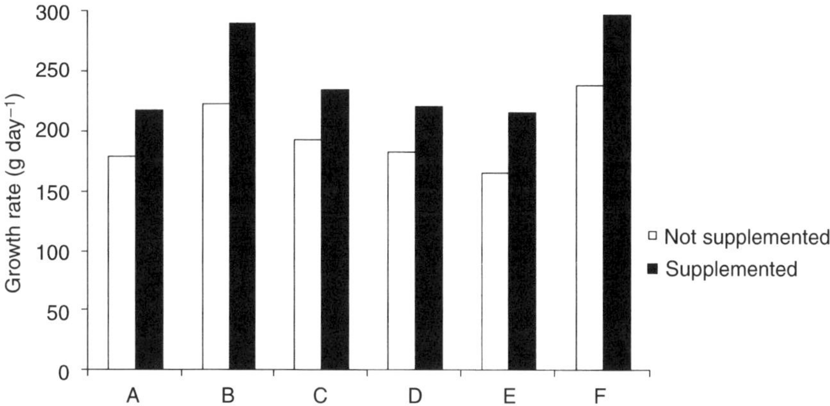

Fig. 2.7. Effects of milk substitute or cow milk supplementation on the growth rate of suckling piglets. Piglets were provided with either milk substitute from days 10 to 20 (A: Dell'Orto *et al.*, 1992; B: Dunshea *et al.*, 1999), days 1 to 21 (C: Azain *et al.*, 1996) and days 14 to 35 (D: Lindberg *et al.*, 1997), or cow milk from days 2 to 20 (E: Sarkar *et al.*, 1981) and days 4 to 28 (F: King *et al.*, 1998).

trated by the findings that food intake (Brooks, 1999a) and growth rate (Miller *et al.*, 1999) in the week after weaning are a significant predictor of subsequent piglet performance. Similarly, Tokach *et al.* (1992) and Azain (1993) reported that piglets gaining well (225–340 g day^{-1}) during the post-weaning diet transition reached slaughter weight some 10–28 days prior to those exhibiting poor gain (0–110 g). Underfeeding is the major factor accounting for the transient change in the small intestine structure, i.e. a shortening of villi and an increase in crypt depth (Pluske *et al.*, 1997). These changes are assumed to impair the ability of the small intestine to digest and absorb nutrients and hence to predispose the piglet to malabsorption and diarrhoea. The question that arises is: to what extent is it possible to make the transition smoother with a proper management and nutrition programme?

Management practices

Weaning is usually associated with mixing of piglets from different litters. Unfamiliar newly reared piglets display aggressive behaviour, leading to the formation of a new social group. The extent to which such behaviour influences the setback at weaning and starting to eat dry food is uncertain. In fact, the duration of fighting does not usually exceed a few hours (Rushen, 1987) and aggressive behaviour is more pronounced in groups of uniform weight than in groups of heterogeneous weight (Rushen, 1987; Francis *et al.*, 1996). In some studies, growth is even superior in groups of unfamiliar piglets or in groups of heterogeneous weight (McGlone *et al.*, 1987; Pluske and Williams, 1996b). It follows that practices aiming at reducing the aggressive behaviour have no marked effect on the weaning setback.

Nutrition programme

Feeding strategies for the weaned piglet, including diet composition, choice of ingredients in relation to their antigenicity and the digestive capacity of the piglet and duration of starter phase, have been reviewed by Hardy (1992), Partridge (1993), Aumaître *et al.* (1995) and Le Dividich *et al.* (1998b). Attention will only be paid to new strategies promoting food intake immediately after weaning. These include the utilization of new animal protein sources and liquid feeding.

PROTEIN SOURCES. Inclusion of dairy products (skimmed milk powder or dry whey) is consistently reported to improve piglet performance (Himmelberg *et al.*, 1985; Tokach *et al.*, 1989; Lepine *et al.*, 1991), mainly through improved diet digestibility. Recently, spray-dried animal plasma (SDPP) has been shown to be an excellent protein source for weanling piglets (Gatnau and Zimmerman, 1991), being superior to dried skimmed milk or soybean protein. During the first 2 weeks after weaning, dietary inclusion of 50–100 g SDPP kg^{-1} improves daily food intake and GR by 30–40% (Gatnau and Zimmerman, 1991; Coffey and Cromwell, 1995). The mechanism by which food intake is increased by SDPP is currently unknown. However, the duration of the response does not usually exceed 2 weeks and the magnitude of the response depends on the piglet health status and environmental conditions, being maximal in continuous-flow nurseries.

LIQUID FEEDING. In changing from liquid to solid food at weaning, the piglet needs to learn separate patterns of feeding and drinking behaviour. It would be anticipated that meeting both food and water at a single delivery point might produce a more regular pattern of feeding more quickly after weaning. Advantages claimed for liquid feeding (fresh or preferably fermented) during the nursing–weaning diet transition include: (i) increased food consumption; (ii) a better control of the gut health (Jensen and Mikkelsen, 1998; Brooks, 1999b); and (iii) minimized villi atrophy (Zijlstra *et al.*, 1996; Pluske *et al.*, 1997). Irrespective of the age at weaning, piglets offered liquid diets during the first week after weaning consumed 75–150% more food (Fig. 2.8) and grew faster than their counterparts offered solid food. Further, relevant studies of Dunshea *et al.* (1999) indicated that feeding liquid diets before and after weaning promoted food intake after weaning. Compared with piglets offered solid food prior to weaning, those offered liquid diets consumed 66% more food during the first week after weaning. However, food intake immediately after weaning is influenced by feed dilution (Table 2.5), being superior in piglets fed a more concentrated diet (250 g l^{-1}) (Geary *et al.*, 1996; Azain, 1997).

Feeding liquid diets during the whole weaning phase improves GR by about 12.3% (Jensen and Mikkelsen, 1998) but, as indicated by studies conducted at the University of Plymouth (Fig. 2.9), most of the advantage is achieved within the first 2 weeks after weaning. In practice, the switch from the liquid diet to solid food takes place gradually during the second week after weaning. Overall, combining the advantages provided by offering supplementary milk during the nursing phase and liquid diets during the nursing–weaning transition enables pigs to be 6.1 kg heavier at 120

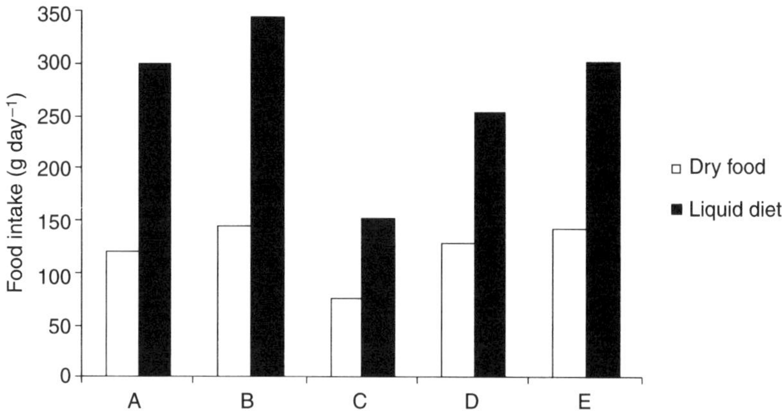

Fig. 2.8. Effect of provision of liquid vs. dry pelleted food on voluntary food intake during the first week after weaning. Piglets were weaned at 7–10 days (A: Azain, 1997), or at 18–23 days (B: Zijlstra *et al.*, 1996; C,D: Dunshea *et al.*, 1999; E: Russel *et al.*, 1996).

Table 2.5. Effect of concentration of the liquid diet on the performance of pigs weaned at 7 days of age.

	Mixing rate (g l^{-1})	
	150	250
Initial bodyweight (kg)	3.4	3.4
Daily weight gain (g)	300	373
Daily food intake (g)	321	400

Duration of the trial, 14 days (Azain, 1997).

days of age than those offered solid food (Dunshea *et al.*, 1997). However, a high level of feeding immediately after weaning is sometimes reported as being a cause of diarrhoea, which is usually attenuated by food restriction (Rantzer *et al.*, 1996). In contrast, an ecopathological approach suggested that high food intake immediately after weaning is a weak predictor of the occurrence of digestive disorders (Madec and Leon, 1999). This indicates the complexity of the early post-weaning period.

Conclusion

The growth potential of the young pig is limited by food intake. In this chapter, examination of the pattern of both energy intake and growth rate and their factors of variation allows the identification of periods of energy and growth deficit. Clearly, the most pronounced periods of deficit are the second half of lactation and the transition from nursing to weaning. However, insufficient intake of colostrum

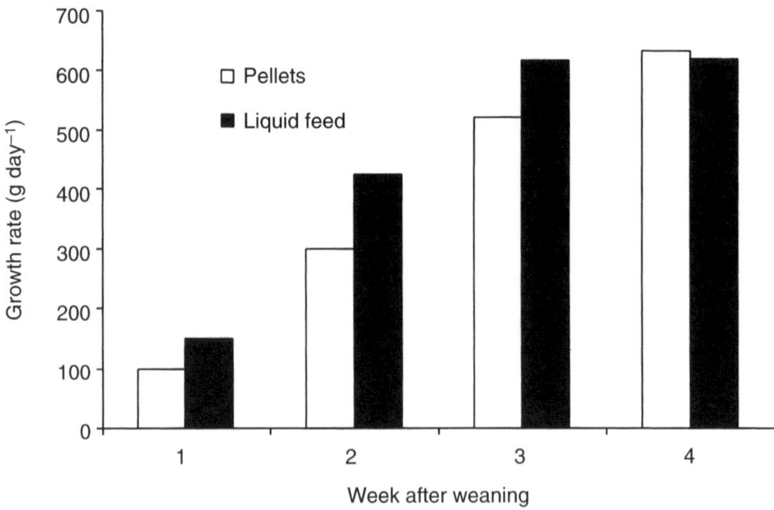

Fig. 2.9. Growth performance recorded in piglets fed dry pellets or liquid diets during the whole post-weaning period (weaning age: 22.6 days). (From Russel *et al.*, 1996.)

and milk in the neonatal period compromises the survival of the piglet and its ability to thrive. Providing supplementary milk to suckling piglets from 10 to 14 days of age and feeding liquid diets during the transition to the post-weaning period appears to be promising in improving piglet performance. The practice needs to be carefully monitored for hygiene reasons and is labour-intensive unless automatic feeding is used. This, and factors initiating and controlling colostrum and milk production by the sow, warrant future extensive research.

References

Allee, G.L. and Hines, R.H. (1972) Influence of fat level and calorie:protein ratio on performance of young pigs. *Journal of Animal Science* 35, 210 (Abstract).

ARC (1981) *The Nutrient Requirement of Pigs.* Commonwealth Agricultural Bureaux, Slough, UK.

Auldist, D.E., Stevenson, F.L., Kerr, M.G., Eason, P. and King, R.H. (1997) Lysine requirements of pigs from 2 to 7 kg live weight. *Animal Science* 65, 501–507.

Auldist, D.E., Morrish, L., Wakeford, C. and King, R.H. (1995) Effect of increased suckling frequency on mammary gland development and milk yield in sows. In: Hennesy, D.P. and Cranwell, P.D. (eds) *Manipulating Pig Production V.* Australasian Pig Science Association, Werribee, Australia, p. 137.

Aumaître, A., Peiniau, J. and Madec, F. (1995) Digestive adaptation after weaning and nutritional consequences in the piglet. *Pig News and Information* 16, 73N–79N.

Azain, M.J. (1993) Impact of starter diet on nursery performance. *Swine Report No. 86.* University of Georgia, pp. 49–54.

Azain, M.J. (1997) Nutrition of the young pig, use of liquid diets. In: *Proceedings of the 13th Annual Carolina Swine Nutrition Conference*, Raleigh, pp. 1–14.

Azain, M.J., Tomkins, T., Sowinski, J.S., Arenston, R.A. and Jewell, D.E. (1996) Effect of supplemental pig milk replacer on litter performance: seasonal variation in response. *Journal of Animal Science* 74, 2195–2202.

Bark, L.J., Crenshaw, T.D. and Leibbrandt, V.D. (1986) The effect of meals intervals and weaning on feed intake of early weaned pigs. *Journal of Animal Science* 62, 1233–1239.

Beyer, M. and Jentsch, W. (1994) Relationship between milk productivity of the sow and growth performance in piglets. In: Souffrant, W.B. and Hagemeister, H. (eds) *Proceedings of the VIIth International Symposium on Digestive Physiology in Pigs*, pp. 226–229.

Boyd, D.R., Kensinger, R.S., Harrell, R. and Bauman, D.E. (1995) Nutrient uptake and endocrine regulation of milk synthesis by mammary tissue of lactating sows. *Journal of Animal Science* 73 (Supplement 2), 36–56.

Brooks, P.H. (1999a) Strategies and methods for the allocation of food and water in the post-weaning period. In: *50th Annual Meeting of the European Association for Animal Production*, Zurich.

Brooks, P.H. (1999b) The potential of liquid feeding systems. In: Lyons, T.P. and Cole, D.J.A. (eds) *Concepts in Pig Science*. Nottingham University Press, Nottingham, UK, pp. 81–98.

Campbell, R.G. (1988) Nutritional constraints to lean tissue accretion in farm animals. *Nutrition Research Reviews* 1, 233–253.

Campbell, R.G., Taverner, M.R. and Mullaney, P.D. (1975) The effect of dietary concentration of digestible energy on the performance and carcass characteristics of early-weaned pigs. *Animal Production* 21, 285–294.

Caugant, A. and Gueblez, R. (1993) Influence of piglet weight at birth on subsequent production traits. *Journées de la Recherche Porcine en France* 25, 123–128.

Cera, R.R., Mahan, D.C. and Reinhart, G.A. (1988) Weekly digestibilities of diets supplemented with corn oil, lard or tallow by weanling swine. *Journal of Animal Science* 66, 1430–1437.

Close, W.H. and Stanier, M.W. (1984a) Effect of plane of nutrition and environmental temperature on the growth and development of the early weaned piglet. 1. Growth and body composition. *Animal Production* 38, 211–220.

Close, W.H. and Stanier, M.W. (1984b) Effect of plane of nutrition and environmental temperature on the growth and development of the early weaned piglet. 2. Energy metabolism. *Animal Production* 38, 221–331.

Coffey, R.D. and Cromwell, G. (1995) The impact of environment and antimicrobial agents on the growth response of early weaned piglets to spray-dried porcine plasma. *Journal of Animal Science* 73, 2532–2539.

Cranwell, P.D., Pierzynowski, S.G., Rippe, C., Pluske, J.R., Power, G.N., Campbell, R.G., Kertan, R.H. and Dunshea, F.R. (1997) Weight and age at weaning influence pancreatic size and enzyme capacity. In: Cranwell, P.D. (ed.) *Manipulating Pig Production VI*. Australasian Pig Science Association, Werribee, Australia, p. 66.

Dell'Orto, V., Savioni, G., Salimei, E. and Navarotto, L. (1992) Automatic distribution of reconstituted milk to pre-weaning piglets: effect on growth performance. *Rivista di Suinicoltura* 33, 61–65.

Dunshea, F.R., Eason, P.J., Morrish, L., Cox, M.C. and King, R.H. (1997) Supplemental milk around weaning can increase live weight at 120 days of age. In: Cranwell, P.D. (ed.) *Manipulating Pig Production VI*. Australasian Pig Science Association, Werribee, Australia, p. 68.

Dunshea, F.R., Kerton, D.J., Eason, P.J. and King, R.H. (1999) Supplemental skimmed milk before and after weaning improves growth performance of pigs. *Australian Journal of Agricultural Research* 50, 1165–1170.

Estienne, M.J., Harper, A.F., Barb, C.R. and Azain, M.J. (2000) Concentrations of leptin in serum and milk from lactating sow differing in body conditions. *Domestic Animal Endocrinology* 19, 4270–4273.

Etienne, M., Dourmad, J.Y. and Noblet, J. (1998) The influence of sow and piglet characteristics and of environmental conditions on milk production. In: Verstegen, M.W.A., Mouhgan, P.J. and Schrama, J.W. (eds) *The Lactating Sow.* Wageningen Pers, Wageningen, pp. 285–299.

Francis, D.A., Chistison, G.I. and Cymbaluk, N.F. (1996) Uniform or heterogeneous weight groups as factors in mixing weanling pigs. *Canadian Journal of Animal Science* 76, 171–176.

Fraser, D., Feddes, J.J.R and Pajor, E.A. (1994) The relationship between creep feeding behavior of piglets and adaptation to weaning: effect of diet quality. *Canadian Journal of Animal Science* 74, 1–6.

Gädeken, D., Oslage, H.J. and Böhme, H. (1985) Energy requirement for maintenance and energy costs of protein and fat deposition in piglets. *Archiv für Tierernährung* 7, 481–494.

Gatnau, R. and Zimmerman, D.R. (1991) Spray dried porcine plasma (SDPP) as a source of protein for weanling pigs. *Journal of Animal Science,* 68 (Supplement 1), 103–104 (Abstract).

Geary, T.N., Brooks, P.T., Morgan, D.T., Campbell, A. and Russel, P.J. (1996) Performance of weaner pigs fed *ad libitum* with liquid feed at different dry matter concentrations. *Journal of the Science of Food and Agriculture* 72, 17–24.

Gerfault, V., Louveau, I., Mourot, J. and Le Dividich, J. (2000) Lipogenic enzyme activity in subcutaneous adipose tissue and skeletal muscle from neonatal pigs consuming maternal or formula milk. *Reproduction, Nutrition and Development* 40, 103–111.

de Goey, L.W. and Ewan, R.C. (1975) Effect of level of feed intake and diet dilution on the energy metabolism of the young pigs. *Journal of Animal Science* 40, 1045–1051.

Halter, H.M., Wenk, C. and Schürch, A. (1980) Effect of feeding level and feed composition on energy utilisation activity and growth performance of piglets. In: Mount, L.E. (ed.) *Energy Metabolism.* Butterworths, London, pp. 395–398.

Hardy, B. (1992) Diets for young pigs. In: Varley, M.A., Williams, P.E.V. and Lawrence, T.L.J. (eds) *Neonatal Survival and Growth.* Occasional Publication No. 15. British Society of Animal Production, Edinburgh, pp. 99–107.

Harrell, R.J., Thomas, M.J. and Boyd, R.D. (1993) Limitations of sow milk yield on baby pig growth. In: *Cornell Nutrition Conference for Feed Manufacturers.* Cornell University, Ithaca, pp. 156–164

Herpin, P. and Le Dividich, J. (1995) Thermoregulation and the environment. In: Varley, M. (ed.) *The Neonatal Pig. Development and Survival.* CAB International, Wallingford, UK, pp. 57–95.

Herpin, P., Le Dividich, J., Hulin, J.C., Fillaut, M., de Marco, F. and Bertin, R. (1996) Effect of level of asphyxia during delivery on viability at birth and early postnatal vitality of newborn pig. *Journal of Animal Science* 74, 2062–2076.

Himmelberg, L.W., Peo, E.R., Lewis, A.J. and Crenshaw, J.D. (1985) Weaning weight response of pigs to simple and complex diets. *Journal of Animal Science* 61, 18–26.

Hodge, R.M.W. (1974) Efficiency of food conversion and body composition of the preruminant lamb and the young pig. *British Journal of Nutrition* 32, 113–126.

Hoy, S. and Puppe, E.B. (1992) Effects of teat order on performance and health in growing pigs. *Pig News and Information* 13, 131N–136N.

Jensen, B.B. and Mikkelsen, L.L. (1998) Feeding liquid diet to pigs. In: Garnsworthy, P.C. and Wiseman, J. (eds) *Recent Advances in Animal Nutrition.* Nottingham University Press, Nottingham, UK, pp. 107–126.

Jentsch, W., Beyer, M. and Hoffman, L. (1992) Energy and nitrogen metabolism in piglets In: *43rd Annual Meeting of European Association for Animal Production*, Madrid.

Jones, G., Edwards, S.A., Traver, S., Jagger, S. and Hoste, S. (1999) Body composition changes in piglets at weaning in response to nutritional modification of sow milk composition and effects on post-weaning performance. In: *50th Annual Meeting of the European Association for Animal Production*, Zurich.

Kavanagh, S., Lynch, P.B., Caffrey, P.J. and Henry, W.D. (1997) The effect of pig weaning weight on post weaning performance and carcass traits. In: Cranwell, P.D. (ed.) *Manipulating Production VI.* Australasian Pig Science Association, Werribee, Australia, p. 71.

Kim, S.W., Hurley, W.L., Han, I.K. and Easter, R.A. (2000) Growth of nursing pigs related to the characteristics of nursed mammary glands. *Journal of Animal Science* 78, 1313–1318.

King, R.H., Mullan, B.P., Dunshea, F.R. and Dove, H. (1997) The influence of piglet body weight on milk production of sow. *Livestock Production Science* 47, 169–174.

King, R.H., Boyce, J.M. and Dunshea, F.R. (1998) Effect of supplemental nutrients on the growth performance of suckling pigs. *Australian Journal of Agricultural Research* 49, 883–887.

King, R., Le Dividich, J. and Dunshea, F.R. (1999) Lactation and neonatal growth. In: Kyriazakis, I. (ed.) *A Quantitative Biology of the Pig.* CAB International, Wallingford, UK, pp. 155–180.

Kornegay, E.T. and Notter, D.R. (1984) Effects of floor space and number of pigs per pen on performance. *Pig News and Information* 5, 23–33.

Le Dividich, J. (1981) Effects of environmental temperature on the growth rates of early-weaned piglets. *Livestock Production Science* 8, 75–86.

Le Dividich, J. and Noblet, J. (1981) Colostrum intake and thermoregulation in the neonatal pig in relation to environmental temperature. *Biology of the Neonate* 40, 167–174.

Le Dividich, J. and Noblet, J. (1986) Effect of dietary energy levels on the performance of individually housed early-weaned piglets in relation to environmental temperature. *Livestock Production Science* 14, 255–263.

Le Dividich, J. and Sève, B. (2000) Effects of underfeeding during the weaning period on growth, metabolism and hormonal adjustments in the piglet. *Domestic Animal Endocrinology* 19, 63–74.

Le Dividich, J., Vermorel, M., Noblet, J., Bouvier, J.C. and Aumaître, A. (1980) Effects of environmental temperature on heat production, energy retention, protein and fat gain in early weaned piglets. *British Journal of Nutrition* 44, 313–323.

Le Dividich, J., Herpin, P. and Rosario-Ludovino, R. (1994). Utilization of colostrum energy by the newborn pig. *Journal of Animal Science* 72, 2082–2089.

Le Dividich, J., Herpin, P., Paul, E. and Strullu, F. (1997) Effect of fat content in colostrum on voluntary colostrum intake and fat utilization in the newborn pig. *Journal of Animal Science* 75, 707–712.

Le Dividich, J., Noblet, J., Herpin, P., van Milgen, J. and Quiniou, N. (1998a) Thermoregulation. In: Wiseman, J., Varley, M.A. and Charlick, J.P. (eds) *Progress in Pig Science.* Nottingham University Press, Nottingham, UK, pp. 229–263.

Le Dividich, J., Tivey, D. and Aumaître, A. (1998b) Gastro-intestinal development and digestive capacity in the young pig. In: Done, S., Thomson, J. and Varley, M. (eds)

Proceedings of the 15th IPVS Congress. Nottingham University Press, Nottingham, UK, pp. 299–308.

Le Huerou-Luron, I., Lafuente, M.J., Thomas, F., Rome, V. and Le Dividich, J. (1999) Effect of milk intake on the development of the digestive function in piglets during the first postnatal week. In: *50th Annual Meeting of the European Association for Animal Production*, Zurich.

Lefaucheur, L., Ecolan, P., Barzic, Y.M. and Le Dividich, J. (2000) Influence of early postnatal nutrition on myofiber differentiation. In: *Proceedings of the International Congress on Myology*, Nice, France, p. 207.

Leibbrandt, V.D., Ewan, R.C., Speer, V.C. and Zimmerman, D. (1975a) Effect of age and calorie:protein ratio on performance and body composition of baby pigs. *Journal of Animal Science* 40, 1070–1076.

Leibbrandt, V.D., Ewan, R.C., Speer, V.C. and Zimmerman, D. (1975b) Effect of age at weaning on baby pigs' performance. *Journal of Animal Science* 40, 1077–1080.

Lepine, A.J., Mahan, M.D. and Chung, J.K. (1991) Growth performance of weanling pigs fed corn–soybean meal diets with or without dried whey at various L-Lysine-HCl levels. *Journal of Animal Science* 69, 2026–2032.

Lindberg, J.E., Neil, M. and Cidh, M.A. (1997) Effect of ad libitum access to milk replacer to piglets on performance of piglets, slaughter pigs and sows. *Proceedings of the British Society of Animal Science* p. 58.

van Lunen, T.A. and Cole, D.J.A. (1998) The effect of dietary energy concentration and lysine/digestible energy ratio on growth performance and nitrogen deposition of young hybrid pigs. *Animal Science* 67, 117–129.

Madec, F. and Leon, E. (1999) The role of management and husbandry in pig health with emphasis on post weaning enteric disorders. In: Cranwell, P.D. (ed.) *Manipulating Pig Production VII.* Australasian Pig Science Association, Werribee, Australia, pp. 200–209.

Mahan, D.C. (1993) Effect of weight, split-weaning, and nursery feeding programs on performance responses of pigs to 105 kilograms body weight and subsequent effects on sow breeding interval. *Journal of Animal Science* 71, 1991–1995.

Mahan, D.C. and Lepine, A.J. (1991) Effect of pig weaning weight and associated nursery feeding program on subsequent performance to 105 kilograms body weight. *Journal of Animal Science* 69, 1370–1378.

Mahan, D.C., Cromwell, G.L., Ewan, R.C., Hamilton, C.R. and Yen, J.T. (1998) Evaluation of the feeding duration of a phase 1 nursery diet to three-week-old pigs and two weaning weights. *Journal of Animal Science* 76, 578–583.

Marion, J. and Le Dividich, J. (1999) Utilization of sow milk energy by the piglet. In: Cranwell P.D. (ed.) *Manipulating Production VII.* Australasian Pig Science Association, Werribee, Australia, p. 254.

Marion, J., Sève, B., Ganier, P., Thibault, J.N. and Le Dividich, J. (1999) The effect of sow milk intake on whole body protein turn-over and its contribution to heat production in the neonatal pig. *Proceedings of the VIIth International Symposium on Protein Metabolism and Nutrition*, Aberdeen, p. 10.

McConnell, J.C., Stuck, M.W., Waldorf, R.C., Byrd, W.P. and Grimes, L.W. (1982) Caloric requirements of early-weaned pigs fed corn–soybean meal-based diets. *Journal of Animal Science* 55, 841–847.

McCracken, K.J. (1989) Post-weaning voluntary food intake of pigs weaned at 2 or 4 weeks of age. In: Forbes, J.M., Varley, M. and Lawrence, T.L.J. (eds) *The Voluntary Food Intake of Pigs.* Occasional Publication No. 13. British Society of Animal Production, pp.101–102.

McCracken, K.J., Eddie, S.M. and Stevenson, W.G. (1980) Energy and protein nutrition of early-weaned piglets. I. Effect of energy intake and energy:protein on growth, efficiency and nitrogen utilization of pigs between 8 and 32 days. *British Journal of Nutrition* 43, 289–304.

McGlone, J.J., Stanbury, W.F. and Tribble, L.F. (1987) Effect of heat and social stressors and within-pen weight variation on pig performance. *Journal of Animal Science* 65, 456–462.

McGlone, J.J., Stansbury, W.F. and Tribble, L.F. (1988) Management of lactating sows during heat stress: effects of water drip, snout coolers, floor type and a high energy-density diet. *Journal of Animal Science* 66, 885–889.

Meunier-Salaün, M.C., Monnier, M., Colléaux, Y., Sève, B. and Henry, Y. (1991) Impact of dietary tryptophan and behavioral type on behavior, plasma cortisol, and brain metabolites of young pigs. *Journal of Animal Science* 69, 3689–3698.

Miller, H.M., Toplis, P. and Slade, R.D. (1999) Weaning weight and daily live weight gain in the week after weaning predict piglet performance. In: Cranwell, P.D. (ed.) *Manipulating Production VII.* Australasian Pig Science Association, Werribee, Australia, p. 130.

Nam, D.S., and Aherne, F.X. (1994) The effects of lysine:energy ratio on the performance of early-weaned pigs. *Journal of Animal Science* 72, 1247–1256.

National Research Council (1998) *Nutrient Requirements of Swine*, 10th edn. National Academy Press, Washington, DC.

Nienaber, J.A., Hahn, G.L. and Shanklin, M.D. (1982) Modelling performance on nursery-age pigs with thermal and nutritional inputs. *Proceedings of the Second International Livestock Environment Symposium.* American Society of Agricultural Engineers, St Joseph, Michigan, pp. 303–310.

Noblet, J. (1997) Digestive and metabolic utilization of dietary energy of pigs' feeds. Comparison of energy systems. In: Garnsworthy, P.C., Wiseman, J. and Haresign, W. (eds) *Recent Advances in Animal Nutrition.* Nottingham University Press, Nottingham, UK, pp. 207–231.

Noblet, J. and Etienne, M. (1986) Effect of energy level in lactating sows on yield and composition of milk and nutrient balance of piglets. *Journal of Animal Science* 63,1888–1896.

Noblet, J. and Etienne, M. (1987) Body composition, metabolic rate and utilization of milk nutrients in suckling piglets. *Reproduction Nutrition Development* 27, 829–839.

Noblet, J. and Etienne, M. (1989) Estimation of sow milk nutrient output. *Journal of Animal Science* 67, 3352–3359.

Noblet, J. and Le Dividich, J. (1982) Effect of environmental temperature and feeding level on energy balance traits of early weaned piglets. *Livestock Production Science* 9, 619–632.

Ogunbameru, R.O., Kornegay, E.T. and Wood, C.M. (1992) Effect of evening or morning weaning and immediate or delayed feeding on postweaning performance of pigs. *Journal of Animal Science* 70, 337–342.

Orgeur, P., Saläun, C., Le Rouix, T., Venturi, E. and Le Dividich, J. (2000) Fostering and very early weaning: a strategy for rearing surplus piglets. *Journées de la Recherche Porcine en France* 32, 143–149.

Owsley, W.F., Orr, D.E. and Tribble, L.F. (1986) Effects of nitrogen and energy source on nutrient digestibility in the young pig. *Journal of Animal Science* 63, 492–496.

Partridge, G.G. (1993) New approaches with pig weaner. In: Garnsworthy, P.C. and Cole, D.J.A. (eds) *Recent Advances in Animal Nutrition.* Nottinhgam University Press, Nottingham, UK, pp. 221–248.

de Passillé, A.M.B., Pelletier, G., Menard, J. and Morrisset, J. (1989) Relationship of weight gain and behaviour to digestive organ weight and enzyme activity in piglets. *Journal of Animal Science* 67, 2921–2929.

Pluske, J.R. and Williams, I.H. (1996a) Split weaning increases the growth of light piglets during lactation. *Australian Journal of Agricultural Research* 47, 513–523.

Pluske, J.R. and Williams, I.H. (1996b) The influence of feeder type and the method of allocation at weaning on voluntary feed intake and growth of piglets. *Animal Science* 62, 115–120.

Pluske, J.R., Williams, I.H. and Aherne, F.X. (1995) Nutrition of the neonatal pig. In: Varley, M.A. (ed.) *The Neonatal Pig. Development and Survival.* CAB International, Wallingford, UK, pp. 187–235.

Pluske, J.R., Hampson, D. and Williams, I.H. (1997). Factors influencing the structure and function of the small intestine in the weaned pig: a review. *Livestock Production Science* 51, 215–236.

Pluske, J.R., Williams, I.H., Zak, L.J., Clowes, E.J., Cegielski, A.C. and Aherne, F.X. (1998) Feeding lactating primiparous sows to establish three divergent metabolic states. III. Milk production and pig growth. *Journal of Animal Science* 76, 1165–1171.

Prunier, A., Messias de Bragança, M. and Le Dividich, J. (1997) Influence of high ambient temperature on the performance of reproductive sows. *Livestock Production Science* 52, 123–133.

Rantzer, D., Svendsen, J. and Weström, B. (1995) Effects of strategies restriction on pig performance and health during the post-weaning period. *Acta Agricultura Scandinavica* 46, 219–225.

Rinaldo, D. (1989) Influence de la température ambiante sur le métabolisme énergétique et tissulaire et le besoin en lysine du porc en croissance: mise en évidence de l'intérêt d'une température élevée. Thèse de l'Université de Rennes I, Rennes, France.

Rinaldo D. and Le Dividich, J. (1991) Assessment of optimal temperature for performance and chemical body composition of the growing pig. *Livestock Production Science* 29, 61–75.

Rousseau, P., Chatelier, C., Dutertre, C. and Léveque, J.C. (1994) Heating systems for piglets nests: comparison between IR lamp and electrically heated floor. Piglets performance, behaviour and energy cost. *Journées de la Recherche Porcine en France* 25, 47–54.

Rushen, J. (1987) A difference in weight reduces fighting when unacquainted newly weaned pigs meet first. *Canadian Journal of Animal Science* 67, 951–960.

Russel, P.J., Geary, T.M., Brooks, P.H. and Campbell, A. (1996) Performance, water use and effluent output of weaner fed ad libitum with either dry pellets or liquid fee and the role of microbial activity in the liquid feed. *Journal of Science and Food Agriculture* 72, 87–94.

Salmon Legagneur, E. and Aumaître, A. (1962) Influence de la quantité de lait et de sa composition sur la croissance du porcelet sous la mère. *Annales de Zootechnie* 11, 181–196.

Sarkar, N.K., Lodge, G.A. and Elliot, J.I. (1981) Effect of a skim milk supplement on growth and body composition of the suckled pigs. *Canadian Journal of Animal Science* 61, 507–509.

Sauber, T.E., Stahly, T.S., Ewan, R.E. and Williams, N.H. (1995) Maximum lactational capacity of sows with a moderate and low genetic capacity for lean tissue growth. *Iowa State University, Swine Research Report,* pp. 60–62.

Schoenherr, W.D., Stalhy, T.S. and Cromwell, G.L. (1989) The effects of dietary fat or fiber addition on yield and composition of milk sows housed in a warm or hot environment. *Journal of Animal Science* 67, 482–495.

Sève, B. (1984) Effets à long terme de la réduction du taux de protéines dans une ration simplifiée de sevrage. *Journées de la Recherche Porcine en France* 16, 347–354.

Sève, B. (1985) Physiological basis of nutrients supply to piglets. *World Review of Animal Production* 26, 8–14.

Sève, B. and Aumaître, A. (1983) Intérêt des protéines de sérum dans les régimes d'allaitement artificiel du porcelet. *Science des Aliments* 3, 53–67.

Sève, B. and Ballèvre, O. (1991) Approches métaboliques du besoin en acides aminés chez le porc en croissance. *Journées de la Recherche Porcine en France* 23, 91–110.

Sève, B. and Bonneau, M. (1986) Long-term effects of early weaning of piglets and initial live weight on performance and body composition. Consequences on the development of the male genital tract. *Journées de la Recherche Porcine en France* 18, 143–154.

Sève, B., Pérez, J.M. and Desmoulin, B. (1985) Effet à long terme du niveau d'alimentation des porcelets entre 10 et 25 kg de poids vif sur les performances et la composition corporelle à l'abattage. *Journées de la Recherche Porcine en France* 17, 419–432.

Sève, B., Reeds, P.J., Fuller, M.F., Cadenhead, A. and Hay, S.M. (1986) Protein synthesis and retention in some tissues of the young pig as influenced by dietary protein intake after early weaning. *Reproduction Nutrition Development* 26, 849–861.

Sève, B., Meunier-Salaün, M.C., Monnier, M., Colléaux, Y. and Henry, Y. (1991) Impact of dietary tryptophan and behavioral type on growth performance and plasma amino acids of growing pigs. *Journal of Animal Science* 69, 3679–3688.

Sève, B., Sawadogo, M., Schaeffer, V., Dufour-Etienne, F. and Bercovici, D. (1999) Interaction entre les apports alimentaires de thréonine et de protéines sur le rendement d'utilisation des acides aminés pour la croissance chez le porcelet au sevrage. *Journées de la Recherche Porcine en France* 31, 267–274.

Shurson, G.C., Hogberg, M.G., Defever, N., Radecki, S.V. and Miller, E.R. (1986) Effect of adding fat to the sow lactation diet on lactation and rebreeding performance. *Journal of Animal Science* 62, 672–680.

Sloat, D.A., Mahan, D.C. and Roehrig, K.L. (1985) Effect of pig weaning weight on post-weaning body composition. *Nutrition Reports International* 31, 627–634.

Stahly, T.S., Cromwell, G.L. and Simpson, W.S. (1980) Effects of level and source of supplmental fat in the lactation diet of sows on the performance of pigs from birth to market weight. *Journal of Animal Science* 51, 352–360.

Stansbury, W.F., McGlone, J.J. and Tribble, L.F. (1987) Effects of season, floor type, air temperature and snout cooler on sow and litter performance. *Journal of Animal Science* 65, 1507–1513.

Stibic J. (1982) Studies on efficiency in piglets as related to nitrogen metabolism. *Zeitschrift für Tierphysiologie, Tierernährung und Futtermittelkunde* 47, 134–148.

Thompson, B.K. and Fraser, D. (1986) Variation in piglets' weights: development of within-litter variation over a 5-week lactation and effect of farrowing crate design. *Canadian Journal of Animal Science* 66, 361–372.

Thompson, B.K. and Fraser, D. (1988) Variation in piglets' weights: weight gains in the first days after birth and their relationship with later performance. *Canadian Journal of Animal Science* 68, 581–590.

Tokach, M.D., Nelssen, J.L. and Allee, G.L. (1989) Effect of protein and (or) carbohydrate fractions of dried whey on performance and nutrient digestibility of early weaned pigs. *Journal of Animal Science* 67, 1307–1312.

Tokach, M.D., Goodbank, R.D., Nelssen, J.L. and Kats, L.J. (1992) Influence of weaning weight and growth during the first week post-weaning on subsequent pig performance. In: *Kansas University Swine Day.* Report of Progress No. 667, pp. 15–17.

Tokach, M.D., Pettigrew, J.E., Johnston, L.J., Overland, M., Rust, M.J.W. and Cornelius, S.G. (1995) Effect of adding fat and (or) milk products to the weanling diet on performance in the nursery and subsequent grow-finishing stages. *Journal of Animal Science* 73, 3358–3368.

Verstegen M.W.A., Moughan, P.J. and Schrama, J.W. (1998) *The Lactating Sow*, 1st edn. Wageningen Pers, Wageningen, 350 pp.

Whittemore, C.T., Aumaître, A. and Williams, I.H. (1978) Growth of body components in young weaned pig. *Journal of Agricultural Science (Cambridge)* 91, 681–692.

Whittemore, C.T., Taylor, H.M., Henderson, R., Wood, J.D. and Brack, D.C. (1981) Chemical and dissected composition changes in weaned pig. *Animal Production* 32, 203–210.

Williams, I.H. (1976) Nutrition of the young pig in relation to body composition. PhD Thesis, The University of Melbourne, Australia.

Yang, H., Kerber, J.A., Pettigrew, J.E., Johnston, L.J. and Walker, R.D. (1997) Evaluation of milk chocolate product as a substitute for whey in the pig starter diets. *Journal of Animal Science* 75, 423–429.

Zhang, Y., Partridge, I.G. and Mitchell, K.J. (1986) The effect of dietary energy level and protein:energy ratio on nitrogen and energy balance, performance and carcass composition of pigs weaned at 3 weeks of age. *Animal Production* 42, 389–395.

Zijlstra, R.T., Whang, K.Y., Easter, R.A. and Odle, J. (1996) Effect of feeding a milk replacer to early weaned pigs on growth, body composition, and small intestine morphology, compared with suckled littermates. *Journal of Animal Science* 74, 2948–2959.

Protein and Amino Acid Requirements of Weaner Pigs

3

M. Cole and M. Sprent

SCA Nutrition Ltd, Maple Mill, Dalton, Thirsk, North Yorkshire Y07 3HE, UK

Introduction

The pre-weaned pig is capable of growing in excess of 300 g a day, but all too often this growth rate is reduced in the weeks following weaning. The digestive capabilities of piglets are still developing at the time of weaning as diets change from one based almost entirely on milk proteins to a more complex one. This has consequences on the lactic acid content in the gut and thus on the bacterial flora, villus height and gut health, often resulting in the impairment of both pig health and growth performance.

In order to ensure that this transitional phase through to 25 kg liveweight is as trouble-free as possible, up to five diets or more may be fed. The raw material composition of these alters as high quality proteins are phased out and cheaper vegetable proteins are introduced. In addition, the nutrient concentration of the diets will diminish as appetite increases.

When formulating diets for pigs of greater than 25 kg liveweight, a nutritionist will generally consider protein (traditionally defined as crude protein, CP) simply as the vehicle to deliver either essential amino acids, or nitrogen from which non-essential amino acids can be synthesized.

In addition to specifying minimum and maximum protein constraints, limits for potentially first-limiting amino acids – lysine, methionine (normally considered with cysteine), threonine and possibly tryptophan – will be defined. Some measure of the availability of the amino acids, such as apparent ileal or true ileal digestibility, may be made. The means of providing this protein will to a large extent be given little consideration apart from limiting the maximum inclusion of certain raw materials that contain performance-limiting anti-nutritional factors, such as rapeseed meal, legumes and soya.

The protein requirement of the weaner pig has to be considered somewhat differently. Raw materials used in diets designed for this age range supply not only nitrogen and essential amino acids but also other proteins that have a secondary role not related to the supply of nutrients. Immunoglobulins, whose inclusion in feed is more to do with their functional properties than their direct nutritional value, would be included in this category.

Published levels for the requirements of amino acids and those used in commercial practice are often somewhat different, particularly for the weaned pig. This chapter will consider both aspects, along with the levels and types of raw materials used to supply protein to the weaned pig.

It is worth noting that formulations for pre-starter and starter feeds are characterized around the world by the use of certain raw materials. Most American nutritionists would include plasma proteins in diets; those in the UK would use cooked cereals and skimmed milk powders; and those in continental Europe would use whey proteins. It is interesting to note that all nutritionists from the regions mentioned are confident that their approach is acceptable.

Amino Acid Levels of Finished Feed

In order to meet the amino acid requirements of the weaner pig, a two-stage approach is adopted to achieve the ideal protein. The requirement for the first-limiting amino acid is calculated and then related to the balance of other essential amino acids to achieve the ideal protein. Relatively little work has been conducted in this area with respect to the weaner pig, compared with the growing and finishing pig.

The National Research Council (NRC) Nutrient Requirements of Swine (1998) included a comprehensive review of data published since 1985 to ascertain the lysine requirement of growing pigs. The published values varied by as much as 30% for the same weight class of pigs. In all the studies conducted between 3 and 35 kg liveweight, the maximum lysine requirement was determined as only 14.9 g kg^{-1}, generally being between 11 and 13 g kg^{-1}.

From a review of literature evaluating the optimum lysine/digestible energy (DE) ratio (g MJ^{-1}), Close (1994) reported requirements ranging from a low of 0.44 to a high of 0.82, for pigs between 20 and 50 kg liveweight. Variation in reported values may in part be a result of poor experimentation. For example, lysine was not measured in the diet, the recommended level was the highest level used or lysine was not the first-limiting amino acid.

Data for the weaned pig are somewhat scarce but van Lunen and Cole (1996) suggested that the requirement for hybrid pigs could be as high as 1.2 g lysine MJ^{-1} DE (Table 3.1) and identified three major factors affecting this ratio: genotype, sex and liveweight.

All producers wish to rear fast-growing pigs and van Lunen and Cole (1996) concluded that the optimum lysine/DE ratio for fast-growing pigs of 20 kg was 1.20. This figure is somewhat higher than the value of 0.84 quoted by the Agricultural Research Council (ARC) (1981) and that of 0.99 g lysine MJ^{-1} DE proposed by

Zhang *et al.* (1984). This may be attributed to van Lunen and Cole (1996) using highly selected genotypes not currently available at the time of the work summarized by ARC (1981) or Zhang *et al.* (1984) for experimentation or commercial production, illustrating the need to identify the genotype being used in the evaluation.

In Northern Europe, diets for pigs up to 30 kg liveweight are typically formulated with a lysine to energy ratio of 1.0. Nutritionists have interpreted the scientific data in different ways. This is apparent if the absolute levels of lysine in the diets in Northern Europe and the USA (Tokach *et al.*, 1997) are compared with those published by review bodies, namely the National Research Council (NRC) (1998) and Institut National de la Recherche Agronomique (INRA) (1984) (Table 3.2). Research values tend to be lower than those used commercially, with the UK generally feeding higher levels than the USA. This is despite the fact that very similar genotypes are being fed worldwide.

Table 3.1. Optimum lysine/digestible energy (DE) (g MJ^{-1}) ratios in the diets of growing pigs (van Lunen and Cole, 1996).

Genotype	Protein deposition rate (g day^{-1})	Weight (kg)	Lysine/DE ratio Castrate	Gilt	Boar
Unimproved	100	< 25	0.78	0.80	0.83
		25–55	0.73	0.75	0.78
Average	125	< 25	0.85	0.85	0.88
		25–55	0.78	0.80	0.83
High	150	< 25	0.88	0.90	0.93
		25–55	0.83	0.85	0.88
Hybrid	175	< 25	1.20	1.20	1.20
		55–55	1.10	1.10	1.10

Table 3.2. Summary of lysine requirements of different classes of weaned pig.

Diet type	Weight band (kg)	Total lysine (g kg^{-1})	Source[a]
Pre-starter	< 5	17–18	Tokach *et al.* (1997)
	3–5	15	NRC (1998)
	< 5	16.5–17.5	UK commercial[a]
Starter	5–7	15–16	Tokach *et al.* (1997)
	5–10	13.5	NRC (1998)
	5–10	14	INRA (1984)
	5–10	16–16.5	UK commercial[a]
Link	7–11.5	13.5–14.5	Tokach *et al.* (1997)
	7–12	15–16	UK commercial[a]
Early grower	11.5–23	12.5–13.5	Tokach *et al.* (1997)
	10–20	11.5	NRC (1998)
	10–25	11	INRA (1984)
	10–25	15	UK commercial[a]
Grower	15–25	10.5	NRC (1998)
	12–25	14–15	UK commercial[a]

[a] UK commercial based on data from SCA.

Published tables exist detailing the amino acid availability, be it apparent or true, ileal or faecal, in many of the commonly used feedstuffs. These have normally been evaluated using pigs over 20 kg, so data for raw materials commonly used in weaner pig diets are lacking. The use of a measure of amino acid availability is a desirable approach but if it is achieved by applying a factor to total values, or if sound feed ingredient values are absent, it is probably of little additional use.

The Balance of Amino Acids

Published balances of amino acids relative to lysine have been well reviewed (van Lunen and Cole, 1996) (Table 3.3). Since ARC (1981), the change that has affected commercial diets the most has been the increase in levels of threonine (usually the second-limiting amino acid in pig feeds) relative to lysine. Fuller *et al.* (1987) suggested a level of 72%, having separated out requirements for maintenance and growth and feeding semi-purified diets. Work by D.J.A. Cole and L. Bong (1989, personal communication) has found the value to be between 65 and 70%. Nutritionists must be cautious in extrapolating the plentiful data generated with older pigs to the newly weaned piglet with its developing digestive system, where data are scant.

Lysine, methionine, threonine and tryptophan are all available as synthetic amino acid sources, with markedly different costs (Table 3.4). With the exception of tryptophan, where cost excludes it from many commercial diets, these amino acids are commonly used as supplements in weaner diets to achieve the ideal balance.

It is unusual to find commercial diets with tryptophan at a level greater than 17% of lysine despite the fact that tryptophan, being a precursor for serotonin, is known to play a role in stimulating feed intake. Recent studies at Moorepark (Forum Feeds, 2000, personal communication) demonstrated that pigs between an average of 8 and 25 kg responded to an increasing level of dietary tryptophan up to 21% of lysine.

Dietary Sources of Protein used in Commercial Post-weaning Diets

Having established the amino acid requirements for weaner feeds, the raw materials used to supply these needs must be considered. Selection of raw material is the key to successful weaner feeding. The newly weaned pig faces an appetite barrier as it makes the transition from liquid to solid feed. Diet palatability and digestibility are crucial factors in overcoming this.

Protein sources commonly used in weaner feeds can be categorized in a number of different ways (Table 3.5). For the purpose of this chapter, they have been broken down into five sections: milk proteins; crystalline amino acids; vegetable protein sources; animal protein sources; and other materials, such as cereals, primarily included in diets for attributes other than their protein content.

From weaning onwards the inclusion of high quality proteins such as fishmeal, milk powders and porcine plasma protein gradually reduces as the levels of soybean meal and other cheaper vegetable proteins increases. Concerns that soya and other vegetable proteins trigger allergic reactions in the young pig limit their inclusion in favour of higher quality, more expensive proteins. Some vegetable protein sources are also thought to be associated with palatability problems.

Typical weaner diets fed in the USA and in the UK from weaning for 7 to 10 days are shown in Table 3.6, which also presents the protein supplied by each raw material. From this it can be seen that the contribution of the cereal fraction to the overall diet should not be underestimated.

One of the fundamental differences between formulations in Europe and the USA is the greater use of processed cereals in the former region. Cereals are cooked primarily to increase the digestibility (gelatinization) of the starch fraction but cooking will also increase the overall digestibility of nutrients in the feeds. The earlier introduction of soybean meal (465 g kg^{-1} crude protein) into US diets will further lower the digestibility of the weaner feeds. This point is worth noting, since the other components of the feeds are markedly different in the diets illustrated in Table 3.6.

Traditionally, weaner feeds in Europe have been formulated using fishmeal and milk proteins as the main protein sources. An abundance of milk powders, be it as skimmed milk powder or some type of whey powder, has led to their widespread use. This has been helped by the payment of subsidies for the denaturation of milk prior to incorporation into animal feed in order that it is removed from the human food chain.

Table 3.3. The ideal amino acid balance of commonly supplemented amino acids in weaner feeds.

Source	Lysine	Methionine and cystine	Threonine	Tryptophan
Cole (1978)	100	50	60	18
Fuller *et al.* (1979)	100	53	56	12
ARC (1981)	100	50	50	15
Fuller *et al.* (1987)				
Maintenance	100	150	142	29
Growth	100	53	69	18
Both	100	56	72	19
van Lunen and Cole (1996)	100	50–55	65–57	18

Table 3.4. The cost (£ sterling, typical prices August 2000) and typical inclusion rates of synthetic amino acids in starter feeds.

Costs and rates	Lysine	Methionine	Threonine	Tryptophan
Cost per tonne	1,000	1,800	2,100	25,000
Typical inclusion rates in weaner feeds (kg t^{-1})	2–5	0.5–2	0.5–1.5	0–1
Cost per 1 g inclusion	1.00	1.80	2.10	25.00

Table 3.5. The cost (£ sterling, approximate prices delivered to mills in UK, August 2000) and protein content of raw materials commonly used in starter diets.

Group and product	Protein (g kg $^{-1}$)	Total lysine (g kg^{-1})	Price per tonne (£)	Typical rate of inclusion (kg t^{-1}) Pre-starter	Starter
Milk proteins					
Skimmed milk	350	29	1600	0–300	0–250
Buttermilk	320	27	1350	0–300	0–250
Whey powder	130	9	330	0–150	0–100
Delactosed whey	230	15	400	0–150	0–100
Whey protein concentrate	350	32.5	950	0–300	0–200
Casein	840	67	3500	0–200	0–100
Vegetable protein					
Hipro soya	475	31.5	175	0	0–50
Full-fat soybeans	350	23.5	150	0–150	0–200
Soya protein concentrate	650	62	550	0–75	0–75
Potato protein	780	62.5	600	0–75	0–75
Soya flour	415	27	190	0–75	0–75
Wheat gluten	800	18	975	0–50	0–25
Animal protein					
Meat and bone	450	21	75	0	0–50
Fishmeal (herring)	710	56	350	75–150	75–150
Egg protein	450–800	30–70	800–1800	75–150	75–150
Bloodmeal	880	85	350	75–150	75–150
Plasma protein	77	61	2200	0–60	0–30
Others					
Wheat	110	3.2	95	0–400	0–400
Barley	105	3.5	90	0–400	0–400
Maize	90	2.5	140	0–400	0–400
Wheatfeed	150	6	55	0	0–50
Porridge oats	120	4.5	155	0–300	0–200

In the USA the availability of large quantities of soya proteins, at prices considerably lower than milk proteins, has seen these incorporated into starter feeds at higher levels. Various processing techniques have been adopted to raise the protein levels of such materials and to remove anti-nutritional factors, such as the storage proteins glycine and β-conglycine.

In recent years new raw materials, such as plasma proteins, have become available and are used extensively in weaner feeds in most countries worldwide. This is not the case in the UK, where the use of animal proteins, excluding milk products, is limited to fishmeal. In the aftermath of the BSE crisis the use of meat and bonemeal has been prohibited in livestock feeds, in order to avoid the possibility of feeding material of the same species back to stock (so-called closed loop feeding).

A peculiar exception to this rule is the use of plasma protein which, whilst obviously derived from pigs, is not prohibited. However, so concerned are most feed companies about its acceptance by the general public within the UK that many do not allow the material on their premises.

Digestibility of Raw Material

Makkink (1993) reviewed the digestibility of different protein sources in various trials and concluded that, in the immediate post-weaning period, the digestibility of milk protein was superior to that of soya, processed in a number of ways, and indeed fishmeal. These findings were confirmed by further work using 15 N-isotope techniques, in which piglets between 35 and 50 days of age were fed diets containing one of the following: skimmed milk powder (SMP) (350 g CP kg^{-1}), soybean meal (526 g CP kg^{-1}), soya protein isolate (895 g CP kg^{-1}) and fishmeal (758 g CP kg^{-1}). The digestibility of the feed was measured as either apparent or true over the whole tract or to the ileum (Fig. 3.1). In addition, levels of endogenous nitrogen loss were recorded at the ileum and over the whole tract (Fig. 3.2).

There were significant differences in the apparent digestibility of the different materials, with SMP being superior to the other three in both instances. Smaller differences were seen with the true N-digestibility. Makkink (1993) concluded that the differences seen in pig performance (growth and food conversion efficiency) when pigs were fed vegetable proteins were as a result of increases in endogenous losses rather than differences in digestibility, compared with SMP. It was thought that the soya protein either stimulates nitrogen secretion by the exocrine glands of the digestive tract or causes excessive loss of gut wall cells by sloughing.

The remainder of this chapter discusses the merits of various proteins used in weaning feeds. Emphasis is given to materials that may be associated with poor properties other than their nutrient content *per se*.

Table 3.6. The difference between US (Kansas State University) and UK commercial weaner feeds (g kg^{-1} diet).

Raw material (RM)		Pre-starter (17 g lysine kg^{-1})				Starter (16 g lysine kg^{-1})			
		UK		USA		UK		USA	
Feedstuff	Crude protein (CP) (g kg^{-1})	RM	CP	RM	CP	RM	CP	RM	CP
Maize	85			335	28.5			400	34
Cooked cereals	85–120	400	42.8			460	46.2		
Fishmeal	710	80	56.8	60	42.6	80	56.8	25	17.8
Hipro soya	456			125	57.0			232.5	106
Full-fat soya	355	150	53.3			175	62.1		
Skimmed milk	370	150	55.5			50	18.5		
Dried whey	130	125	16.3	250	32.5	200	26	200	26
Plasma	780			67	52.3			25	19.5
Bloodmeal	878			16.5	14.5			25	22

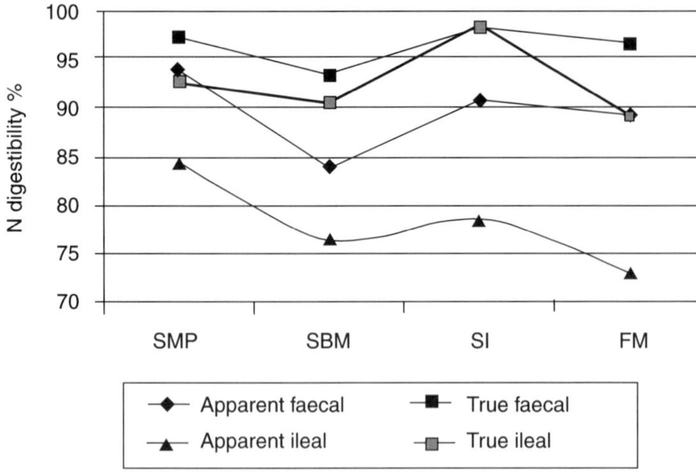

Fig. 3.1. Apparent and true ileal and faecal digestibility of raw materials (after Makkink, 1993). SMP, skimmed milk powder; SBM, soybean meal; SI, soya protein isolate; FM, fishmeal.

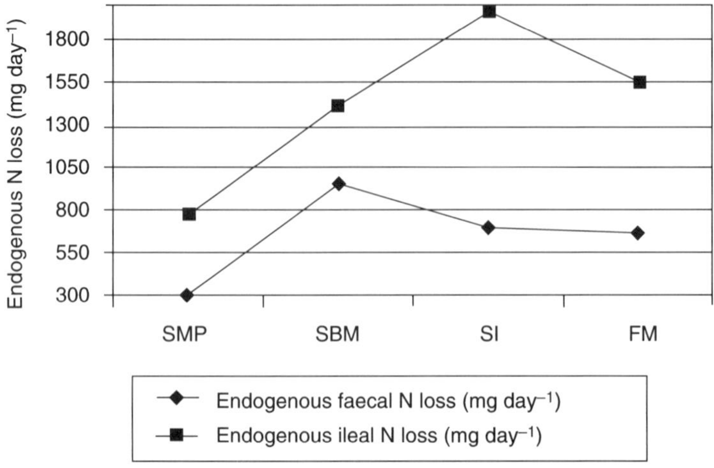

Fig. 3.2. Endogenous faecal and ileal N losses (mg day^{-1}) (after Makkink, 1993). SMP, skimmed milk powder; SBM, soybean meal; SI, soya protein isolate; FM, fishmeal.

Milk Products

It should be remembered that, besides supplying protein, milk products contain other components believed to be equally important in feeding the newly weaned pig, most notably lactose. There is debate as to whether responses to milk products are caused by lactose or not. Patterson (1987) found that growth performance deteriorated significantly when soya or fish meals replaced the protein element of skimmed milk powder, whereas lactose replacement with glucose or starch was satisfactory. Tokach *et al.* (1989) placed equal importance on the protein (lactalbumin) and lactose components of whey when considering improved performance.

As the pig is weaned it is deprived of sow milk, a diet in which the protein fraction is rich in casein relative to whey proteins (1.0:0.61) (Elsley, 1970). Milk products also contain non-protein material that is believed to be important in the diet of newborn piglets. Milk contains both non-protein nitrogen and true proteins. True proteins are divided into casein and whey proteins. Casein can be defined as the solids that precipitate at pH 4–5, and the whey fraction is that which remains in solution at the same pH. It is the casein fraction that is particularly rich in amino acids, whereas whey contains many beneficial factors that have a role other than simply providing a source of nutrients. These include immunoglobulins, growth factors, lactoferrin and lactoperoxidase (Table 3.7).

Europe has traditionally had a plentiful supply of milk products, mostly originating as by-products of industries such as cheese manufacture as well as from surpluses of liquid milk production. Milk can be separated to produce skimmed milk powder or used for the manufacture of cheese, resulting in the production of whey powder (Fig. 3.3).

Skimmed milk powder

The payment of an EU subsidy for the denaturation of skimmed milk powder for incorporation into animal feed made the product particularly attractive for use in piglet feeds, particularly through the 1970s and 1980s. Since then its use has declined, as companies have looked to less expensive alternative products.

Table 3.7. The different components of milk fractions.

Fraction	Components
True protein	Casein
	Alpha
	Beta
	Kappa
	Whey
	Lactalbumins
	Immunoglobulins
	Lactoferrin
	Lactoperoxidase
	Growth factors
Non-protein nitrogen	Urea
	Nitrates
	Free amino acids

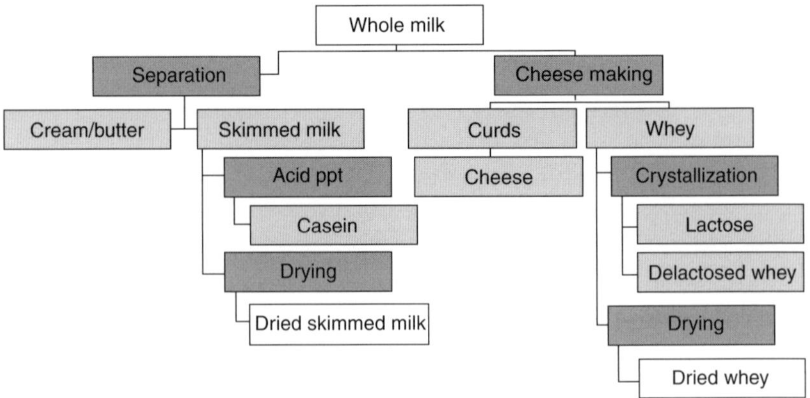

Fig. 3.3. The various sources of milk proteins for use in piglet feeds.

In the USA, where no subsidies are paid for the incorporation of milk powder into animal feed, unadulterated rather than denatured skimmed milk powder is often used in weaner feeds. Skimmed milk powder contains 350 g CP kg^{-1}, primarily casein protein. The protein composition of sow milk obviously varies with stage of lactation but on average is about 60 g kg^{-1} (Gallagher *et al.*, 1997). The essential amino acid components of porcine and bovine casein are fairly similar, with the possible exception of methionine, which in porcine casein is less than 50% of that in bovine casein (Table 3.8). It would appear logical, therefore, that skimmed milk powder should play some role in formulating a diet for feeding after sow milk.

Despite the fact that studies have continually demonstrated superior performance of pigs fed diets based on skimmed milk, its use has declined in recent years due to cost.

Whey proteins

Dried whey products are included in starter feeds as a source of lactose and protein and were traditionally classified as sweet, acid or delactosed, depending upon the method of production. Typically the protein content is around 120 g kg^{-1}, considerably lower than the 350 g kg^{-1} found in denatured skimmed milk powder. This, coupled with the price being approximately one-third that of skimmed milk, has resulted in levels of inclusion being higher at between 200 and 300 g kg^{-1} in weaner feeds.

Not only is whey a variable product inasmuch as it is a by-product of the cheese industry, but also the drying of whey can be aggressive, with proteins being damaged during the process. The lactose forms a complex with the protein, which reduces the biological value of the product. Exposure to 85°C for 30 min was shown to denature over 60% of whey protein, whereas at 65°C only 15% was denatured (Richert *et al.*, 1974). Since the biologically active proteins, such as immunoglobulins, are the most heat sensitive, the application of severe processing techniques results in these being lost first (Smithers *et al.*, 1996).

By feeding pigs two different types of whey, either feed or food grade, for 4 weeks after weaning (Table 3.9), Mahan (1984) demonstrated that different grades of whey could affect piglet performance. The only noticeable difference in the two materials was the mineral content, with the ash being 3% higher in the feed grade material, yet the use of the edible grade material conferred benefits of over 13% in growth rate. Mahan (1992) has since concluded that the main response seen to whey powder in post-weaning diets was due to the contribution of lactose.

Alternative Protein Sources

Various products have become available in recent years for use in weaner pig feeds. Many of these are by-products of the human food industry but some, such as porcine plasma, have been developed specifically for use in piglet feeds.

Spray-dried porcine plasma

Spray-dried porcine plasma first appeared on the market in the early 1990s and its use in the USA was reported to have a dramatic effect on both feed intake and growth rates. Spray-dried plasma protein comprises the albumin and globulin fractions of

Table 3.8. The amino acid composition of porcine and bovine casein.

	Composition in casein (g 100 g^{-1} protein)	
Amino Acid	Porcine (Nakae and Kataoka, 1973)	Bovine (Devendra, 1980)
Lysine	8.7	8.1
Methionine	1.3	3.2
Threonine	4.8	4.3
Tryptophan	1.5	1.3
Leucine	10.2	10.0
Isoleucine	5.8	6.6
Valine	8.1	7.5

Table 3.9. Effect of grade of dried whey on performance (Mahan, 1984).

	Dried whey source	
	Feed grade	Edible grade
Start weight (kg)	6.03	6.12
Day 28 weight (kg)	13.56	14.69
Daily liveweight gain (DLWG) (g)	263	304
Daily feed intake (DFI) (g)	531	562
Food conversion ratio (FCR)	2.00	1.87

blood and contains 680 g CP kg^{-1} and 61 g lysine kg^{-1}. Blood is collected in refrigerated tanks and prevented from coagulating by the addtion of sodium citrate. The plasma fraction is separated from the blood cells by centrifugation and stored at −4°C until the product is preheated (25 min at 93°C) and then spray-dried. The finished product is a fine-grained powder.

Maxwell (1996) summarized early experiments with the use of spray-dried porcine plasma (Table 3.10). In these trials, all of which were conducted in the USA, significant improvements were seen in daily liveweight gain, primarily as a result of feed intake.

Nelssen *et al.* (1997) reported that the only protein source necessary for inclusion in weaner feeds, albeit early weaner feeds used in segregated early weaning (SEW) programmes, was porcine plasma. The importance of plasma declines in the weeks after weaning; typically, it is included in US commercial rations at up to 60 g kg^{-1} in the first diet after weaning and up to 30 g kg^{-1} in the second feed.

Little independent work has been conducted on the use of porcine plasma in the UK. This stems from the fact that most UK feed compounders do not use porcine plasma or even allow it into their mills. The practice of feeding animal products back to the same species of livestock is not deemed acceptable by the food retail chain. However, recent studies with porcine plasma found that its inclusion consistently increased performance in the first week after weaning (Table 3.11). In both studies, the growth rates in the second and third weeks were superior for pigs fed non-plasma diets.

In recent studies at a US test farm, porcine plasma was included in a UK-style pre-starter diet at a rate of 50 g kg^{-1} and in a starter feed at 25 g kg^{-1} (both diets contained skimmed milk powder and heat-processed cereals). No improvements were seen in pig liveweight up to 3 weeks afer weaning (Fig. 3.4).

Whey protein concentrates

In recent years the human food industry has had a requirement for high-protein whey products with particular functional properties for use in many non-fat or reduced-calorie products. The development of ultra-filtration processes, in which whey is moved through a series of porous membranes, has led to the production of dried concentrated whey proteins. These typically have a protein content of between 340 and 800 g CP kg^{-1} and, unlike many of the conventionally dried lower-protein whey products, still have intact many of the biologically active proteins such as immunoglobulins, lactoferrins and lactoperoxidase. It has been hypothesized that the use of such whey proteins may offer an alternative to spray-dried porcine plasma.

In a series of experiments, Grinstead *et al.* (2000) evaluated the use of whey protein concentrates in weaner feeds in the first 14 days after weaning, with pigs weaned at 19 days of age at an average weight of 5.8 kg. The skimmed milk fraction of the diet was replaced with either spray-dried porcine plasma or whey protein concentrate containing 780 g CP kg^{-1} (Table 3.12).

Table 3.10. Early experiments evaluating spray-dried porcine plasma for 3-week-old newly weaned pigs (all refs cited by Maxwell, 1996).

		Improvement over pigs fed the control diet (%)		
Experiment[a]	Protein in control diet	DLWG	DFI	FCR
Hansen *et al.* (1990)	Skimmed milk	+42.0	+37.2	−3.6
Gatnau (1991)	Soybean meal	+101.6	+75.6	+12.0
DeRodas *et al.* (1995)	Skimmed milk	+28.6	+24.2	+1.2
Gatnau *et al.* (1991)	Soybean meal	+81.9	+34.2	+59.5
Kats *et al.* (1994)	Skimmed milk	+15.2	+27.9	−10.9
Gatnau *et al.* (1990)	Skimmed milk	+50.0	+54.0	+29.3
Central Soya (1989)	Skimmed milk	+22.5	+18.6	+3.1
Central Soya (1989)	Skimmed milk	+12.4	+8.6	+3.6
Central Soya (1989)	Skimmed milk	+14.7	+1.9	−1.9

[a] Each test was 2 weeks duration after weaning.

Table 3.11. The effect of inclusion of porcine plasma in diets for weaner pigs (Toplis and Miller, 1999; Slade and Miller, 2000).

	Liveweight gain (g day^{-1})			
	Trial One		Trial Two	
Period	Control	Plasma (40 g kg^{-1})	Control	Plasma (75 g kg^{-1})
Week 1	138	194	120	154
Week 2	406	409	407	317
Week 3	473	448	417	379

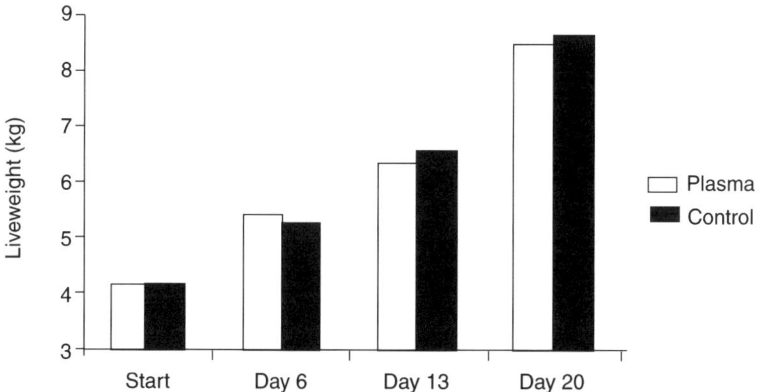

Fig. 3.4. The effect of inclusion of spray-dried porcine plasma into UK-type weaner feeds (SCA Nutrition USA, 1999).

Table 3.12. Efect of replacement of skimmed milk powder with either spray-dried porcine plasma (SPP) or whey protein concentrate (WPC) in starter diets on weaner pigs (Grinstead *et al.*, 2000).

	Days 0–7		Days 7–14		Overall	
Group diet	DLWG (g)	DFI (g)	DLWG (g)	DFI (g)	DLWG (g)	DFI (g)
Control	190	173	363	383	277	278
SPP						
25 g kg^{-1}	219	203	359	357	289	280
50 g kg^{-1}	239	222	316	359	278	290
WPC						
27 g kg^{-1}	196	190	366	368	281	279
54 g kg^{-1}	195	185	384	376	289	281

Inclusion of whey protein concentrate had no effect on daily liveweight gain or feed intake compared with the control containing skimmed milk powder, whereas inclusion of spray-dried porcine plasma resulted in an increase in both daily feed intake and daily gain from day 0 to day 7. It was concluded that the use of whey protein concentrate could partially or totally replace animal plasma in weaner feeds, but that the optimum inclusion of either protein source should be based upon weight, age and health status of pigs. In practice this would be particularly difficult for feed compounders to achieve when producing a commercial range of weaner feeds.

In a further US study (Richert *et al.,* 1991), it was found that the replacement of the skimmed milk powder of an early weaning diet with whey protein concentrate did not affect the growth performance of pigs weaned at 3.8 kg liveweight during the 14 days in which the treatment feeds were offered (Table 3.13). During this period, daily piglet growth rate on both these milk-based diets was some 30 g less than when piglets were fed diets containing spray-dried porcine plasmas.

In trials conducted by SCA Nutritions, the inclusion of lower-protein whey protein concentrates (350 g CP kg^{-1}) in place of skimmed milk powder did not result in improved performance in diets free from hipro soya and raw cereals (Table 3.14).

Egg protein

Several sources of egg protein material are commercially available to the nutritionist for inclusion in weaner feeds. Some of these are classed as hatchery waste, others as pure whole egg, spray-dried to form products with crude protein levels between 450 and 800 g kg^{-1}. The cost of egg protein varies but is often considerably less than for porcine plasma or skimmed milk powder.

In addition to eggs being a source of high quality protein, they are also reported to contain up to 150 mg antibodies kg^{-1} – some 20 times higher than those found in cow colostrum (Shipp *et al.*, 1999). Shipp *et al.* (1999) suggested that antibodies present in eggs could help to protect the young pig from dietary pathogens, and that the feeding of 'hyperimmune' eggs containing elevated levels of antibodies to young piglets might confer some degree of protection against disease, thus improving growth performance. Little evidence exists to support the use of this type of product in feed.

Researchers at the University of Nebraska (Norin *et al.*, 1998), working with pigs below 5 kg liveweight and weaned at between 11 and 14 days of age, found that inclusion of egg protein (120 g kg^{-1} diet) in place of spray-dried porcine plasma and fed for the first 2 weeks post weaning, resulted in a drop in daily gain in the first 7 days after weaning (Table 3.15). In the following week, performance was comparable with piglets fed either a complex starter feed or one containing spray-dried porcine plasma at 60 g kg^{-1}. No reduction in performance was seen when spray-dried egg powder was used at 60 g kg^{-1} in combination with porcine plasma at 30 g kg^{-1}.

In separate trials at Kansas State University, the inclusion of spray-dried egg albumin, containing 810 g CP kg^{-1}, included in the diet at 70 g kg^{-1} in place of 50 g porcine plasma kg^{-1}, resulted in a decrease in performance when fed to piglets weaned at less than 6 kg liveweight. When used at 35 g kg^{-1} in conjunction with plasma protein, performance was comparable with the control (a simple corn/soybean meal diet) (Table 3.16).

Table 3.13. Comparison of the growth performance of pigs fed diets containing either skimmed milk powder (SMP), whey protein concentrate (WPC) or spray-dried porcine plasma (SPP) (Richert *et al.*, 1991).

	Days 0–14		Days 14–18	
Diet material	DLWG (g)	DFI (g)	DLWG (g)	DFI (g)
SMP	242	245	435	688
WPC	244	252	454	697
SPP	278	299	432	690

Table 3.14. Effect of replacing skimmed milk powder (SMP) with whey protein concentrate (WPC) in starter feeds (SCA Nutrition, 1999).

Diet material	DLWG (g)	DFI (g)
SMP	332	363
WPC		
(350 g crude protein kg^{-1})	305	349

Table 3.15. Effect of inclusion of spray-dried egg protein on the growth performance of segregated early-weaning (SEW) weaned piglets (Norin *et al.*, 1998).

	Days 0–7		Days 7–14	
Diet	DLWG (kg)	DFI (kg)	DLWG (kg)	DFI (kg)
Complex	0.24	0.33	0.35	0.50
Simple + porcine plasma (60 g kg^{-1})	0.20	0.30	0.35	0.50
Simple + porcine plasma (30 g kg^{-1}) + egg protein (60 g kg^{-1})	0.20	0.30	0.33	0.49
Simple + egg protein (120 g kg^{-1})	0.15	0.26	0.35	0.53

Table 3.16 Effects of spray-dried egg albumin on the growth performance of early-weaned pigs (Llata *et al.*, 1998).

	Days 0–14	
Diet	DLWG (kg)	DFI (kg)
Control	0.28	0.38
Spray-dried porcine plasma (50 g kg^{-1})	0.34	0.43
Spray-dried egg albumin (70 g kg^{-1})	0.25	0.40
Spray-dried egg albumin (35 g kg^{-1}) + porcine plasma (25 g kg^{-1})	0.28	0.39

Potato proteins

In recent years potato protein has appeared as a protein source for the weaned pig. Potatoes are refined in a process that removes the starch, to produce a material with a protein content in excess of 800 g kg^{-1}.

Sève (1977) found that the inclusion of potato proteins at a level of 150 g kg^{-1} in place of fish and milk protein did not affect either growth rate or food conversion efficiency. Other workers have found that, at levels of between 30 and 50 g kg^{-1}, potato proteins can be included in feeds without detrimental effects on performance. However, problems have traditionally been encountered with the acceptance of potato proteins due to the presence of steroidal glycoalkaloids, a bitter group of compounds that can lead to palatability problems.

Kerr *et al.* (1998) conducted a series of experiments investigating the inclusion of conventional and low glycoalkaloid potato proteins in place of spray-dried animal plasma and skimmed milk powder. Pigs fed the low glycoalkaloid potato protein performed better than those fed the conventional source. It was concluded that low glycoalkaloid potato proteins could partially replace porcine plasma in diets fed from weaning to 14 days and could totally replace menhaden fishmeal or bloodmeal in diets fed from 7 to 14 days after weaning.

Conclusion

The amino acid requirements of the young pig are well defined, even if those used in commercial feeds tend to be somewhat different from published research. The nutritionist has to consider factors other than the nutrient contribution of raw materials when formulating diets for the weaner. Factors such as the digestibility of the feed, the presence or otherwise of biologically active proteins (e.g. immunoglobulins) and the absence of anti-nutritional factors have to be considered.

Responses to protein sources have to be considered in relation to the make-up of the overall diet. Porcine plasma would appear to be of benefit in simple diets containing raw cereals and soybean products. Where milk proteins are present, the response is diminished.

Whatever the weaner pig is fed, it is important that the developing digestive tract of the young piglet is not irreparably damaged by the use of poor quality raw materials, since this will impair digestive capabilities and hence growth performance through to slaughter.

Despite access to common data, nutritionists around the world characterize feeds by the inclusion of different raw materials. In the USA, porcine plasma is of primary importance; in the UK skimmed milk powder has traditionally been incorporated in feed; whilst in mainland Europe whey proteins are preferred.

Acknowledgement

The authors would like to thank Forum Feeds for the supply of previously unpublished material.

References

Agricultural Research Council (ARC) (1981) *The Nutrient Requirements of Pigs.* Commonwealth Agricultural Bureau, Slough, UK.

Close, W.H. (1994) Feeding new genotypes: establishing amino acid/energy requirements. In: Cole, D.J.A., Wiseman, J. and Varley, M.A. (eds) *Principles of Pig Science.* Nottingham University Press, Nottingham, UK.

Cole, D.J.A. (1978) Amino acid nutrition of the pig. In: Haresign, W. and Lewis, D. (eds) *Recent Advances in Animal Nutrition.* Butterworths, London, pp. 59–72.

Devendra, C. (1980) Milk production in goats compared to buffalo and cattle in humid tropics. *Journal of Dairy Science* 63, 1755–1767.

Elsley, F.W.H. (1970) Nutrition and lactation in the sow. In: *Proceedings of the 17th Easter School in Agricultural Science, University of Nottingham.* Butterworths, London, pp. 393–411.

Fuller, M.F., Livingston, R.M., Baird, B.A. and Atkinson, T. (1979) The optimal amino acid supplementation of barley for growing pigs. 1. Response of nitrogen metabolism to progressive supplementation. *British Journal of Nutrition* 41, 321–331.

Fuller, M.F., McWilliam, R. and Wang, T.C. (1987) The amino acid requirements of pigs for maintenance and growth. *Animal Production* 44, 476–488.

Gallagher, D.P., Cotter, P.F. and Mulvihill, D.M. (1997) Porcine milk protein: a review. *International Dairy Journal* 7, 99–118.

Grinstead, G.S., Goodband, R.D., Dritz, S.S., Tokach, M.D., Nelssen, J.L, Woodworth, J.C. and Molitor, M. (2000) Effects of a whey protein product and spray-dried animal plasma on growth performance of weanling pigs. *Journal of Animal Science* 78, 647–657.

INRA (1984) *L'alimentation des animaux monogastriques (porc, lapin, volailles).* Institut National de la Reserche Agronomique, Paris.

Kerr, C.A., Goodband, R.D., Tokach, M.D., Nelssen, J.L., Dritz, S.S., Richert, B.T. and Bergström, J.R. (1998) Evaluation of enzymatically modified potato starches in diets for weanling pigs. *Journal of Animal Science* 76, 2838–2844.

de la Llata, M., Goodband, R.D., Tokach, M.D., Nelssen, J.L., Dritz, S.S., Grinstead, G.S. and Woodworth, J.C. (1998) Effects of spray dried egg albumin on growth performance of early-weaned pigs. In: *1998 Swine Industry Day Report of Progress* (SRP819). Kansas State University (Internet pubn ozxnet.ksu.edu/library/).

van Lunen, T.A. and Cole, D.J.A (1996) Energy–amino acid interactions in modern pig genotypes. In: Garnsworthy, P.C., Wiseman, J. and Haresign, W. (eds) *Recent Advances in Animal Nutrition.* Nottingham University Press, Nottingham, UK, pp. 233–261.

Mahan, D.C. (1984) Dried whey is not always dried whey. Part 1. Ohio Swine Research and Industry Report 84(1), 30–33.

Mahan, D.C. (1992) Efficacy of dried whey and its lactalbumin and lactose components at two dietary lysine levels on post weaning pig performance and nitrogen balance. *Journal of Animal Science* 70, 2182–2187.

Makkink, C.A. (1993) Of piglets, dietary proteins and pancreatic proteases. PhD Thesis, Department of Animal Nutrition, Wageningen, The Netherlands.

Maxwell, C. (1996) Nutrition and management of the early-weaned pig. In: Lyons, T.P. and Jacques, K.A. (eds) *Under the Microscope. Proceedings of Alltech's 15th Annual Symposium.* Nottingham University Press, Nottingham, UK, pp. 203–222.

Nakae, T. and Kataoka, K. (1973) Comparative studies on the milk constituents of various mammals in Japan. VII. Amino acid composition of milk proteins from various mammals. *Japanese Journal of Dairy Science* 22, A20–A28.

National Research Council (1998) *Nutrient Requirements of Swine,* 10th edn. National Academy Press, Washington, DC.

Nelssen, J.L., Dritz, S.S., Tokach, M.D. and Goodband, R.D. (1997) Nutritional programmes for early weaned pigs. In: *Congresso Brasileiro Especialistas em Suinos de 20 a 23 de Outubro 1997,* pp. 126–135.

Norin, S.L., Miller, P.S., Lewis, A.J. and Reese, D.E. (1998) Protein sources for early weaned (SEW) pigs. In: *1998 Nebraska Swine Report,* pp. 21–23

Patterson, D.C. (1987) The use of milk products in the diet of early weaned pig. In: *Annual Report 1986–1987, Agricultural Research Institute of Northern Ireland,* pp. 39–46.

Richert, B.T., Hancock, J.D., Hines, R.H. and Burton, K.S. (1991) Use of whey protein concentrate, dried buttermilk and porcine plasma protein to replace skim milk powder in diets for weanling pigs. *Journal of Animal Science* 70 (Supplement 1), 231.

Richert, S.H., Morr, C.V. and Cooney, C.M. (1974) Effect of heat and other factors upon foaming properties of whey protein concentrates. *Journal of Food Science* 39, 42–48.

de Rouchey, J.M., Tokach, M.D., Nelssen, J.L., Goodband, R.D., Dritz, S.S., James, B.W. and Webster, M.J. (2000) Effects of irradiation of spray dried blood meal and animal plasma on nursery pig growth performance. In: *2000 Swine Industry Day Report of Progress* (SPR858). Kansas State University (Internet pubn oznet.ksu.edu/library/), p.57.

Sève, B. (1977) Utilisation d'un concentre de proteine de pommes de terre dans l'aliment de sévérage du porclet a 10 jours et a 21 jours. *Journées de la Reserche Porcine en France 9*, 205–210.

Shipp, T.E., Godfresden-Kisie, J.A. and Kichura, T.S. (1999) Hyperimmune spray dried eggs as a feed supplement for weanling pigs. *Feed Mix* 7(6), 30–33.

Slade, R.D. and Miller, H.M. (2000) Early post weaning benefits of porcine plasma remerge in later growth performance. In: *Proceedings of the British Society of Animal Science Winter Meeting 2000*, p. 120.

Smithers, G.W., Ballard, F.J., Copeland, A.D., DeSilva, K.J., Dionysuis, D.A., Francis, G.L., Goddard, C., Grieve, P.A., McIntosh, G.H., Mitchell, I.R., Pearce, R.J. and Regester, G.O. (1996) New opportunities from the isolation and utilisation of whey proteins. *Journal of Dairy Science* 79 (8), 1454–1459.

Tokach, M.D., Nelssen, J.L. and Allee, G.L. (1989) Effect of protein and/or carbohydrate fractions of dried whey on performance and digestibility of early weaned pigs. *Journal of Animal Science* 67, 1307–1312.

Tokach, M.D., Dritz, S.S., Goodband, R.D. and Nelssen, J.L. (1997) *K State Starter Pig Recommendations*. In: *1997 Swine Industry Day Report of Progress* (SRP795). Kansas State University (Internet pubn oznet.ksu.edu/library/).

Toplis, P. and Miller, H.M. (1999) Post weaning benefits of porcine plasma. In: *Proceedings of the British Society of Animal Science Winter Meeting 1999*, p. 173.

Zhang, Y., Partridge, I.G., Keal, H.D. and Mitchell, K.G. (1984) Dietary amino-acid balance and requirements for pigs weaned at 3 weeks of age. *Animal Production* 39, 441–448.

Starch Digestion in Piglets

4

J. Wiseman, J. Pickard and L. Zarkadas

Division of Agriculture and Horticulture, School of Biosciences, University of Nottingham, Sutton Bonington Campus, Loughborough, Leicestershire LE12 5RD, UK

Introduction

Starch is the major energy-yielding component of the diet of post-weaned piglets and contributes more than double the digestible energy compared with that from dietary fat. Starch in the diets of post-weaned piglets is almost invariably provided from cereals, where it comprises between 600 and 800 g kg^{-1} cereal dry weight (although the specific content is affected by a number of factors, including cultivar, environment and growing season). Accordingly, knowledge of the structure of cereal starch and how this might be influenced by the numerous processing techniques to which cereals are subjected is of considerable importance. The effects on digestive physiology and subsequent digestibility are fundamental to the ability of starch to contribute to the dietary energy requirements of piglets.

Basic Structure of Starch

Cereal starch comprises two glucose polymers, amylose and amylopectin (Fig. 4.1), which have different structures and properties. Although environmental conditions may influence the relative proportion of the two, varietal genotype is the major determinant.

Amylose is essentially a linear polymer composed of glucopyranose ('glucose') units linked through α-D-(1–4)-glycosidic linkages. It has between 1000 and 4400 glucose residues, giving it a molecular mass between 1.6×10^5 and 7.1×10^5 Da. Amylose consists of a double helix of these linear polymers which, although essentially amorphous in nature, does demonstrate crystallinity and may also be crystallized further on cooling subsequent to contact with heat and water (a process referred to as

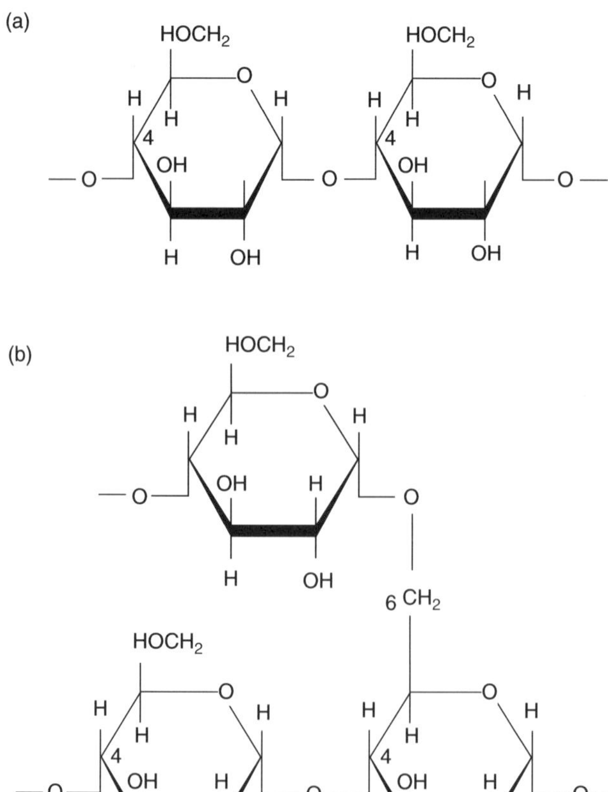

Fig. 4.1. Basic structure of (a) amylose and (b) amylopectin.

retrogradation) into two forms, depending upon the packing density and quantity of water associated with them: 'A' crystals are packed more tightly and accordingly have smaller amounts of water present than 'B' crystals. The amylose contents of most cereal starches range between 200 and 350 g kg^{-1}, but mutants have been used in breeding programmes to produce cultivars (mainly in polyploid species such as maize and barley) with abnormally high or low amylose contents.

Amylopectin is also a polymer of glucose, with a branch every 20–25 glucopyranoses in which a chain of α-D-(1–4)-glucopyranose units is linked to the C-6 hydroxymethyl position of a glucose residue through an α-D-(1–6)-glycosidic linkage (MacGregor and Fincher, 1993; d'Appolonia and Rayas-Duarte, 1994). Amylopectin has a molecular mass of the order of 10^8 Da (d'Appolonia and Rayas-Duarte, 1994) and is usually less crystalline and more amorphous than amylose. The molecule is generally considered to consist of three types of chain arranged as clusters (Eliasson and Gudmundsson, 1996): (A) linear side-chains linked only via their reducing ends to the rest of the molecule; (B) those to which A chains are attached; and (C) chains that carry the only reducing group of the molecule. Cereal starches give an A pattern; tuber, stem and amylomaize starches give a B pattern; and bean and root starches a C pattern, which is a mixture of A and B. The degree of branching may be represented by the ratio of A to B chains.

Amylopectin is insoluble in cold water but forms stable gels on heating. It is thought to be responsible for the solubility of starch granules (Reddy *et al.*, 1984). High amylopectin types are generally described as waxy, as the appearance of the endosperm of the first mutants discovered suggested a waxy composition.

Granule Organization

Starch is deposited in discrete, roughly spherical, insoluble granules (Greenwood, 1970, 1979) within cellular organelles called amyloplasts (Tester, 1997). The morphology of the starch granules is characteristic of the species, although each granule within a population possesses individuality. The size and shape of starch granules differ among several grains; the molecular weight and fine structure of both amylose and amylopectin may vary with cultivar and the stage of maturity of the plant. Starch granules are covered with a protein layer (French, 1973). The starch granule is heterogeneous (Greenwood, 1970) and consists of two different parts: crystalline regions and amorphous or gel-like regions.

The polymers in the granule are hydrogen bonded and aligned in an ordered radial manner (Young, 1984), as shown by the birefringent properties of ungelatinized starch granules when viewed under polarized light (French, 1984; Blanshard, 1987). Under crossed polarizers, birefringence (in the form of a Maltese cross) is indicative of a high degree of order within the structure (French, 1984; Galliard and Bowler, 1987). Amylopectin is most probably the principal crystalline component of the granule as amylose can be leached from normal granules without affecting the X-ray pattern. The A and B patterns are thought to indicate crystals formed by double helices in amylopectin that occur in the outer chains of amylopectin molecules, where they form regions. The crystalline parts of starch granules are responsible for many of the physical characteristics of the granules. Surface characteristics of the starch granules are also affected by the minor components of starches, including proteins and lipids.

Changes to Starch Structure During Processing

Mechanical damage

Milling or fine grinding (in contrast to heating leading to gelatinzation, described below) can result in mechanical damage to starch granules and contents. There are several consequences of this, including: an increased capacity to absorb water, from 0.5-fold starch dry mass when intact to 3–4-fold when damaged (gelatinized granules absorb as much as 20-fold); an increased susceptibility to amylolysis; loss of organized structure (as seen, for example, through loss of X-ray pattern and birefringence); and increased solubility, leading to leaching of mainly amylopectin. In gelatinized granules, amylose is preferentially leached (Craig and Starck, 1984).

At the molecular level, the disorganization of granules appears to be accompanied by fragmentation of amylopectin molecules during damage, whereas gelatinization achieves loss of organization without either polymer being reduced in size.

Heating

There has been much interest in the changes that take place in starch following heating. This interest has come principally from the fields of human nutrition and medicine (e.g. the influence that such changes might have on digestibility, postprandial plasma glucose increases associated with diabetes and links to 'dietary fibre') and food technology (structural changes associated with texture and keeping quality). Three broad categories of starch have been defined: rapidly digested, slowly digested and resisant to digestion. The last is itself divided into three subcategories: (i) starch which, through its physical structure, is inaccessible to digestive enzymes; (ii) starch which, because of its granular structure, is resistant to enzyme attack; and (iii) starch that is retrograded following heating at high temperatures (e.g. Berry, 1986; Englyst *et al.*, 1992) (Tables 4.1 and 4.2). Interestingly, mild to modest heat treatments are considered beneficial in terms of pig nutrition, as it is likely that the reduction in crystallinity that occurs leads to an increase in digestibility as access by hydrolytic enzymes is increased. Excessive heat treatment may reduce digestibility, which is considered detrimental to the piglet but an advantage to the human (glucose absorption will be reduced, associated with an increase in hindgut fermentation that will lower the pH of contents; Brown, 1996).

Gelatinization

The susceptibility of starch granules to enzyme action is of considerable importance. It is generally accepted that digestibility of starch is affected by many factors, including the nature and physical form of the starch, the presence or absence of inhibitors and the physical distribution of starch in relation to the dietary fibre components. It is well known that the physical and chemical properties of starch – such as the amylose to amylopectin ratio, the crystalline structure of amylose and the starch granule size – may influence its digestibility by endogenous α-amylase. Heat treatment results in changes to starch crystallinity and/or gelatinization. The susceptibility of starch to enzymes can be increased by gelatinization or by any other process that destroys the granular structure of starch (Holm *et al.*, 1985).

At room temperature, starch granules are not totally impermeable to water and in fact water uptake can be detected microscopically by a small increase in granule diameter. The swelling is reversible and the wetting and drying can be cycled repeatedly without permanent change. If the temperature of a suspension of starch in excess water is raised progressively, a condition is reached, around 60°C, at which irreversible swelling begins, and continues with increasing temperature (French, 1984).

Further heating leads to irreversible uncoiling or dissociation of double-helical regions, the intermolecular hydrogen bonds in the crystalline areas of the starch granules break down and so the birefringence of the starch is lost. The crystallinity of the granule is destroyed by the disappearance of the crystalline structure of amylopectin. Thus there is increased viscosity, by reduction of the mobile phase surrounding the granules. Accompanying leaching of starch polymers into this phase can further increase viscosity.

The temperature at which this phenomenon happens is called the 'gelatinization temperature', which is defined as the temperature at which 98% loss of birefringence occurs. In the case of wheat starch, the onset of gelatinization takes place at 48–52°C and is complete at 59–62°C (Gough and Pybus, 1971; Tester and Morrison, 1990; Cooke and Gidley, 1992; Liu and Lelievre, 1993), though variations do occur. In general, and for the major cereal grains, gelatinization temperatures range between 60–70°C; for maize, wheat and barley these are 62–72°C, 65–67°C and 59–64°C, respectively (Greenwood, 1970). Larger granules gelatinize at a lower temperature and small granules at a higher one. The lower the moisture content, the higher is the temperature required to produce gelatinization.

Table 4.1. Classification of starch based on digestibility (Englyst *et al.*, 1992).

Type of starch	Example	Digestion in small intestine
Rapidly digested starch (RDS)	Gelatinized starch (heated)	Rapid, complete
Slowly digested starch (SDS)	Most raw cereals	Slow but complete
Resistant starch (RS)		
Physically inaccessible starch	Partially milled grain	Resistant
Resistant starch granules	Raw potato	Resistant
Retrograded starch	Overheated/cooked	Resistant

Table 4.2. Starch fractions (g kg^{-1} DM) in cooked feed ingredients (after Patil *et al.*, 1998).

	Type of starch			
Grains	Rapidly digested	Slowly digested	Resistant	Total starch
Wheat	299	73	136	508
Barley	249	121	182	552
Maize	371	156	252	779
Sorghum	292	139	361	792

Retrogradation

Suspensions of gelatinized granules containing more than 30 g starch kg^{-1} form a viscous or semi-solid starch paste which, on cooling, sets to a gel (Osman, 1967). When starch gels are held for prolonged periods, retrogradation develops. Applied to starch, this means a return from a solvated, dispersed amorphous state to an insoluble, aggregated or crystalline condition. Retrogradation is due largely to crystallization of amylose, which is much more rapid than that of amylopectin: yields of resistant starch were extremely low with waxy (high amylopectin) maize (Berry, 1986). Although regarded as crystalline, retrograded gels are susceptible to amylolysis; however, a fraction known as 'resistant starch' is resistant to enzyme attack, behaves as 'dietary fibre' and is most abundant in autoclaved amylomaize starch suspensions (Morrison, 1988). This specific fraction of starch was considered by Berry (1986) to be responsible for increases in dietary fibre content. Higher yields (90 g kg^{-1}) of resistant starch have been noticed in autoclaved wheat starch gels compared with only approximately 5 g kg^{-1} for uncooked wheat starch (Berry, 1986).

Starch Digestion in Pigs

It is not proposed to consider starch digestion in detail, as this has been the subject of a number of comprehensive reviews. Complete enzymic hydrolysis of starch to D-glucose is accomplished by a number of intestinal enzymes working in combination (Robyt, 1984). Perhaps the principal enzyme initially responsible for hydrolysing starch is α-amylase secreted by salivary glands and by the pancreas. The enzymes from the two sources appear to be identical in terms of mode of action, with pH optima between 5.5 and 8.0. Although salivary amylase has been thought to be inactivated in the acidic conditions operating in the stomach, in humans it may survive in combination with the substrate to be effective in the duodenum, as reviewed by Gray (1992). If this is confirmed in piglets, then it would be questionable whether total reliance on pancreatic amylase as an indication of overall activity of this enzyme is valid, particularly as the development of pancreatic amylase in the human infant occurs somewhat later than in the piglet.

The enzyme attacks the α-D-(1–4) non-terminal linkages in both linear and branched glucans but not α-D-(1–6) or the α-D-(1–4) bonds within four glucose residues adjacent to an α-D-(1–6) branched linkage due to stearic hindrance. These two bonds are found in the branched-chain polymer amylopectin. The major breakdown products of starch (amylose and amylopectin) digestion are maltose, maltotriose, maltotetraose and much smaller amounts of glucose; products specific to amylopectin are α-limit dextrins, oligosaccharides of four or more glucose molecules and containing at least one branched α-D-(1–6) linkage (e.g. Manners, 1979; Fogarty, 1983; Robyt, 1984). These products are subsequently hydrolysed – by a range of amyloglucosidases and α-limit dextrinase (specifically for the α-limit dextrins) located predominantly in the brush border – to glucose, which is then transported across the intestinal membrane.

Development of Enzyme Activity

Effective formulation of starter diets requires accurate knowledge of the physiological structure and function of the piglet digestive system. Weaning is associated with considerable changes in proteolytic and disaccharidase activities. The development of the digestive system and its enzymatic activity (both intestinal and pancreatic) in young pigs has been extensively studied (Aumaître and Corring, 1978; Corring *et al.*, 1978; Kidder and Manners, 1980; Efird *et al.*, 1982a, b; Lindemann *et al.*, 1986; Owsley *et al.*, 1986; Kelly *et al.*, 1991a,b).

According to Corring *et al.* (1978), activities of lipase, amylase, trypsin and chymotrypsin undergo a two-step development: increasing steadily from birth to the third week and more rapidly from the fourth to the eighth week of age. It was also stated that there is a high correlation between pancreas:bodyweight ratio and growth, which increases in the first week and is more rapid from the fourth week of age. Similarly, Efird *et al.* (1982a) observed dramatic changes in the development of the digestive system from birth to the third week and at weaning. Lindemann *et al.* (1986) concluded that increases with age (Fig. 4.2) were due to increases in both tissue weight and enzyme activity per gram of tissue.

Activity of maltase is low at birth and increases gradually with age. Aumaître and Corring (1978) reported that maltase activity increased considerably to the eighth week. Maltase II activity was significantly higher during the post-weaning period for pigs fed uncooked cereal-based diets (McCracken, 1984). Similar observations have been reported by Miller *et al.* (1986), who weaned pigs at 6 weeks of age. The influence of previous dietary history on development of enzyme systems has been extensively studied. Kelly *et al.* (1991a) stated that creep feeding of piglets prior to weaning was not associated with the subsequent increase in activity of maltase enzymes and reported that total activity steadily increased during the post-weaning

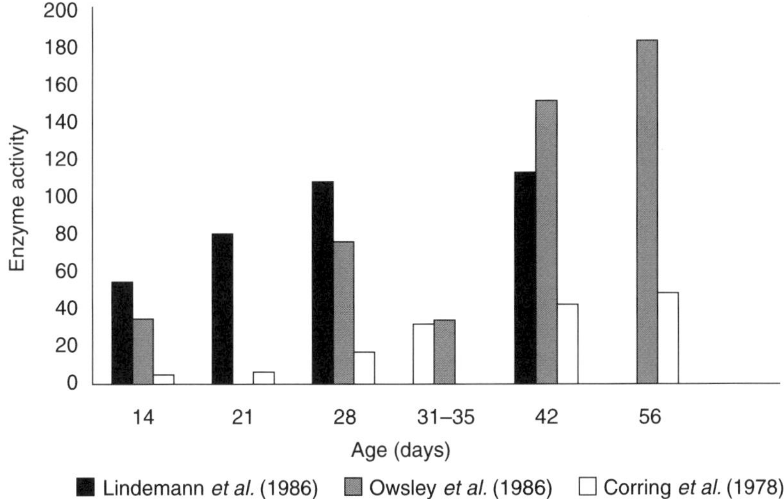

Fig. 4.2. Development of α-amylase activity in piglets.

period; after 7 days it was twice that found at 3 days after weaning. Kelly *et al.* (1991b) found that total maltase activity was significantly higher for continuously fed pigs. In a study with 4-week-old weaned pigs fed a cereal-based diet, maltase activity was found to be higher for the untreated (containing no exogenous protease) groups (Rooke *et al.*, 1998).

The increase in amylase activity from 5 to 8 weeks of age is probably due to the requirement for amylase in order to digest dietary starch. Corring *et al.* (1978) reported that, at 3 weeks of age, specific activity was higher (× 2.8) than at birth and thereafter increased more rapidly until the sixth week (× 27.5 between 0 and 42 days of age). This is in agreement with the results of Owsley *et al.* (1986), who measured activities in the pancreas and small intestine, and reported that the total activity of amylase, trypsin and chymotrypsin increased significantly until 56 days of age.

There has been considerable interest in the role of feed intake (both the quantity and nature) in stimulating digestive activity in piglets. Owsley *et al.* (1986) fed weaned piglets from 29 days to 56 days on maize–soya-based diets, alone or containing either 200 g dried whole whey or 50 g lard kg^{-1}. The different diet treatments had no effect on pancreatic amylase activity in the intestinal contents. However, amylase activity (expressed as units kg^{-1} bodyweight) was significantly higher for the pigs fed diets without whey. It was concluded that the lack of the appropriate substrates in the intestine may inhibit the release of pancreatic enzymes.

In early-weaned pigs, it has been suggested that low feed consumption, growth check and diarrhoea in the immediate post-weaning period are related to the limited capacity of the gastrointestinal tract to digest and effectively use weaner diets (Hampson *et al.*, 1984; Hampson, 1994). However, food intake tends to be positively correlated with enzymic activity (McCracken and Kelly, 1984). Reduced digestive and absorptive capacity may not be as limited as once thought, since weaned pigs actually have considerable 'spare' digestive capacity for cereal-based weaner diets (Kelly *et al.*, 1991b).

Effects of Processing on Starch Digestibility and Performance in Piglets

A large number of trials have been undertaken to assess the consequences of processing on the degree of amylolysis. Although *in vitro* measurements of starch degradation suggest that processing can have major effects (Bjorck *et al.*, 1984a,b) changes to starch structure *per se* in the outer segments of granules may not be entirely responsible for differences recorded and starch:protein together with starch:lipid complexes may be important in this respect. Mechanical damage itself may result in improvements in starch digestibility. Flaking has been shown to improve this in barley, but the determinations were on an *in vitro* basis (Osman *et al.*, 1970) (Table 4.3).

Many of the studies reported refer simply to named processes in an attempt to comment upon the effectiveness or otherwise of a specific process in changing the nutritional value of starch-containing diets for piglets (Sauer *et al.*, 1990). Such analyses are somewhat limited, as there is no assessment of the consequences in terms of changes to starch structure. There have been attempts to characterize the degree of alteration of starch structure using enzyme systems. For example, van der Poel *et al.* (1989) (Table 4.4) employed *in vitro* enzyme hydrolysis with amyloglucosidase to define starch 'availability' from maize. Despite the range in values, there were no subsequent differences in total tract digestibility or performance in piglets immediately after weaning.

Other techniques have been employed to assess the degree of change in starch structure consequent on heating. The use of X-ray diffraction in the study of solid materials is common; it distinguishes between ordered and disordered states – amorphous substances such as liquids or gases produce diffuse X-ray diffraction patterns consisting of one or more haloes on a diffractogram, whilst crystalline substances yield patterns of numerous sharp spots or lines. This makes it suited to the investigation of polymers that are only partially crystalline, such as starch.

Polymers are not usually fully crystalline and the crystals present are not perfect. They inherently contain a degree of disorder due to the long chain nature of the material and the thermal motion of the atoms in the chains of the polymer. The effects of these defects contribute to background scattering and lead to broadening of diffraction line profiles. In order to interpret responses, two pieces of information are required from the X-ray diffraction data: (i) in which polymorphic form of starch the crystals had grown (achieved by visual inspection of the diffraction pattern); and (ii) the amount of crystallinity in the sample (computation of the data).

Table 4.3. Effect of flaking on *in vitro* coefficient of starch digestion of pressure-cooked barley (after Osman *et al.*, 1970).

Pressure (kg cm^{-2})	Poor flaked	Intermediate flaked	Flat flaked
1.4		0.290	0.338
2.8	0.304	0.436	0.470
4.2	0.384	0.411	0.483

Table 4.4. Availability of starch (proportion) in maize after infrared treatment at two different initial moisture levels (after van de Poel *et al.*, 1989).

Treatment [a]	Moisture (g kg^{-1})	Available starch (g kg^{-1} DM)	Available starch (g 100 g^{-1} total starch)
Untreated	127	52	7.3
Infrared + flaked	135	210	29.3
	268	109	15.2
Infrared + flaked + stored[b]	135	195	27.2
	268	103	14.4

[a] Dwell time = 46 s, exit temperature = 110–120°C.
[b] Stored for 15 min in a well-insulated container.

To measure the level of crystallinity, the method of Hermans and Weidinger (1948) as cited by Marsh (1986) was applied to wheat samples in a recent study at the University of Nottingham (L. Zarkadas and J. Wiseman, unpublished data). The analysis involved the drawing of a smooth demarcation line on a powder diffractogram, which had been corrected for incoherent scattering, between the background scattering and the crystalline scattering. The ratio of the area enclosed by this line and the diffraction curve above the line to the total area of the diffractogram due to coherent scattering was taken as a measure of the absolute crystallinity. By using an interactive graphics program a smooth curve could be drawn between the amorphous and crystalline coherent scattering curves. The details of the analysis are shown schematically in Fig. 4.3.

The percentage of the gelatinized starch in the cereal samples was calculated as:

$$GS = (R - C)/R$$

where GS = gelatinized starch, R = crystallinity of the raw sample, and C = crystallinity of the cooked sample.

Wheat was subjected to various processing conditions designed to give a wide range of gelatinized starch content and incorporated into diets for weaned piglets. Total tract digestibility coefficients of gross energy of diets, whilst not being significantly different between treatment within trials, indicated that there was a marginal improvement with increased gelatinization when all trials were combined (Fig. 4.4). However, there were no significant performance differences.

Whilst over-processing may not be too common, it is possible that excessively high temperatures may be associated with a reduction in the nutritional value of cereal starches. The enzyme α-amylase is unable to penetrate some gelatinized starch granules (Würsch *et al.*, 1986). Even at low moisture levels, branched structures (trans-glucosidation) formed during heat treatment of starch-containing food items are not degradable with amylolytic enzymes and represent an irreversible chemical modification (Siljeström *et al.*, 1986). Moreover, formation of amylose–lipid complexes, observed during heat treatment of wheat, may impair amylolysis (Holm *et al.*, 1983, 1988). Formation of these complexes may also decrease the amylose content, thereby reducing further the enzymic hydrolysis of wheat starch (Hoover and Vasanthan, 1994). Differences in the rate of α-amylase attack may affect the rate of digestion of the starch and the speed of movement of digesta through the digestive tract. These factors may affect gut fill and therefore affect feed intake.

Extract viscosity of wheat samples did increase with increasing degree of gelatinization (Fig. 4.5) and there may be a possible correlation with digestibility (Fig. 4.4), though it would apear that *in vitro* viscosity values of wheat associated with changes in starch structure cannot be used confidently as a predictor of piglet responses.

Conclusions

Although based on a simple monosaccharide (glucose), starch has an extremely complex physical and organizational structure. In addition to genetic and environmental effects influencing the growth of the cereal grain, there are many variables associated with processing (time, temperature and moisture) which can have a dramatic influence on this structure. Even so, it appears that the digestion of starch *in vivo* is a comparatively simple process and the piglet should have an adequate complement of the necessary enzymes from a comparatively early age – accepting that this may not be fully developed at weaning. However, the weaning process itself (removal of

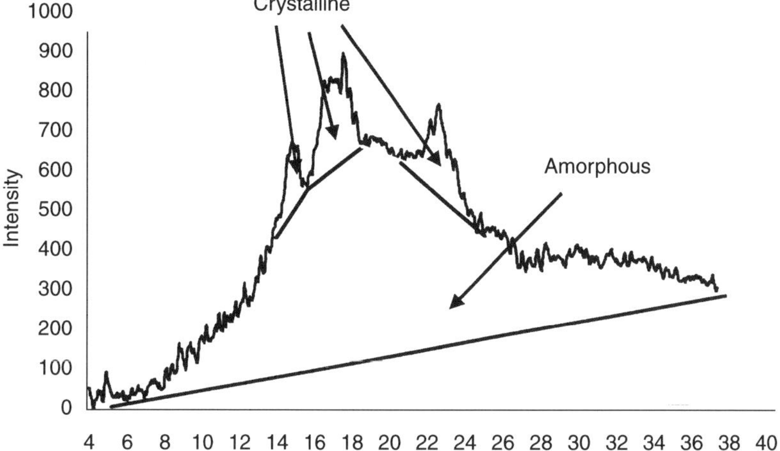

Fig. 4.3. Assessment of degree of crystallinity of starch through X-ray diffraction; gelatinization = (R − C)/R, where R = crystallinity of the raw sample and C = crystallinity of the cooked sample.

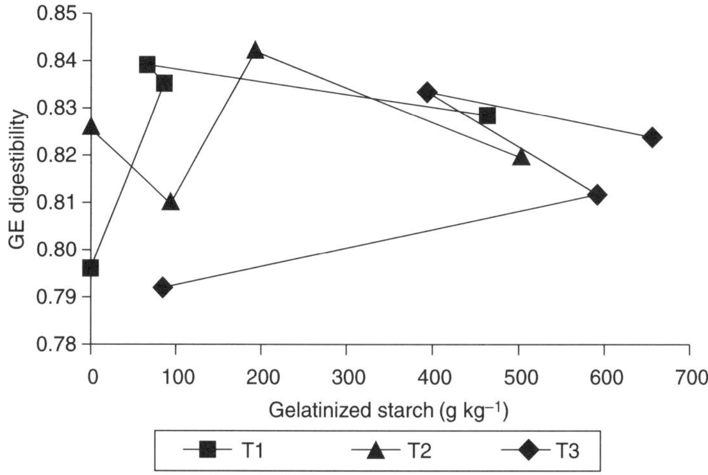

Fig. 4.4. Relationship between content of gelatinized starch in heat-treated wheat and coefficient of digestibility of gross energy in diets containing 700 g wheat kg^{-1}, determined with piglets of approximately 15 kg liveweight (L. Zarkadas and J. Wiseman, University of Nottingham, unpublished data).

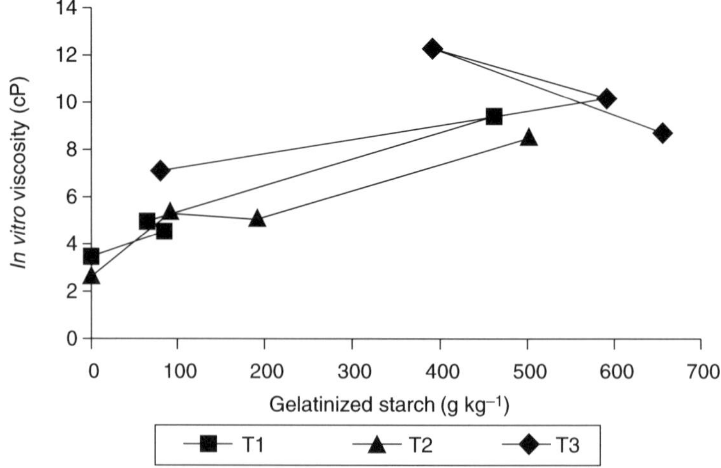

Fig. 4.5. Relationship between content of gelatinized starch in heat-treated wheat and *in vitro* extract viscosity (L. Zarkadas and J. Wiseman, University of Nottingham, unpublished data).

the animal from the dam, placing it in an unfamiliar environment and offering it a novel dry food) is associated with considerable stress and nutritional risk – the so-called post-weaning growth check, which may even lead to death (usually rapidly).

Another area of interest concerns the fate of starch that has escaped amylolysis in the small intestine and will pass into the large intestine. Huang *et al.* (1998), in an assessment of micronizing hull-less barley, established that there were differences in starch and gross energy digestibility at the ileum but not over the total tract (Table 4.5). This indicates that digestibility in the large intestine in unprocessed cereals may be significant. There is considerable debate as to whether this is beneficial (generation of volatile fatty acids, which will contribute to lowering of pH and are absorbed as an energy-yielding substrate) or detrimental (proliferation of pathogens, e.g. coliforms) to the piglet.

It is generally considered that, rather than inadequacies in digestive physiology, a reduction in voluntary feed intake during the first few days after weaning is the main reason for these problems. Bark *et al.* (1986) noticed a failure of early-weaned pigs to consume an adequate amount of food during the first 3 days after weaning in an unfa-

Table 4.5. Coefficients of ileal and total tract digestibility of starch determined with piglets between 9 and 13 kg liveweight (Huang *et al.*, 1998).

	Diets based on		
Starch digestibility	Hull-less barley	Micronized hull-less barley	Maize starch
Ileal	0.79	0.973	0.979
Total tract	0.996	0.999	1.000

miliar environment. Such a reduction (or, indeed, cessation) of intake has been associated with rapid deterioration in gastrointestinal function which, when combined with pathological challenges, will lead to the characteristic scouring seen in weaned piglets. It is therefore possible that digestive disorders (with accompanying poor digestibility) are an effect and not a cause of the problem of poor feed intake. It is imperative that top-quality processed raw materials are employed in the formulation of diets for weaned piglets. However, when considering cereal starch, this may be more to do with issues such as promoting palatability rather than an improvement in digestibility.

References

d'Appolonia, B.I. and Rayas-Duarte, P. (1994) Wheat carbohydrates: structure and functionality. In: Bushuk, W. and Bushuk, V.F. (eds) *Wheat: Production, Properties and Quality.* Cambridge University Press, Cambridge, UK, pp. 107–127.

Aumaître, A. and Corring, T. (1978) Development of digestive enzymes in the piglet from birth to 8 weeks. II. Intestine and intestinal disaccharidases. *Nutrition and Metabolism* 22, 244–255.

Bark, L.J., Crenshaw, T.D. and Leiberbrandt, V.D. (1986) The effect of meal intevals and weaning on feed intake of early weaned pigs. *Journal of Animal Science* 62, 1233–1239.

Berry, C.S. (1986) Resistant starch: formation and measurements of starch that survives exhaustive digestion with amylolytic enzymes during the determination of dietary fibre. *Journal of Cereal Science* 4, 301–314.

Bjorck, I., Asp, N.-G., Birkhed, D., Eliasson, A.-C., Sjoberg, L.-B. and Lundquist, I. (1984a) Effects of processing on starch availability *in vitro* and *in vivo.* I. Extrusion cooking of wheat flours and starch. *Journal of Cereal Science* 2, 91–103.

Bjorck, I., Asp, N.-G., Birkhed, D. and Lundquist, I. (1984b) Effects of processing on starch availability *in vitro* and *in vivo.* II. Drum drying of wheat flour. *Journal of Cereal Science* 2, 165–178.

Blanshard, J.M.V. (1987) Starch granule structure and function: a physicochemical approach. In: Galliard, T. (ed.) *Critical Reports on Applied Chemistry,* Vol. 13, *Starch: Properties and Potential.* Published for the Society of Chemical Industry by John Wiley & Sons, Chichester, UK, pp. 16–54.

Brown, I. (1996) Complex carbohydrates and resistant starch. *Nutrition Reviews* 54, S115-S119.

Central Soya Feed Research (1989) *Pig Tests* 1132, 1098, 1124.

Cooke, D. and Gidley, M.J. (1992) Loss of crystalline and molecular order during starch gelatinisation: origin of the enthalpic transition. *Carbohydrate Research* 227, 103–112.

Corring, T., Aumaître, A. and Durand, G. (1978) Development of digestive enzymes in the piglet from birth to 8 weeks. I. Pancreas and pancreatic enzymes. *Nutrition and Metabolism* 22, 231–243.

Craig, S.A.S. and Starck, J.R. (1984) The effect of physical damage on the molecular structure of wheat starch. *Carbohydrate Research* 125, 117–125.

Efird, R.C., Armstrong, W.D. and Herman, D.L. (1982a) The development of digestive capacity in young pigs: effects of weaning regimen and dietary treatment. *Journal of Animal Science* 55, 1370–1379.

Efird, R.C., Armstrong, W.D. and Herman, D.L. (1982b) The development of digestive capacity in young pigs: effect of feeding different diets on the exocrine pancreatic secretion of nitrogen, amino acids and enzymes in growing pigs. *Journal of the Science of Food and Agriculture* 62, 229–234.

Efird, R.C., Armstrong, W.D. and Herman, D.L. (1982c) The development of digestive capacity in young pigs. Effect of age and weaning system. *Journal of Animal Science* 55, 1380–1394.

Eliasson, A.-C. and Gudmundsson, M. (1996) Starch: physicochemical and functional aspects. In: Eliasson, A.-C. (ed.) *Carbohydrates in Food.* Marcel Dekker, New York, pp. 431–503.

Englyst, H.N., Kingman, S.N. and Cummings, J.H. (1992) Classification and measurement of nutritionally important starch fractions. *European Journal of Clinical Nutrition* 46, 533–550.

Fogarty, W.M. (1983) Microbial amylases. In: *Microbial Enzymes and Biotechnology*. Applied Science Publishers, London, pp. 1–92.

French, D. (1973) Chemical and physical properties of starch. *Journal of Animal Science* 37, 1048–1061.

French, D. (1984) Organisation of starch granules. In: Whistler, R.C., BeMiller, J.N. and Pachall, E.F. (eds) *Starch: Chemistry and Technology*. Academic Press, London, pp. 183–247.

Galliard, T. and Bowler, P. (1987) Morphology and composition of starch. In: Galliard, T. (ed.) *Critical Reports on Applied Chemistry*, Vol. 13, *Starch: Properties and Potential.* Published for the Society of Chemistry by John Wiley & Sons, Chichester, UK, pp. 55–78.

Gatnau, R. and Zimmerman, D.R. (1991) Spray dried porcine plasma (SDPP) as a source of protein for weanling pigs in two environments. *Journal of Animal Science* 69 (suppl. 1), 103.

Gatnau, R., Paul, R. and Zimmerman, D. (1990) *Spray Dried Porcine Plasma as a Source of Immunoglobulins for Newborn Pigs.* Iowa State University Swine Research Report, p.13.

Gatnau, R., Zimmerman, D.R., Diaz, T. and Johns, J. (1991) Determination of optimum levels of spray dried porcine plasma (SDPP) in diets for weanling pigs. *Journal of Animal Science* 69 (suppl. 1), 369.

Gough, B.M. and Pybus, J.N. (1971) Effect on the gelation temperature of wheat starch granules of prolonged treatment with water at 50°C. *Stäerke* 23(6), 210–212.

Gray, G.M. (1992) Starch digestion in non-ruminants. *Journal of Nutrition* 122, 172–177.

Greenwood, C.T. (1970) Starch and glycogen. In: Pigman, W. and Horton D. (eds) *The Carbohydrates: Chemistry and Biochemistry*, 2nd edn, Vol. 11B. Academic Press, New York and London, pp. 471–481.

Greenwood, C.T. (1979) Observations of the structure of the starch granule. In: Blanshard, J.M.V. and Mitchell, R.J. (eds) *Polysaccharides in Foods.* Butterworths, London, pp. 129–138.

Hampson, D.J. (1994) Post-weaning *Escherichia coli* diarrhoea in pigs. In: Gyles, C.L. (ed.) Escherichia coli *in Domestic Animals and Humans.* CAB International, Wallingford, UK, pp. 171–191.

Hampson, D.J., Kidder, D.E. and Hampson, E.M. (1984) Attempts to modify post-weaning changes in the piglet gastrointestinal tract. *Proceedings of the Nutrition Society* 43, 19A.

Hansen, J.A., Goodband, R.D., Nelssen, J.L. and Weeden, T.L. (1990) Effects of substituting spray dried plasma protein for milk products in starter pig diets. In: *1990 Swine Industry Day Report of Progress* (SRP610), Kansas State University (Internet Publication – oznet.ksu.edu/library/), p. 30.

Harbers, L.H. (1975) Starch granule structural changes and amylolytic patterns in processed sorghum grain. *Journal of Animal Science* 41, 1496–1501.

Holm, J., Bjorck, I., Ostrowska, S., Eliasson, A.-C., Asp, N.-G., Larsson, K. and Lundquist,

I. (1983) Digestibility of amylose–lipid complexes in-vitro and in-vivo. *Starch/Starke* 35, 294–297.

Holm, J., Bjorck, I., Asp, N.-G., Sjoberg, L.-B. and Lundquist, I. (1985) Starch availability in vitro and in vivo after flaking, steam-cooking and popping of wheat. *Journal of Cereal Science* 3, 193–206.

Holm, J., Bjorck, I. and Eliasson, A.-C. (1988) Effects of thermal processing of wheat on starch. I. Physico-chemical and functional properties. *Journal of Cereal Science* 8, 249–260.

Hoover, R. and Vasanthan, T. (1994) Effect of heat-moisture treatment on the structure and physicochemical properties of cereal, legume and tuber starches. *Carbohydrate Research* 252, 33–53.

Huang, S.X., Sauer, W.C., Pickard, M., Li, S. and Hardin, R.T. (1998) Effect of micronisation on energy, starch and amino acid digestibility in hull-less barley for young pigs. *Canadian Journal of Animal Science* 78, 81–87.

Kats, L.J., Nelssen, J.L., Tokach, M.D., Goodband, R.D., Hansen, J.A. and Laurin, J.L. (1994) The effect of spray dried porcine plasma on growth performance in the early weaned pig. *Journal of Animal Science* 72, 2075.

Kelly, D., Smyth, J.A. and McCracken, K.J. (1991a) Digestive development of the early-weaned pig. 1. Effect of continuous nutrient supply on the development of the digestive tract and on changes in digestive enzyme activity during the first week post-weaning. *British Journal of Nutrition* 65, 169–180.

Kelly, D., Smyth, J.A. and McCracken, K.J. (1991b) Digestive development of the early-weaned pig. 2. Effect of level of food intake on digestive enzyme activity during the immediate post-weaning period. *British Journal of Nutrition* 65, 181–188.

Kidder, D.E. and Manners, M.J. (1980) The level and distribution of carbohydrates in the small intestine mucosa of pigs from 3 weeks of age to maturity. *British Journal of Nutrition* 43, 141–153.

Lindemann, M.D., Cornelius, S.G., Kandelgy, S.M., Moser, R.L. and Pettigrew, J.E. (1986) Effect of age, weaning and diet on digestive enzyme levels in the piglet. *Journal of Animal Science* 62, 1298–1307.

Liu, H. and Lelievre, J. (1993) A model of starch gelatinisation linking differential scanning calorimetry and birefringence measurements. *Carbohydrate Polymers* 20, 1–5.

MacGregor, A.W. and Fincher, G.B. (1993) Carbohydrates of the barley grain. In: MacGregor, A.W. and Bhatty, R.S. (eds) *Barley: Chemistry and Technology.* American Association of Cereal Chemists, St Paul, Minnesota, pp. 73–130.

Manners, D.J. (1979) The enzymic degradation of starch. In: Blanshard, J.M.V. and Mitchell, J.R. (eds) *Polysaccharides in Food.* Butterworths, London, pp. 75–91.

Marsh, R.D.L. (1986) A study of the retrogradation of wheat starch systems using X-ray diffraction. PhD Thesis, University of Nottingham, UK.

McCracken, K.J. (1984) Effect of diet composition on digestive development of early-weaned pigs. *Proceedings of the Nutrition Society* 43, 109A.

McCracken, K.J. and Kelly, D. (1984) Effect of diet and post-weaning food intake on digestive development of early-weaned pigs. *Proceedings of the Nutrition Society* 43, 110A.

Miller, B.G., James, P.S., Smith, M.W. and Bourne, F.J. (1986) Effect of weaning on the capacity of pig intestinal villi to digest and absorb nutrients. *Journal of Agricultural Science (Cambridge)* 107, 579–589.

Morrison, W.R. (1988) Lipids. In: Pomeranz, Y. (ed.) *Wheat: Chemistry and Technology*, 3rd edn, Vol. 1. American Association of Cereal Chemists, St Paul, Minnesota, pp. 373–439.

Osman, E.M. (1967) Starch in the food industry. In: Whistler, R.L. and Paschall, E.E. (eds) *Starch: Chemistry and Technology*, Vol. 2. Academic Press, New York, pp. 163–215.

Osman, H.F., Theurer, B., Hale, W.H. and Mehen, S.M. (1970) Influence of grain processing on *in vitro* enzymatic starch digestion of barley and sorghum grain. *Journal of Nutrition* 100, 1133–1139.

Owsley, W.F, Orr, D.E. and Tribble, L.F. Jr (1986) Effects of age and diet on the development of the pancreas and the synthesis and secretion of pancreatic enzymes in the young pig. *Journal of Animal Science* 63, 497–504.

Patil, A.R., Murray, S.M., Hussein, H.S. and Faheyg, C. Jr (1998) Quantification of starch fractions in selected starchy feed ingredients. *Journal of Animal Science* 76, 168 (Abstract).

van der Poel, A.F.B., den Hartog, L.A., van den Abeele, T., Boer, H. and van Zuilichem, D.J. (1989) Effect of infrared irradiation or extrusion processing of maize on its digestibility in piglets. *Animal Feed Science and Technology* 26, 29–43.

Reddy, N.R., Pierson, M.D., Sathe, S.K. and Salunkhe, D.K. (1984) Chemical, nutritional and physiological aspects of dry bean carbohydrates: a review. *Food Chemistry* 13(1), 25–68.

Robyt, J.F. (1984) Enzymes in the hydrolysis and synthesis of starch. In: Whistler, R.L., BeMiller, J.N. and Paschall, E.F. (eds) *Starch: Chemistry and Technology*, 2nd edn. Academic Press, London, pp. 87–123.

de Rodas, B.Z., Sohn, K.S., Maxwell, C.V. and Spicer, L.J. (1995) Plasma protein for pigs weaned at 19 and 24 days of age: effect on performance and plasma insulin-like growth factor 1, growth hormone, insulin and glucose concentrations. *Journal of Animal Science* 73, 3657.

Rooke, J.A., Slessor, M., Fraser, H. and Thomson, J.R. (1998) Growth performance and gut function of piglets weaned at four weeks of age and fed protease-treated soya bean meal. *Animal Feed Science and Technology* 70, 175–190.

Sauer, W.C., Modsenthin, R. and Pierce, A.B. (1990) The utilisation of pelleted, extruded and repelleted diets by early weaned pigs. *Animal Feed Science and Technology* 31, 269–275.

Siljeström, M., Westerlund, E., Borck, I., Holm, J., Asp, N.-G. and Theander, O. (1986) The effect of various thermal processes on dietary fibre and starch content of whole grain wheat and white flour. *Journal of Cereal Science* 4, 315–323.

Tester, R.F. (1997) Starch: the polysaccharide fractions, In: Frazier, P.J., Donald, A.M. and Richmond, P. (eds) *Starch: Structure and Functionality*. The Royal Society of Chemistry, London, pp. 163–171.

Tester, R.F. and Morrison, W.R. (1990) Swelling and gelatinisation of cereal starches. I. Effects of amylopectin, amylose and lipids. *Cereal Chemistry* 67, 551–557.

Würsch, P., Delvedovo, S. and Koellrevtter, B. (1986) Cell structure and starch nature as key determinants of the digestion rate of starch in legume. *American Journal of Clinical Nutrition* 43, 25–29.

Young, A.H. (1984) Fractionation of starch. In: Whistler, R.L., BeMiller, J.N. and Paschall, E.F. (eds) *Starch: Chemistry and Technology*, 2nd edn. Academic Press, London, pp. 249–283.

5

Non-starch Polysaccharides in the Diets of Young Weaned Piglets

J.R. Pluske, J.C. Kim, D.E. McDonald, D.W. Pethick and D.J. Hampson

Division of Veterinary and Biomedical Sciences, Murdoch University, Murdoch WA 6150, Australia

Introduction

Carbohydrates comprise between 60% and 80% of the digestible dry matter (DM) in the majority of feedstuffs for weaner pigs. The carbohydrate ingested by piglets can be divided into one fraction (starch) that is targeted by the endogenous enzymes of the gastrointestinal tract (α-amylase), and another component ('dietary fibre', DF) that is digested predominantly in the terminal ileum and large intestine. The major constituent of DF in diets for young pigs influencing production, gut function and possibly enteric disease susceptibility is derived from non-starch polysaccharides (NSPs). Traditionally the 'indigestible' carbohydrate content of feedstuffs has been expressed as 'crude fibre', a term that provides no information regarding possible physicochemical effects of NSP *in vivo* in the animal or its likely digestibility, simply because it is the residue remaining from extractions of the major plant cell wall components.

Non-starch polysaccharides are a large variety of polysaccharide molecules, often in association with phenolic lignified polymers, protein and starch, that have glucosidic bonds other than the α-(1→4), (1→6) bonds of starch. The building blocks of plant cell wall polysaccharides are the pentoses (arabinose and xylose), hexoses (glucose, galactose and mannose), 6-deoxyhexoses (rhamnose and fucose) and uronic acids (glucuronic and galacturonic acids). These monomers are chemically linked to each other to build various NSPs in the plant cell walls of both cereals and legumes. The major NSPs of plant cell walls, therefore, comprise cellulose (linear β-glucan chains), non-cellulosic polysaccharides (arabinoxylans, mixed-linked β-glucans, mannans, galactans, xyloglucan) and pectic polysaccharides (polygalacturonic acids, which may be substituted with arabinan, galactan and arabinogalactan) (Theander *et al.*, 1989; Lewis, 1993; Bach Knudsen, 1997; Choct, 1997). The NSPs, in turn, can be divided into the 'soluble' fraction and the 'insoluble' fraction. The term 'soluble' refers to solubility in water or weak alkali solutions. It is generally accepted that the large majority of NSPs exhibiting anti-nutritive properties in monogastric animals, or at least the broiler chicken, are water-soluble and give rise to viscous aqueous solutions even when present at relatively low levels (Annison, 1993).

The major polysaccharides present in grains and legumes are presented (as average values) in Table 5.1. However, and from a physiological perspective, it is important to consider also the *variation* that exists within a particular cereal, because this may determine some of the physicochemical properties of the grain *in vivo* that, in turn, are likely to have an effect on digestibility and performance. Further discussion of the implications of this variation will be discussed later in this chapter.

'Fibre' Recommendation in Weaner Pig Diets

One of the difficulties associated with reviewing the literature in this field, and hence in making dietary recommendations, is the lack of any requirements *per se* for NSP in *any* class of pig, let alone the weaner, where the task would be more difficult. The National Research Council (NRC, 1998), for instance, does not have a requirement as such for NSP (or any fibrous constituent for that matter) in young pig diets. However, Bolduan *et al.* (1988) commented that the crude fibre content of a starter diet should be about 50 g kg^{-1} in order for the hindgut to be 'activated' after weaning, and that this could be supplied by adding 500 g barley kg^{-1}, 200–300 g wheat bran kg^{-1}, 100 g lucerne meal kg^{-1} or 50 g straw meal kg^{-1} to a starter feed. It is unlikely that commercial starter diets would contain such inclusion levels of these particular ingredients.

Another issue associated with NSP in weaner diets is the selection of a suitable 'control' diet against which to test the addition of NSP sources, and hence assess the effects of NSP on production. Ideally a 'control' diet would be one containing as little NSP as possible. Virtually all weaner diets, however, contain cereals, legumes and (or) oilseeds and so, by default, contain considerable levels of NSP. Studies comparing 'low' NSP diets with 'high' NSP diets in weaner pig nutrition, therefore, are rare. Ingredients low in NSP are maize, sorghum and rice, and diets can be formulated with these cereals and animal protein products that are very low in total and soluble NSP. Diets such as these provide a useful comparison to assess the contribution of NSP to performance and gut physiology.

Pluske *et al.* (1998) fed either a sorghum-based diet (41 g total NSP kg^{-1}, 8 g soluble NSP kg^{-1}) or a 'conventional' wheat/barley/lupin-based diet (176 g total NSP kg^{-1}, 40 g soluble NSP kg^{-1}) to growing pigs between 20 and 55 kg liveweight. These authors found that the sorghum-fed pigs grew 11% faster, ate 2% more food and converted feed to liveweight gain 10% better than pigs fed the wheat/barley/lupin-based diet. Empty bodyweight gain could not be ascertained, but the digesta weight in the large intestine (caecum plus colon) of the sorghum-fed pigs was 29% less than that in the conventionally fed pigs (1.7 vs. 2.4 kg, $P < 0.001$). As a consequence the dressing percentage in the sorghum-fed pigs was 5% higher ($P < 0.05$) compared with their counterparts fed the wheat/barley/lupin-based diet. As will be discussed later, it is evident that NSPs are likely to have an influence on carcass gain.

It is well described that, at least in growing pigs, the 'fibrous' components of feedstuffs usually have a negative influence on energy and protein digestibility, and sometimes growth and feed efficiency. For example, the Agricultural Research Council (1967) concluded that in growing pigs every 10 g increase in crude fibre kg^{-1} feed depressed energy digestibility by 1.3%, utilization of metabolizable energy (ME) by 0.9%, feed efficiency by 3% and 'growth' by 2%. Similarly, Just *et al.* (1983) reported that a 1% increase in the crude fibre content of the diet depressed gross energy digestibility by 1.3% and utilization of ME by 0.9%. More recent work (Zijlstra *et al.*, 1999) shows that in growing pigs the digestible energy (DE) content in wheat was more negatively related (higher coefficient of determination) to total xylose ($r = -0.78$, $P < 0.001$) than other chemical measurements. However, from the point of view of practical chemical analysis, correlations found between DE content and neutral detergent fibre (NDF) and acid detergent fibre (ADF) concentrations were just as valid ($r = -0.70$, $P < 0.01$).

Given that the newly weaned pig lacks endogenous enzymes capable of hydrolysing NSP and is assumed to have an undeveloped microflora for carbohydrate digestion in the ileum and large intestine, there is a certain degree of hesitation when it comes to including 'high' levels of NSP in young pig diets. In the absence of any guidelines regarding requirements, how much NSP is too much? And is some NSP actually 'good' for the weaned pig?

Aumaître *et al.* (1995) suggested that 'appropriate' fibrous feeds could be added to weaner diets to supply fermentative substrates to the flora of the large intestine, thereby promoting physiological and functional development. In turn, a shift to acid fermentation based on NSP could decrease the formation of amines

Table 5.1. Typical NSP contents (g kg^{-1} dry matter) and major polysaccharides in grains and legumes.

Ingredient	Soluble NSP	Insoluble NSP	Total NSP	Major NSP
Wheat[a]	25	94	119	Arabinoxylan
Barley (hulled)[b]	45	122	167	β-Glucan
Barley (hull-less)[a]	50	74	124	
Rye[a]	42	110	152	Arabinoxylan
Oat (hulled)[a]	40	192	232	β-Glucan
Oat (hull-less)[a]	55	63	116	
Soybean[c]	27	16	192	Galacturonans, arabinans and galactose
Pea[c]	25	322	347	Rhamnogalacturonan, glucan
Lupin[c]	46	320	366	Rhamnogalacturonan, arabinose and galactose
Lupin kernel[d]	27	218	245	

[a]Bach Knudsen (1997); [b]Englyst (1989); [c]Choct (1997); [d]Annison *et al.* (1996).

(e.g. cadaverine, putrescine, histamine, tryptamine) in the colon and reduce diarrhoea. The hypothesis raised by these observations is that an 'optimum' amount of appropriate NSP in a starter diet could not only prevent digestive disturbances, but also contribute to the adaptation of the digestive function of the large intestine. But what is this amount? The question also remains whether the inclusion of NSP in the diet of pigs immediately after weaning at the expense of more highly digestible (and palatable?) ingredients is advantageous from an economic, nutritional and disease perspective. Ball and Aherne (1987) stated that a low level of faecal energy digestibility was significantly related to a higher incidence of diarrhoea in weaned pigs. If one assumes that the major contributor to this energy 'indigestibility' was the NSP fraction, does this then mean that the likelihood of post-weaning diarrhoea is increased if more NSP is included in the diet? Or is it the nature of the NSP that is more important?

Influence of NSP on Weaner Pig Performance

Research gathered since the 1980s has shown, in general, that the newly weaned pig can adapt remarkably quickly to a marked change in diet. During lactation the suckling pig has a large intestinal volume of 30–40 ml kg^{-1} liveweight, but this volume increases rapidly to 90–100 ml kg^{-1} liveweight in the first 7–14 days following weaning (Bolduan *et al.*, 1988). The large intestine, therefore, provides considerable 'spare capacity' for microbial fermentation of undigested carbohydrates that, in turn, can provide energy (albeit less efficiently) to the animal.

In this regard, interest has been expresssed in the use of certain by-products that can be included cheaply yet promote similar performance to existing and more expensive feedstuffs. One of the products used in some parts of the world is sugarbeet pulp, the residue remaining after the extraction of sucrose from sugarbeet roots. Sugarbeet pulp contains high levels (> 600 g kg^{-1} DM) of NSP that is largely of pectic origin, and hence is rapidly fermented in the distal ileum and hindgut of the pig. This contrasts with the NSP content of 100–200 g kg^{-1} DM present in cereal grains such as barley and wheat, which is less fermentable. Growing pigs and sows can digest and perform adequately on diets containing appreciable levels of sugarbeet pulp. However, sugarbeet pulp may be unsuitable for young pigs, because reduced voluntary food intake and the underdeveloped microflora of the large intestine have generally been considered the major factors limiting the use of fibrous foods in such diets for weaned pigs (Gill *et al.*, 2000). Furthermore, where growth and digestibility studies have been conducted using young pigs, the types of fibrous ingredients and the basal diets used are often incompatible with those used commercially (Mosenthin *et al.*, 1999).

With reference to sugarbeet pulp, Gill *et al.* (2000) used a total of 240 piglets weaned at 28 days of age in a 3 × 2 factorial arrangement of treatments to assess the suitability of this feedstuff for young pigs. The experiment comprised three different base diets (wheat; barley; wheat/unmolassed dried sugarbeet pulp) in the presence or absence of appropriate enzyme complements. There were no consistent and significant effects of diet type on food intake, liveweight gain or food conversion ratio (FCR) (except in week 1, where pigs fed the wheat-based diet performed significantly better). Although the general response to enzyme supplementation, averaged for all three diets, was positive for weight gain and FCR, within-week differences were not statistically significant. However, over the 28-day trial, pigs given enzyme-supplemented diets showed improved FCR compared with pigs not offered an enzyme (1.56:1 vs. 1.50:1, $P < 0.05$). These data suggest that growth rate and FCR were not adversely affected when wheat in a simple diet was replaced either completely by barley or partly by unmolassed sugarbeet pulp.

Other authors (Edwards *et al.*, 1991; Vervaeke *et al.*, 1991; Longland *et al.*, 1994) have reported similar results with sugarbeet pulp. For example, Longland *et al.* (1994) compared the performance of pigs weaned at 21 days of age (5.51 kg) and fed either a cereal-based (wheat and barley) diet or a diet where 150 g kg^{-1} cereal was replaced with sugarbeet pulp. Between 21 and 57 days of age, Longland *et al.* (1994) reported no difference in performance indices between pigs fed the two diets, nor were there any major influences on the apparent faecal digestibility of dietary nitrogen or gross energy (GE) (Table 5.2). Pigs given 150 g sugarbeet pulp kg^{-1} were capable of digesting twice the total intake of NSP compared with pigs given the cereal-based diet (164 vs. 75 g NSP kg^{-1} diet), and this occurred within 2 weeks of weaning. The apparent digestibilities of arabinose, mannose, glucose and uronic acids were greater ($P < 0.05$) in the pigs fed sugarbeet pulp than in those fed the cereal-based diet, suggesting that the polysaccharides in cereal plant cell walls are more resistant than those contained in beet pulp (Vervaeke *et al.*, 1991).

A repeatable effect of feeding NSP to weaned piglets is an increase in gut fill and increases in the weight of the large intestine (caecum plus colon) caused by enhanced microbial fermentation and production of volatile fatty acids. The studies of Edwards *et al.* (1991), Longland *et al.* (1994) and Gill *et al.* (2000) showed no adverse effects on liveweight gain and FCR of feeding fermentable NSP to young pigs. However, if body gain is expressed in terms of carcass gain then carcass growth

Table 5.2. Performance data and faecal digestibilities of nitrogen (N) and gross energy (GE) in diets containing 0 (Control) or 150 g sugarbeet pulp (SBP) kg^{-1} (after Longland *et al.*, 1994).

		Age of pigs (days)				
	Diet	28	35	42	57	Average
Food intake (g day^{-1})	Control	354	468	603	763	544
	15% SBP	337	431	570	740	519
Gain (g day^{-1})	Control	248	331	352	512	370
	15% SBP	216	338	381	512	366
FCR (g food g^{-1} gain)	Control	1.42	1.45	1.75	1.51	1.47
	15% SBP	1.57	1.28	1.50	1.46	1.42
Digestibility						
N (days 32 and 56)	Control		0.70		0.71	
	15% SBP		0.67		0.74	
GE (days 32 and 56)	Control		0.78		0.76	
	15% SBP		0.76		0.77	

rate (i.e. empty bodyweight gain, or the gain of the animal corrected for gut fill) may well have been compromised. A key question, therefore, is whether feeding higher levels of NSP after weaning has any long-term effects on carcass yield at slaughter, in terms of both carcass acceptability and profitability.

McDonald *et al.* (1998) established an experimental model based on feeding white rice to the weaned pig to examine the effects of NSP and resistant starch (i.e. starch not digested by endogenous α-amylase) on aspects of digestive physiology and post-weaning colibacillosis. McDonald *et al.* (1998) weaned piglets at 21 days of age and fed them one of four diets on an *ad libitum* basis. The reference diet (R) was hydrolysed rice fortified with sources of animal protein. In the remaining diets, R was substituted with either: (i) a high amylose maize starch (resistant starch, or RS) at 22% of the diet (RS diet); (ii) an NSP isolate derived from Australian sweet lupins (31%; L diet); or (iii) both RS (14%) and the lupin isolate (20%; RS+L diet). The lupin NSP isolate contained both soluble (7% by weight) and insoluble (34% by weight) NSP. Pigs were fed for 10 days, at which time they were euthanased and indicators of intestinal development were measured.

The addition of either RS or lupin NSP increased fermentation in the large intestine of piglets, as shown by a decreased pH of digesta (for both treatments) and an increased size of the large intestine (lupin NSP only) (Table 5.3). Pigs fed the diet based on rice and animal protein grew faster ($P < 0.05$), ate more food ($P < 0.05$) and had drier faeces ($P < 0.01$) (data not shown). Based on carcass growth rate, it appeared that diets promoting growth of large intestinal tissue did so at the expense of carcass gain.

Furthermore, changes in organ size in response to feeding high levels of NSP are likely to have an impact on energy metabolism, as visceral organs have a high rate of energy expenditure relative to their size. For example, Yen *et al.* (1989) found that hepatic–portal-vein-drained organs, including the large intestine, consumed a disproportionately high amount (approximately 25%) of whole-body maintenance expressed as O_2 consumption, although it represents < 4% of bodyweight. Given this, it is possible that some of the declines in efficiency of feed use seen when weaner pigs are fed 'high fibre' diets may be related to increased basal heat production in addition to reduced energy (and amino acid) digestibility (Varel and Yen, 1997). However, whether such events early in the pig's phase of growth have detrimental effects on carcass composition later is largely unknown.

Physicochemical Properties of NSP

Effects of NSP on gut physiology

Numerous workers have shown that soluble NSP depresses the activity of certain pancreatic enzymes *in vitro*, namely, amylase, lipase, trypsin and chymotrypsin (Mosenthin *et al.*, 1999). An *in vitro* human study found that cellulose and xylan reduced the activity of amylase, lipase, trypsin and chymotrypsin to less than half their original activity (Dunaif and Schneeman, 1981). The reduction in enzyme activity was due to non-specific binding of the enzymes by the NSP polymers.

However, the inhibitory effects of NSP on the activity of the intestinal enzymes may not have significant effects on the digestibility of food in the small intestine, because of the very large excess ('spare capacity') of enzyme activity present in pancreatic secretions (Selvendran *et al.*, 1987).

To support this notion, Lizardo *et al.* (1997) fed 25-day-old weaned piglets one of four wheat-based diets that contained 0 or 120 g sugarbeet pulp kg^{-1} as the major source of NSP, with additional protein being supplied by either soybean meal or soluble fish protein concentrate. These authors showed that when pigs were euthanased at 56 days of age, feeding 120 g sugarbeet pulp kg^{-1} sustained the activity of intestinal and pancreatic enzymes, though no data showing digestibility were presented. Similarly, Jensen *et al.* (1998) found no differences in pancreatic enzyme activities when 28-day-old weaned pigs were fed barley-based diets with or without β-glucanase addition.

Larger amounts of water are found within the gut when the content of dietary NSP is increased, due to the hydrophilic nature of NSP and increased endogenous secretions. For example, wheat bran increased secretion of pancreatic juice by 115%, protein by 40%, chymotrypsin by 59%, trypsin by 53%, lipase by 78% and amylase by 70% (Low, 1989). Moreover, NSPs elevate the secretory output from the salivary glands, stomach, liver, pancreas and intestinal wall. These cause an increased excretion of water, proteins, lipid and electrolytes. In particular, soluble NSPs increase the volume of food in the gut by entrapping a large volume of water. This process will increase intestinal secretion, because the greater volume of food elevates intestinal secretions (Low, 1989).

Prolonged intake of soluble NSP shows significant adaptive changes in the digestive system. The changes in the gut are characterized by enlargement of the digestive organs and increased secretion of digestive juices (Low, 1989; Choct, 1997). As evidence of adaptive changes, Younosjai *et al.* (1978) and Southon *et al.* (1985) demonstrated that a high NSP diet given to rats induced higher rates of protein synthesis in the jejunum and ileum, and more rapid mucosal cell division.

Apart from the above direct effects of NSP on gut physiology, soluble NSP may also modify gut physiology by interacting with microflora in the small and

Table 5.3. The influence of feeding NSP and resistant starch (RS) to weaner pigs on growth rate, gut physiology and colon size (after McDonald *et al.*, 1998).

	Diet				
	Reference (R)	RS	Lupin (L)	RS + L	*P*-value
Growth rate (g day^{-1})	89.3 ± 15.6^{a}	73.1 ± 21.3^{a}	4.6 ± 14.8^{b}	59.7 ± 25.1^{ab}	*
Full colon (g)	124.7 ± 13.5^{a}	188 ± 5.5^{ac}	219.9 ± 13.5^{bc}	235.5 ± 40.9^{bc}	**
pH colonic contents	6.65 ± 0.1^{a}	5.84 ± 0.1^{b}	5.67 ± 0.2^{b}	5.51 ± 0.1^{b}	***
Empty bodyweight gain (g day^{-1})	17.1 ± 9.0^{a}	11.8 ± 22.5^{ac}	−59.6 ± 7.6^{b}	−10.1 ± 16.7^{ac}	**

Values are mean ± standard error of the mean (SEM).
Empty bodyweight: weight of pig at slaughter minus weight of the contents of the gastrointestinal tract.
a b c Means in same row without common superscripts are significantly different; * $P < 0.05$, ** $P < 0.01$, *** $P < 0.001$.

large intestine of pigs (Langhout *et al.*, 2000). Insoluble NSP may diminish the overall bacterial activity in the intestinal tract by decreasing the time available for fermentation in the gut; also, bacteria may adhere to the insoluble NSP structure (Smits and Annison, 1996). However, viscous soluble NSPs significantly elevate fermentation in the terminal part of the small intestine. Since soluble NSPs increase the average retention time of digesta in the gastrointestinal tract (Smits and Annison, 1996; Choct, 1997), an excellent environment is created for anaerobic microflora, due to decreased oxygen tension. Increased retention time may increase the amount of undigested material in the small intestine, which gives the anaerobic microflora more time and substrates to colonize the proximal small intestine. Consequently, greater retention time provides more chance for enhanced bacteria adhesion to the mucosal surface, which is an essential process in a number of bacterial infections. Production of toxins and deconjugation of bile salts, which are essential for the digestion of fat, may be increased by proliferation of some anaerobic organisms, because most of the soluble NSP sources are fermentable (Smits and Annison, 1996; Choct, 1997).

An increase in bacterial activity in the small intestine may cause a systemic effect on the gut secretions and morphology of the small intestine. As a result, poor digestibility of nutrients may result from reduced nutrient absorption through the affected gut walls. In addition, digestible carbohydrates such as starch and glucose are converted through microbial action to volatile fatty acids, which represents inefficiency of nutrient utilization by monogastric animals (Smits and Annison, 1996; Choct, 1997).

Effects of NSP on nutrient digestion

Non-starch polysaccharides have been recognized as 'anti-nutritive' due to their negative influence on digestion and absorption of starch, protein and lipid in the gut of some monogastric animals, particularly the broiler chicken. The soluble fractions are considered of major importance in determining the nutritive value of feedstuffs for monogastric animals.

The soluble NSP may increase digesta viscosity, and the increased bulk and viscosity of the intestinal contents will decrease the rate of diffusion of substrates and digestive enzymes (Ikegami *et al.*, 1990; Classen and Bedford, 1991). In other studies, the convective transport of glucose was impaired in an *in vitro* viscous environment (Smits and Annison, 1996) and an arabinoxylan-rich extract from rye decreased the rate of dialysis of glucose (Fengler and Marquardt, 1988). Choct and Annison (1992a, b) showed decreased starch digestion using wheat pentosan in poultry. Supplementation with appropriate enzymes has been shown to increase energy utilization in wheat-fed broiler chickens (Annison, 1992) and in pigs (Graham *et al.*, 1989; Baidoo *et al.*, 1998). In addition, NSP depressed protein digestion in broiler chickens (Choct and Annison, 1990, 1992a) and in pigs (Bedford *et al.*, 1992; Baidoo *et al.*, 1998).

Numerous authors have demonstrated that endogenous nitrogen loss, at least in part from mucins, is significantly increased with increased digesta viscosity (Larsen *et al.*, 1993, 1994; Mariscal-Landín *et al.*, 1995). Furthermore, high viscosity stimulates epithelial cell proliferation and may contribute to some loss of epithelial cells when an animal is given soluble NSP (Gee *et al.*, 1996). It was found that soluble NSP with high water-holding capacities disrupted protein digestion and absorption in pigs, while insoluble NSP with high water-holding capacity had no influence on protein digestion and absorption (Leterme *et al.*, 1998). Using 11-week-old pigs (14 kg) fed different sources of NDF, Schulze *et al.* (1995) determined that NDF caused increases in ileal nitrogen flow (reduced digestibility) as a result of increases in both endogenous and undigested dietary ileal nitrogen losses. These authors recommended that diets containing in excess of 200 g NDF kg^{-1} should be supplemented with approximately 10 g ileal digestible protein per kilogram of diet to compensate for these losses.

Absorption

All nutrients that move into the intestinal epithelial cell must pass through the unstirred water layer (UWL) that is adjacent to the intestinal mucosa, and through the lipid membrane of the microvillus surface (Wilson *et al.*, 1971; Wilson and Dietschy, 1974; Smithson *et al.*, 1981). The thickness of UWL is a rate-limiting physiological barrier for nutrient absorption in the gut of monogastric animals. It increases when gel-forming gums and pectin are included in the diet (Johnson and Gee, 1981; Flourie *et al.*, 1984), and an *in vitro* study demonstrated that dietary fibre increased the thickness of UWL by interacting with the glycocalyx of the intestinal brush border, which in turn led to increased mucus production (Satchithanandam *et al.*, 1990). With the thickened UWL, nutrient transport may decrease by increasing resistance to the passive diffusion, resulting in decreased nutrient absorption (Johnson and Gee, 1981). Although there are data suggesting that soluble NSP is gel-forming in the small intestine of the weaned pig, it is possible that subtle changes in the UWL do occur with certain NSPs that reduce absorption. For example, Rainbird *et al.* (1984) and Ellis *et al.* (1995) reported reduced glucose absorption in growing pigs fed guar gum.

Improving the Utilization of NSP in Weaner Diets

In view of certain restrictions on the use of antibiotic growth promoters and some minerals in weaner diets, there is renewed interest in the use of supplementary enzymes to reduce the (purported) anti-nutritive effects of NSP and improve their utilization. Concern, at least in Australia and certain parts of Europe, is predominantly with anti-nutritive effects of NSP contained in wheat and barley but other cereals (e.g. rye, maize, sorghum and triticale), legumes (see Gdala, 1998, for review on dietary fibre in legume seeds) and oilseeds (Bell, 1993) also contain varying levels of NSP. Considerable research effort has been invested in recent times to enhance the utilization of NSP in starter pig diets by the use of added enzymes, and in general results are equivocal. In particular, the association between viscosity (one of the main features causing anti-nutritive effects of NSP in broiler diets) and enzyme use is less pronounced in the young pig. For example, Jensen *et al.* (1998) reported that the addition of β-glucanase to a hull-less barley-based diet increased β-glucan digestibility concurrent with reduced viscosity along the small intestine; however, this did not influence starch and nitrogen digestibility at the ileal level or influence production.

Nevertheless, Bedford *et al.* (1992) reported that the addition of β-glucanase to a barley-based diet fed to weanling pigs improved performance, though a xylanase added to a rye-based diet had no effect. Bedford *et al.* (1992) remarked that the digesta from weaned pigs was more 'watery' compared with that found in chickens, suggesting that diffusional constraints, if any, would be less severe in pigs. In fact, these authors reported *increased* viscosities in the presence of an added pentosanase in rye-fed weanling pigs, but levels overall were far less than those reported for birds. Pigs differ physiologically from young chicks in that the digesta generally tends to have a lower DM (higher water) content. The DM content of the digesta in poultry is approximately 80% (Bedford *et al.*, 1991) compared with approximately 90% for pigs (Baas and Thacker, 1996). Since viscosity induced by β-glucan has been reported to be logarithmically related to concentration (Bedford and Classen, 1992), simple dilution can essentially eliminate the viscosity problem and the associated constraints on luminal diffusion (Campbell and Bedford, 1992).

Cereals

Barley

Typical hulled barley contains 170 g total NSP kg^{-1}, in which 30% is soluble in water; hull-less barley, however, contains 120 g total NSP kg^{-1}, in which 40% is water-soluble (Englyst, 1989; Bach Knudsen, 1997). The major polysaccharide in grain barley is mixed-linked β-glucan, present at levels of 35–45 g kg^{-1} DM in hulled barley and 40–70 g kg^{-1} DM in hull-less barley (Baidoo and Liu, 1998). The soluble and insoluble β-glucan content in hulless barley is consistently higher than in hulled barley. About 50% of both hulled and hulless barley β-glucans is the soluble fraction. Barley also contains considerable amounts of arabinoxylans (hulled: 79 g kg^{-1} DM; hull-less: 48 g kg^{-1} DM) and trace amounts of mannose and galactose (> 5 g kg^{-1} DM) (Bach Knudsen, 1997; Choct, 1997). Large variation in the NSP content of barley is evident according to variety (Baidoo and Liu, 1998), climatic factors, maturity at harvest, and use of fertilizer (Campbell and Bedford, 1992; Jeroch and Dänicke, 1995).

Hulled barley is generally avoided in diets for newly weaned pigs as it does not support maximum growth. The development of hull-less cultivars of barley with a higher DE content may result in increased usage by young pigs but the higher β-glucan levels in hull-less barley may interfere with digestion and absorption

(Åman and Graham, 1989), necessitating the use of exogenous enzymes. Li *et al.* (1996) reported that β-glucanase supplementation to a hull-less barley-based diet (58 g β-glucans kg^{-1} DM in the barley and 38 g β-glucans kg^{-1} in the complete diet) fed to young pigs increased digestion and absorption of DM, energy and amino acids in the small intestine. The enzyme in turn decreased microbial fermentation in the small intestine. This shift in the disappearance of energy from the large to the small intestine due to the addition of β-glucanase also results in improved efficiency of energy utilization (Just *et al.*, 1983). Baidoo *et al.* (1998), working with Canadian varieties of hull-less barley, reported similar benefits in terms of nutrient digestibility when an appropriate enzyme complement was added to diets for weaner pigs. Baidoo *et al.* (1998) also reported improvements in average daily gain (8%, $P < 0.05$) and FCR (8%, $P = 0.92$) when pigs were fed hull-less barley between 9 and 60 kg liveweight.

The addition of β-glucanase to hulled (Graham *et al.*, 1986a, b; Fadel *et al.*, 1988) and hull-less (Li *et al.*, 1996; Baidoo *et al.*, 1998) barley-based diets has been reported to improve the pre-caecal digestibility of mixed-linked β-glucans. This is presumably caused by a combination of the linear β-glucan structure, enhanced microbial activity and residual β-glucanase activity from the grain. This relatively high degree of degradation may play a role in rendering nutrients available for digestion in the small intestine, and may in part explain why the response to added β-glucanase in barley-based diets for pigs is much less than that seen in chickens. In addition, it is thought that β-glucan degradation increases with age (Graham *et al.*, 1986b), though the type of barley (hulled vs. hull-less) has an important influence on the pigs' response to enzyme use.

Wheat

There is less published evidence regarding enzyme use in wheat-based diets, though commercial enzyme preparations targeting wheat NSP are routinely used in some parts of the world with success. Wheat contains, on average, around 120 g total NSP kg^{-1} DM, of which approximately 25 g kg^{-1} is the soluble (extractable) fraction. The major polysaccharides in wheat are arabinoxylans (16 g soluble and 60 g insoluble kg^{-1} DM) and cellulose (20 g kg^{-1} DM). The mixed-linked β-glucans (10 g kg^{-1} DM), mannose (3 g kg^{-1} DM), galactose (4 g kg^{-1} DM) and uronic acids (5 g kg^{-1} DM) exist as minor components (Bach Knudsen, 1997). However, the NSP content of wheat is variable, according to factors such as cultivar type, agronomic conditions and climate (Longstaff and McNab, 1986; Coles *et al.*, 1997). For example, Austin *et al.* (1999) reported a 56% difference in the total NSP content of 12 UK wheats (range: 82.6–129.2 g kg^{-1} DM), with arabinoxylans (arabinose plus xylose) contributing between 36.8 and 81.0 g kg^{-1} DM and β-glucans between 5.6 and 7.2 g kg^{-1} DM. In a cohort of Western Australian wheats, J.C. Kim and J.R. Pluske (unpublished data) found a 26% difference in total NSP contents (range: 84–105 g kg^{-1} DM), with arabinoxylans contributing between 60.8 and 81.4 g kg^{-1} DM. In a Canadian study involving 15 wheat samples, Zijlstra *et al.* (1999) reported a range in total NSP from 105 to 166 g kg^{-1} DM, with arabinoxylans ranging from 69 to 106 g kg^{-1} DM.

In contrast to barley, total-tract degradation of wheat NSP appears to be less. In growing pigs, Bach Knudsen and Canibe (2000) reported quantitative ileal recovery of NSP when animals were fed wheat-flour-based diets with or without wheat bran, with some breakdown of soluble NSP in the large intestine. Similar results would be expected in weaned pigs but the reasons for the variation often observed in the field in wheat-based diets fed to weaner pigs, and for the (sometimes) equivocal effects of in-feed arabinoxylanases, are not known.

Part of the cause may well be the concentration of arabinoxylans in wheat, though data to support this notion, at least in the young pig, are scarce. In a study using 12 UK wheat samples fed to broiler chickens to investigate causes for variation in AME contents, Austin *et al.* (1999) found no correlation between AME content of the wheat samples to: (i) total water-soluble NSP content; (ii) soluble arabinoxylan content; or (iii) (1→3, 1→4)-β-glucan released from the grain. No correlation was found either with the viscosity *in vitro* of aqueous extracts, a result supported in wheat-fed pigs by Lewis *et al.* (1998, 1999). Instead, Austin *et al.* (1999) reported that the degree of branching of arabinoxylan in the wheat samples was positively associated with AME content, while the amount of soluble NSP was negatively correlated to AME content. This indicated, at least in this study, that arabinoxylan structure is at least of equal importance to NSP concentration as a factor determining AME content in broiler chickens.

Further evidence for this notion comes from the work of Annison *et al.* (1995), who investigated the nutritive activity of soluble arabinoxylans extracted from rice bran in broiler chickens. In numerous studies the addition of isolated wheat arabinoxylans to broiler diets has been shown to be detrimental with respect to nutrient digestibility (Choct and Annison, 1992a,b). These studies demonstrated that the viscous nature of solutions of wheat arabinoxylans is an important factor in their anti-nutritive activity, but in the study where the soluble NSP was extracted from rice bran, Annison *et al.* (1995) were unable to find any anti-nutritive properties. Differences in their activity between the two arabinoxylan sources may be explained by differences in their chemical and physical properties. The data in Table 5.4 indicate that, although NSP preparations from rice bran and wheat are predominantly arabinoxylans, the arabinose:xylose ratio in rice bran is greater than that in wheat (1.23 vs. 0.58). This indicated that the β-xylan main chain of the rice bran polysaccharide carries many more arabinose side chains than the wheat β-xylan main chain.

In turn, this has implications for the viscosity-forming properties of polysaccharides. An important factor in the formation of viscous polysaccharide solutions is the strength of the interaction between molecules. In solutions of arabinoxylans the strongest interactions occur between the unsubstituted sections of the xylan main chain that can come into close contact through the formation of inter-chain hydrogen bonds (Morris, 1986). The presence of arabinose side chains inhibits these interactions. Solutions of wheat NSP preparations are much more viscous than those of rice bran (Table 5.4) and this is probably a direct consequence of reduced between-polysaccharide chain interactions caused by the more highly branched structure of the rice bran arabinoxylan (Annison *et al.*, 1995). Therefore, the difference in structures and hence physicochemical properties between these two arabinoxylan preparations is the most likely reason for their difference in nutritive activity. In addition, and as emphasized by Austin *et al.* (1999), the likely confounding influences of endogenous enzyme release associated with grain storage should be considered when investigating relationships between physical and chemical components and production indices.

The implication of the findings presented by Annison *et al.* (1995) and Austin *et al.* (1999) is that, despite the increasing sophistication of analytical techniques available to quantify NSP, it is evident that the physicochemical properties of NSPs still remain extremely important determinants of their (potential) anti-nutritive activity. Measurements *in vitro*, therefore, may not necessarily reflect the physiological events occurring in the gut *in vivo* that will dictate the response, for example, of a supplementary enzyme in a diet. Increased understanding of the physicochemical properties of polysaccharide hydrolysis *in vivo* may provide further understanding of why variation exists in enzyme efficacy, or just feeding value of NSP *per se*.

Legumes

Legumes, such as peas, beans, soybean and lupins, and oilseeds and (or) oilseed meals, such as canola (rapeseed meal), are commonly included in weaner pig diets. These feedstuffs contain considerable quantities of NSP (see Table 5.1 and Gdala, 1998) and may themselves exert anti-nutritive effects. Whilst soybean meal is recognized as virtually the 'gold standard' in vegetable protein sources available for pig diets, in some countries such as Australia there are cheaper sources of protein. In Western Australia, studies conducted with legumes such as yellow lupins (*Lupinus luteus*; Mullan *et al.*, unpublished data) and lathyrus (*Lathyrus cicera*; Trezona *et al.*, 2000) found comparable performances when these feedstuffs replaced soybean meal. Similarly, the use of canola meal in places such as Canada as a replacement for soybean meal in weaner pig diets is well documented (Bell, 1993).

Table 5.4. Chemical and physical properties of soluble NSP isolates from wheat and rice bran (after Annison *et al.*, 1995).

	Isolate composition							
Isolate source	CHO[a]	Ara[b]	Xyl[b]	Man[b]	Gal[b]	Glc[b]	A/X[c]	Vis, cP[d]
Wheat	0.72	0.35	0.60	–	–	0.05	0.58	64
Rice bran	0.63	0.40	0.32	0.03	0.17	0.08	1.23	1.6

[a]Proportion of isolated material that is carbohydrate.
[b]Ara = arabinose; Xyl = xylose; Man = mannose; Gal = galactose; Glc = glucose. The molar proportions of the sugars are shown.
[c]A/X = arabinose:xylose ratio.
[d]Vis = viscosity (cP, centipoise) of a 10 g l^{-1} solution on 0.1 M NaCl at 25°C.

Do legumes and oilseeds exert anti-nutritive properties? The effects of certain proteins in soybean meal such as glycinin and β-conglycinin on gut structure and function in the young pig are well documented (see review by Dréau and Lallès, 1999) but, in general, little research has been conducted in young pigs to answer this question. Gdala *et al.* (1997) included different lupin varieties in maize starch–casein-based diets, with or without α-galactosidase supplementation, and fed them to cannulated pigs weighing between 11 and 24 kg, to examine carbohydrate and amino acid digestibility. Gdala *et al.* (1997) found large and significant improvements in oligosaccharide (raffinose, stachyose and verbascose) digestibility in the presence of α-galactosidase, and also reported improvements in the ileal digestibility of several essential amino acids. In contrast the NSPs (comprising 307–375 g kg^{-1} DM of the lupin varieties used) were poorly digested (11–16%). Oligosaccharides in legumes such as lupins are not viscous-forming (van Barneveld *et al.*, 1995), suggesting that increased osmolarity in the small intestine caused by the residues of the oligosaccharides may reduce nutrient absorption (Wiggins, 1984).

Effects of Processing on NSP Structure

There is some commercial evidence that feeding pelleted wheat-based diets causes an increased occurrence of an enteric condition described as non-specific colitis (Partridge, 1998). The condition has similar clinical signs to porcine intestinal spirochaetosis caused by *Brachyspira pilosicoli*, but it is unclear how and whether these two conditions are related. Where non-specific colitis has been reported, the use of an in-feed arabinoxylanase has been found to reduce this type of enteric condition (Partridge, 1998).

There is some evidence to support the claim that high temperature processing can change the structure of NSP. Pluske *et al.* (1996) showed chemically that extrusion increased the soluble:insoluble NSP ratio in wheat. Working with barley-based diets but in 78 kg pigs, Fadel *et al.* (1988) showed that extrusion caused a shift from insoluble NSP to soluble NSP and insoluble β-glucans to soluble β-glucans (Table 5.5). This, in turn, significantly improved digestibility of starch at the terminal ileum by 16%, dry matter by 12% and energy by 12%. The shift from insoluble to soluble NSP after extrusion cooking resulted in increased digestion of soluble NSP at the ileum (54% more) and increased fermentation of insoluble NSP in the lower tract (56% more). Whether such changes in digestibility would occur in wheat-based diets is questionable.

'Gut Health' and Non-starch Polysaccharides: Is there a Connection?

The resident microbial flora

The diverse collection of microorganisms colonizing the healthy gastrointestinal tract of pigs (i.e. the microbiota) plays an essential role in the overall well-being of the animal (see Pluske *et al.*, 1999, for review). The most dominant bacteria in the stomach and small intestine of pigs are enterobacteria, streptococci and lactobacilli, while the greater diversity of bacteria in the hindgut includes bacterial genera such as *Bacteroides*, *Prevotella*, *Eubacterium*, *Lactobacillus*, *Fusobacterium*, *Peptostreptococcus*, *Selenomonas*, *Megasphaera*, *Veillonella* and *Streptococcus* (Table 5.6) (Jensen, 1999). The population level of the microbiota in various regions of the gastrointestinal tract depends on the doubling time of the microorganism under the physicochemical conditions in the part of the tract under question, and the emptying rhythm of the particular section (Jensen, 1999). The greater stasis time in the lower small intestine and hindgut results in increased multiplication and hence higher microbial numbers. The density of the microbial population in the caecum and colon amounts to approximately 10^{10}–10^{11} viable bacteria per gram of digesta and has more than 500 different species (Moore *et al.*, 1987).

Jensen (1999) has listed the bacteria isolated to date in various segments of the gastrointestinal tract of the pig, and it is interesting to note that *Bifidobacterium* spp. comprise fewer than 1% of the total population of bacteria in the pig gut. This is significant, because there are many reports (largely from the human literature) that inclusion of certain feedstuffs and feed additives that increase bifidobacteria numbers is beneficial for 'gut health'. In the pig, it is questionable whether a feed additive that increases bifidobacteria numbers when it represents only 1% of the total population of bacteria in the pig gut will ever enhance 'gut health'. In contrast, *Streptococcus intestinalis* is the most dominant bacteria in the pig gut, forming approximately 23% of the total population of the cultivable bacteria.

Why the interest in 'gut health'?

Antimicrobial agents are the main tools used for the control of enteric infections, but unfortunately problems are arising over the use of antimicrobials in the pig industry. Their long-term use eventually selects for the survival of resistant bacterial

Table 5.5. The NSP composition of raw and extruded barley-based diets fed to pigs (adapted from Fadel *et al.*, 1988).

	Content of pig diets (g kg^{-1} dry weight)	
	Raw	Extruded
Total NSP	151	153
Insoluble NSP	129	124
Soluble NSP	22	29
Ratio soluble:insoluble NSP	0.17	0.23
β-Glucans		
Total	28	26
Insoluble	19	15
Soluble	2	11
Ratio soluble:insoluble NSP	0.10	0.73

Table 5.6. The predominant bacteria isolated from various regions of the gastrointestinal tract of pigs (% in each region, with number of species in parentheses) (after Jensen, 1999).

	Total [1679]		Ileum [579]		Caecum [529]		Colon [571]	
Bacteria	%	(no.)	%	(no.)	%	(no.)	%	(no.)
Enterobacteria	24.5	(2)	53.4	(2)	10.4	(2)	5.8	(2)
Streptococci	22.8	(2)	32.2	(2)	18.8	(2)	16.8	(2)
Bacteroides	17.7	(19)	0.7	(3)	30.7	(10)	23.5	(13)
Eubacteria	6.3	(20)	0.7	(3)	8.5	(10)	10.3	(12)
Lactobacilli	5.9	(8)	3.4	(7)	5.6	(4)	8.9	(6)
Peptostreptococci	5.1	(15)	0.5	(1)	3.2	(5)	11.9	(10)
Fusobacteria	4.6	(18)	0.2	(1)	9.0	(13)	5.1	(11)
Selenomonads	3.3	(5)	1.2	(1)	5.9	(4)	3.0	(3)
Ruminococci	1.8	(1)	0.0	(0)	3.4	(1)	2.2	(1)
Clostridia	1.5	(3)	2.7	(1)	1.5	(2)	0.3	(3)
Scarcina	1.3	(5)	3.1	(2)	0.2	(1)	0.4	(2)
Megasphaera	1.0	(1)	0.0	(0)	0.5	(1)	2.5	(1)
Butyrivibrios	0.8	(4)	0.0	(0)	1.3	(2)	1.1	(3)
Propionibacteria	0.4	(1)	0.0	(0)	0.9	(1)	0.4	(1)
Bifidobacteria	0.2	(1)	0.5	(1)	0.0	(0)	0.0	(0)
Veillonellae	0.1	(1)	0.0	(0)	0.2	(1)	0.0	(0)
Not characterized	0.5	(7)	0.2	(1)	0.4	(2)	1.1	(4)
No. of species	113		25		61		75	

Numbers in square brackets represent number of isolates.

species or strains, and genes encoding this resistance can be transferred to other formerly susceptible bacteria. A variety of bacterial pathogens of pigs are now showing resistance to a range of antimicrobial drugs. Not only is this reducing the number of antimicrobials available to control bacterial diseases in pigs, but this resistance also poses risks to human health. Alternative methods of control of bacterial pathogens that do not require antimicrobials include: modifying management practices to limit exposure to pathogens and to minimize stress; using and improving available vaccines; undertaking selective breeding of animals for resistance to infectious diseases; and improving the pig's immune responses through the use of cytokines and other immunomodulatory agents (Hampson *et al.*, 2001). Strategies for selective destruction of bacterial pathogens in the gastrointestinal tract include the oral application of organic acids, or inorganic chemicals such as zinc oxide, and the use of specific bacteriophages or bacteriocins. One form of this approach is to feed pigs specialized strains of certain probiotic bacteria, especially *Lactobacillus* spp., which are selected because they are considered to promote gut health and exclude pathogens. Such probiotics might have most promise for use in controlling infections in young pigs – for example, in the period after weaning, when the resident intestinal microflora is not yet stable – but data supporting this notion are equivocal.

A variation on the probiotic approach has been to feed specific dietary components that act as substrate for natural populations of 'protective' bacteria, such that these proliferate and more effectively exclude pathogens. For example, in the case of *Clostridium difficile* infection, different DF sources have been investigated to optimize inhibition by the resident microbial flora through its production of specific short-chain fatty acids (May *et al.*, 1994). Similarly, so-called prebiotic dietary supplements, such as fructose-containing oligosaccharides, have been used to increase numbers of *Bifidobacterium* spp. selectively in the large intestine (though mainly in rats and humans), the presence of which in turn is thought to result in an inhibition of colonization by certain pathogens (Gibson *et al.*, 1995).

The main source of growth substrate for the gastrointestinal microflora comes from the diet, although endogenous secretions can also be utilized by different classes of bacteria. Simple sugars tend to act as the main growth substrate in the upper part of the gastrointestinal tract, whilst in the large intestine, where the main bacterial biomass is located, NSPs serve as the major bacterial substrate. Different forms of fibre in the diet can broadly influence the composition and metabolic activity of the large intestinal microflora in pigs (Pluske *et al.*, 1999; Reid and Hillman, 1999). Currently, even where addition of appropriate substrate is known to stimulate proliferation of specific groups of resident bacteria, little is known about the way in which these bacteria interact with pathogenic species of bacteria. This lack of information makes it difficult to predict how a given dietary component could be used to influence a given enteric pathogen indirectly.

Besides influencing the normal gastrointestinal microflora, diet could also influence colonization by pathogens through other routes. For example, it could act by modulating the amount of specific substrate available for the pathogen at a given site, by influencing viscosity of the intestinal contents and hence altering accessibility of receptor sites and (or) affecting intestinal motility, and by direct or indirect effects on the intestines (Brunsgaard, 1998). Similar changes may occur in specific colonization sites or bacterial receptors on the enterocytes. The diet might also influence intestinal function: for example, components in boiled rice inhibit secretion in the small intestine and hence reduce the magnitude of secretory diarrhoea due to pathogens such as enterotoxigenic *Escherichia coli* (Mathews *et al.*, 1999).

At Murdoch University we have been interested in the interactions between dietary components, but in particular NSP from cereal grains, on specific enteric diseases of pigs. We have taken this approach rather than the more general approach of 'gut health' due to the specific nature of the diseases in question, all of which are serious industry problems. Furthermore, an exact definition of 'gut health' appears to be lacking. Does it refer to a shift in bacterial populations or to one in population numbers? Much of this ambiguity can be attributed to the lack of quantitative information on microbial ecology of the gut in the young pig in response to dietary interventions. The following sections summarize some of the diseases we have studied and our findings in more detail.

Post-weaning colibacillosis

The growth checks and diarrhoea that regularly occur in many piggeries in the first 5–10 days after weaning are a serious industry problem. Colonization of the small intestine by enterotoxigenic strains of *E. coli* in this period results in severe secretory diarrhoea – post-weaning colibacillosis (PWC). Besides mortalities and the requirement for antimicrobial medication, the associated growth checks result in overall increases in the time pigs take to reach market weight (Hampson *et al.*, 2001).

PWC is a multifactorial condition and there are various dietary influences on the disease. For example, diets containing 'high' concentrations of protein (210 g kg^{-1}) have been shown to predispose to the condition (Prohaszka and Baron, 1980). Some highly digestible and milk-based weaner diets have been associated with increased post-weaning diarrhoea, whilst conversely it has been suggested that the inclusion of 'fibre' sources to weaner diets will reduce the incidence and severity of PWC (Bertschinger *et al.*, 1978; Bolduan *et al.*, 1988). The mechanisms of protection are uncertain, but may relate to reduced availability of substrate for the *E. coli* within the lumen of the small intestine.

The single most important control for microbial fermentation in the gastrointestinal tract is the amount and type of substrate available to the microbiota. Given that the NSPs (in terms of chemical and structural composition) are important energy substrates for microbial fermentation, then it might be possible to manipulate the microbiota – for example, to ameliorate PWC – by feeding different levels and types of NSP to young pigs.

In one study, weaner pigs were fed a highly digestible cooked white rice/animal protein diet, or the same diet supplemented with 100 g guar gum kg^{-1} as a source of soluble NSP, and their growth rates, gastrointestinal physiology and faecal excretion of haemolytic *E. coli* were recorded (McDonald *et al.*, 1999). Pigs fed the basic rice diet for 2 weeks after weaning were heavier, and had lighter large intestines and less fermentation at this site, than the pigs fed the diet containing guar gum. When pigs on the two diets were experimentally challenged with enterotoxigenic *E. coli* (serovar O8; G7; K88:K87), significantly more of these organisms were recovered from the small intestine of the pigs on the diet supplemented with guar gum than on the other diet (Table 5.7). Pigs fed a commercial wheat/lupin-based diet had significantly more of the pathogens isolated than did the pigs on the basal rice diet (data not shown).

In a follow-up study, McDonald *et al.* (2001) conducted an experiment aimed at assessing the effect of a soluble NSP source (pearl barley, or hulled barley with the husk removed), with or without β-glucanase supplementation, on the performance, gastrointestinal physiology and intestinal proliferation of enterotoxigenic *E. coli* in 21-day-old weaned pigs experimentally inoculated with *E. coli* (serovar O8; G7; K88: K87). Pigs were infected at 48, 72 and 96 h after weaning and were allowed *ad libitum* access to their feed. All pigs were euthanased between 7 and 9 days after weaning.

Pigs fed the rice-based diet grew faster, had a greater gain in empty bodyweight and had a reduced large intestinal weight (expressed as a proportion of empty bodyweight) than pigs fed the pearl barley-based diet (Table 5.8). Higher concentrations of volatile fatty acids and a lower pH of digesta in the large intestine (data not shown) indicated greater fermentative activity in pigs fed the pearl barley-based diet. Pigs offered the rice-based diet showed a reduction in empty bodyweight gain associated with the enterotoxigenic *E. coli* infection, and showed less proliferation of *E. coli* in the small and large intestine than their barley-fed counterparts (Fig. 5.1). Addition of a β-glucanase enzyme to the pearl barley-based diet failed to reduce viscosity (data not shown) or influence bacterial counts (Fig. 5.1).

The addition of pearl barley to the diet altered the physicochemical properties in the intestines, raising the ileal viscosity and altering the site of microbial fermentation. The energy expended in adapting the intestinal tract for digestion of NSP caused a depression in carcass growth, and this was exacerbated by PWC. The data

Table 5.7. Performance data, gastrointestinal physiology and recovery of haemolytic *E. coli* from the intestinal tract of non-infected and experimentally infected weaner pigs fed either a diet based on cooked white rice plus animal protein or the same diet supplemented with guar gum (after McDonald *et al.*, 1999).

	Uninfected			Experimentally infected		
	Rice	Rice + GG	*P*-value	Rice	Rice + GG	*P*-value
No. of pigs	12	17		12	16	
Start weight (kg)	7.8 ± 0.30	7.8 ± 0.24	NS	6.9 ± 0.22	6.9 ± 0.59	NS
Finish weight (kg)	8.9 ± 0.36	8.4 ± 0.28	*	7.3 ± 0.26	6.6 ± 0.20	**
Daily gain (g)						
Liveweight	141 ± 21	85 ± 15	*	54 ± 26	−52 ± 16	***
EBW	93 ± 18	12 ± 12	***	41 ± 20	−53 ± 17	***
Large intestine (g)						
Full	221 ± 15	296 ± 21	**	218 ± 15	239 ± 27	NS
Empty	95 ± 4	124 ± 9	*	91 ± 9	90 ± 6	NS
pH						
Caecum	6.1 ± 0.08	5.2 ± 0.03	**	5.8 ± 0.14	5.4 ± 0.15	NS
Faeces	6.6 ± 0.05	6.0 ± 0.13	**	6.9 ± 0.07	6.5 ± 0.11	*
VFA pool in colon						
(mmol per pig)	7.5 ± 0.17	16.2 ± 0.26	*	ND	ND	–
Small intestine						
Mean CFU g^{-1}	–	–	–	1.3×10^4	8×10^9	*
No. positive sites	–	–	–	0.53	1.63	**
Colon						
Mean CFU g^{-1}	–	–	–	1.4×10^7	3.7×10^{10}	NS
No. positive sites	–	–	–	1.71	1.69	NS

Maximum number of positive sites = 3.
Data are means ± SEM; NS, not significant; * $P < 0.05$, ** $P < 0.01$, *** $P < 0.001$.
Diets:
Rice = cooked white rice (744 g kg^{-1}) plus animal protein sources (197 g kg^{-1})
Rice + GG = cooked white rice (637 g kg^{-1}) plus animal protein sources (200 g kg^{-1}) plus guar gum (100 g kg^{-1}).
Abbreviations:
EBW, empty bodyweight (i.e. weight of pig at slaughter minus weight of contents of gastrointestinal tract)
VFA, volatile fatty acids
CFU, colony-forming units
ND, not determined.

Table 5.8. Performance, large intestinal weights, number of haemolytic *E. coli* colonies and ileal viscosity in non-infected and infected pigs fed either a rice-based diet or one containing pearl barley (after McDonald *et al.*, 2000a).

	Non-infected		Infected		*P*-value	
	Rice	Barley	Rice	Barley	Diet	Infection
Total weight gain (kg)	1.4 ± 0.36	0.9 ± 0.76	1.4 ± 0.36	−0.6 ± 0.12	*	**
EBW gain (g day^{-1})	74 ± 30.3	26 ± 6.5	−28 ± 10.9	−56 ± 16.1	**	****
Large intestine as % pig weight	2.7	3.8	2.6	3.2	****	NS
Ileal VFA (mmol per pig)	18.2	9.1	11.5	13.5	Interaction = 0.053	
Viscosity in ileum (cP)	2.1 ± 0.31	2.8 ± 0.56	1.6 ± 0.22	2.3 ± 0.29	**	**

Data are means ± SEM; NS, not significant; * $P < 0.10$, ** $P < 0.05$, *** $P < 0.01$, **** $P < 0.001$.
Diets:
Rice = cooked white rice (702 g kg^{-1}, dietary soluble NSP content 4 g kg^{-1}) + animal protein sources (197 g kg^{-1})
Barley = pearl barley (500 g kg^{-1}; dietary soluble NSP content 5 g kg^{-1}) + rice (275 g kg^{-1}) + animal protein sources (200 g kg^{-1}).
EBW = empty bodyweight (i.e. weight of pig at slaughter minus weight of contents of gastrointestinal tract).

from these two experiments suggest that the presence of soluble NSP in weaner diets is detrimental for piglet growth and causes proliferation of enterotoxigenic *E. coli* in the small intestine. It also indicates that there are benefits in feeding a highly digestible rice-based diet to weaners. The mechanism(s) involved in the protection from PWC is not certain, but may be related to the reduced availability of substrate for the bacteria in the small intestine of pigs fed the rice-based diet compared with the wheat-based diet. Certainly the addition of guar gum increases the viscosity of the digesta in the small intestine, and may lead to stasis in the unstirred layer immediately above the epithelium, allowing the bacteria trapped here to proliferate and attach to the epithelium (Cherbut *et al.*, 1990). If dietary soluble NSP is confirmed as a predisposing factor in PWC, then careful selection of ingredients to minimize these components and/or treatment with exogenous enzymes may be helpful in controlling the condition.

The two types of soluble NSP used by McDonald *et al.* (1999, 2001) were highly fermentable and viscous in nature, which raises the question of whether fermentability, viscosity or combinations of both are likely to influence the small intestinal porcine microbiota. To investigate further the potential detrimental effects of increased intestinal viscosity in weaner pigs on proliferation of haemolytic *E. coli*, McDonald *et al.* (2000a) fed experimental diets supplemented with 40 g carboxymethylcellulose (CMC) kg^{-1}, a water-soluble synthetic viscous polysaccharide resistant to microbial fermentation, to 21-day-old weaned pigs for 13 days. At this time they were euthanased and the effects of two types of CMC, either low-viscosity (50–200 cP *in vitro*) or high-viscosity (400–800 cP *in vitro*), on gastrointestinal development, growth performance, faecal dry matter and proliferation of haemolytic *E. coli* were monitored (Table 5.9).

Dietary CMC increased the luminal viscosity in a viscosity-dependent manner along the entire length of the small intestine and in the caecum, and resulted in increased intestinal weights. Faecal dry matter content was significantly reduced in pigs fed CMC. Feeding high-viscosity CMC reduced empty bodyweight gain but not liveweight gain. Pigs fed the rice-based diet remained healthy, whereas those fed either low- or high-viscosity CMC developed diarrhoea within 7 days of weaning and continued to do so until they were euthanased on day 10. Pigs fed the low- or high-viscosity CMC diets shed more haemolytic *E. coli* (serovar 0141; K88) daily than pigs fed the rice-only based diet.

Porcine intestinal spirochaetosis

Porcine intestinal spirochaetosis (PIS) is a chronic diarrhoeal disease of weaner and grower/finisher pigs, resulting from colonization by the anaerobic intestinal spirochaete *Brachyspira (Serpulina) pilosicoli* (Trott *et al.*, 1996; Ochiai *et al.*, 1997). As with the closely related *Brachyspira hyodysenteriae,* the agent of swine dysentery, *B. pilosicoli* colonizes the caecum and colon. Unlike *B. hyodysenteriae*, which is chemotactic to mucus and moves deep into the crypts, *B. pilosicoli* remains largely in the lumen of the intestine, or may attach by one cell end to the epithelium adjacent to the intestinal lumen (Hampson and Trott, 1995).

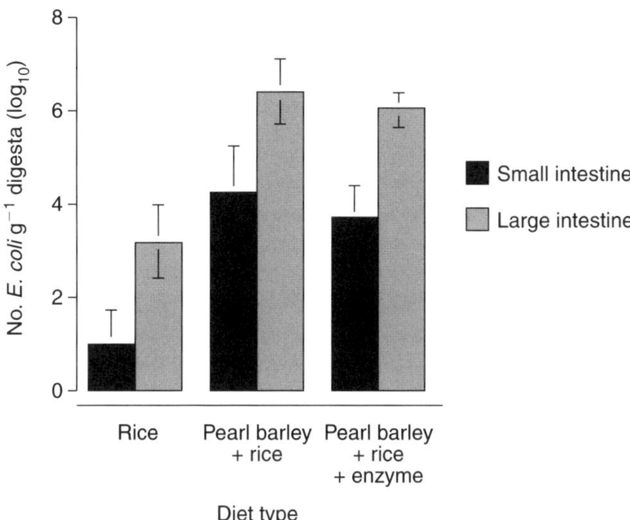

Fig. 5.1. Bacterial counts in the small and large intestine of piglets slaughtered 7 days after weaning (weaning occurred at 21 days of age). Pigs fed diets 'Pearl barley + rice' and 'Pearl barley + rice + enzyme' had significantly ($P < 0.05$) higher numbers of *E. coli* in both the small and large intestines than pigs fed the rice-based diet (D.E. McDonald, unpublished data).

Table 5.9. The effect of diet on intestinal viscosity, gastrointestinal physiology, performance and expression of diarrhoea in weaner pigs (after McDonald *et al.*, 2000a).

	Diet type			Standard error of the mean	
	Rice	Rice+LVCMC	Rice+HVCMC	(SEM)	*P*-value
Viscosity (cP)					
Duodenum	1.42[a]	6.14[b]	8.93[b]	2.631	< 0.001
Ileum	1.40[a]	6.04[b]	7.91[b]	2.205	< 0.001
Caecum	1.67[a]	4.09[ab]	6.18[b]	2.433	*
Weight of small + large intestines (g)	647[a]	882[b]	883[b]	90.8	< 0.001
Daily gain (g)					
Liveweight	291[a]	393[b]	325[ab]	63.8	*
EBW	105[a]	149[a]	29[b]	60.1	*
No. of pigs with diarrhoea after weaning					
Day 7	0/8[a]	5/8[b]	7/7[b]	–	*
Day 8	1/8[a]	3/8[b]	7/7[b]	–	*
Day 9	0/8[a]	4/8[b]	7/7[b]	–	*
Day 10	0/8[a]	4/8[b]	5/7[b]	–	*
Mean % of cultured bacteria that were *E. coli*					
Day 7	13.6	61.4	84.7	?	?
Day 8	14.3	62.5	61.4	?	?
Day 9	25.0	58.8	70.0	?	?
Day 10	0	15.0	29.3	?	?

[a,b] Means in rows not having the same superscripts differ significantly.
Diet types:
Rice=cooked white rice + animal protein supplement
Rice+LVCMC = cooked white rice/animal protein supplement + low viscosity CMC (40 g kg^{-1})
Rice+HVCMC = cooked white rice/animal protein supplement + high viscosity CMC (40 g kg^{-1}).
EBW = empty bodyweight (liveweight minus weight of full gastrointestinal tract).

In view of the close similarity between *B. pilosicoli* and *B. hyodysenteriae*, their very similar habitats in the large intestine, and reports from the field of dietary influences on PIS, an investigation was made into whether the cooked rice diet that protects from swine dysentery might also protect from PIS (Hampson *et al.*, 2000). In this study, two groups of weaner pigs were fed either a standard commercial wheat/lupin weaner diet (n = 8) or the rice-based diet described above (n = 6) for 3 weeks after weaning. All pigs were then challenged orally over 3 days with 10^{10} active mid-log phase cells of a Western Australian field strain of *B. pilosicoli* (strain 95/1000). The pigs were killed 3–4 weeks post inoculation (pi). All animals became colonized with *B. pilosicoli* strain 95/1000, but this occurred significantly later (mean of 10 days pi compared with 3 days), and lasted for significantly less time (mean of 5 days compared with 16 days), in the pigs fed the cooked white rice/animal protein diet compared with those fed the wheat/lupin based diet. One pig fed the wheat/lupin diet developed an acute and severe erosive colitis with severe watery diarrhoea within 3 days pi and was euthanased. All the other pigs on both diets developed mild transient diarrhoea, lasting only 2–3 days. Small areas of mild patchy colitis were observed grossly at post-mortem, but no spirochaete attachment to the epithelium was detected. This study demonstrated, as we have shown previously but in growing pigs with *B. hyodysenteriae*, that colonization by *B. pilosicoli* can be influenced by diet. In this case the rice-based diet did not prevent colonization, but only retarded it.

In a further experiment, McDonald *et al.* (2000b) studied the influence of feeding high-viscosity CMC on intestinal proliferation of *E. coli* and also that of *B. pilosicoli*. Weaner pigs were inoculated with *B. pilosicoli* 8 days after weaning and euthanased 14 days later. The results were suggestive of an increased duration of colonization with *B. pilosicoli* in pigs fed the diet with CMC, although it was evident that proliferation of the gut with *E. coli* also occurred with feeding CMC.

The basis of diet and intestinal disease

How might the type of diet fed alter the colonization and (or) proliferation of an enteric pathogen? There are numerous reasons for this but very little factual information. A factor that needs to be considered when discussing this issue is the interactions between the resident bacteria in the gut in response to the diet. For example, swine dysentery is a mixed synergistic infection that, in association with other anaerobic bacteria in the hindgut, can either enhance or reduce colonization by *B. hyodysenteriae* and hence influence the clinical expression of this disease. Our work in this area (Durmic *et al.*, 1998) has shown that synergistic bacteria (i.e. *Fusobacterium necrophorum*, *Fusobacterium nucleatum*), which have previously been found to facilitate colonization by *B. hyodysenteriae*, were found only amongst isolates from pigs fed a diet that predisposed them to swine dysentery (see Pluske *et al.*, 1999, for review). Despite some inconsistencies between studies in bacterial populations, diet type and clinical expression of this disease, these data provide some preliminary evidence that different carbohydrate sources fermented to varying degrees in the hindgut can alter the large intestinal microbiota that, in turn, may predispose some pigs to disease.

In terms of how this relates to dietary manipulation of 'gut health', especially in the light of the number of bacteria isolates, the number of species sequenced in the pig to date (see Table 5.6) and the exhausting nature of such work, the practical implications and possible applications of such findings appear somewhat intangible at present. Nevertheless, research at Murdoch University into stomach ulcers, using a model with weaner pigs, has demonstrated that interactions between diet, disease and specific bacterial species can occur.

Weaner pigs fed finely ground wheat developed quite severe ulceration of the pars oesophagea after 2–3 weeks, whilst pigs fed the same wheat that had been subjected to high pressure and temperature extrusion did not develop lesions (Accioly *et al.*, 1998). In this model system, when polymerase chain reaction tests were applied to stomach samples from the pigs, *Helicobacter* spp. were found only in a few pigs on both diets,

Table 5.10 Occurrence of ulcers and colonization by *Helicobacter* spp. and *Campylobacter* spp. in weaner pigs fed diets containing finely ground wheat, or the same wheat after extrusion (after Phillips, 1999).

Diet	Pig	Ulcers	*Helicobacter* spp.	*Campylobacter* spp.
Raw wheat-based	1	–	+	+
	2	+	+	+
	3	+	–	+
	4	+	–	+
	5	+	–	–
	6	+	–	–
	7	+	–	–
	8	+	–	–
Extruded wheat-based	9	+	–	+
	10	–	+	+
	11	–	+	+
	12	–	–	+
	13	–	–	+
	14	–	–	+
	15	–	–	+
	16	–	–	+

whilst *Campylobacter* spp. were significantly more common in the pigs fed extruded wheat and lacking ulcers (Table 5.10; Phillips, 1999). This suggests a link between diet, *Campylobacter* spp. and protection from ulcers. A similar highly significant association between the presence of *Campylobacter* spp. and lack of ulcers was seen in a recent abattoir survey of pigs from nine farms, though details of the diets fed on these farms is not yet available. These results suggest either that *Campylobacter* spp. may protect from ulcers, or that they are part of the normal microflora of the stomach that is displaced in the presence of ulcers, or by dietary factors that predispose to ulcers.

Although *Helicobacter heilmannii* is suspected as being involved in the aetiology of ulcers in pigs, other evidence does not support this. In the above survey, for example, a positive association between ulcers and *Helicobacter* spp. was found on only one of the nine farms. Similarly, a recent study in gnotobiotic swine failed to produce ulceration of the pars oesophagea when the animals were inoculated with *H. heilmannii* and fed a carbohydrate-enriched liquid diet (Krakowa *et al.*, 1998). In contrast, pigs fed this diet and inoculated with *Lactobacillus* and *Bacillus* spp. did develop ulcers. In this case it was suggested that fermentation by these latter bacterial species was encouraged by the presence of readily available dietary substrate, and that the acidic short-chain fatty acids produced as end-products of the fermentation were damaging to the epithelium. Whether or not *H. heilmannii* is a primary pathogen in the stomach, or whether other bacteria contribute to damage to the epithelium of the stomach, both possibilities provide links between diet, enteric bacteria and disease. Similarly, the links between *Campylobacter* spp., diet and an absence of ulceration hint at possible protective interactions. Knowledge about such links will provide new opportunities for the control of ulceration of the pars oesophagea in pigs.

Conclusions

This chapter has described a number of effects and implications of NSP in diets for weaner pigs. Whilst no specific recommendations appear to have been made for NSP inclusion levels and types in weaner diets, it is by virtue of the ingredient mix available and the economics of feeding that diets for weaners will contain NSP from cereal, legume and (or) oilseed sources. Although the newly weaned pig appears to have considerable capacity to digest NSP, especially from sugarbeet pulp that is high in fermentable NSP, one consequence would appear to be the diversion of nutrients from carcass gain to growth of the intestines, in particular the large intestine. However, whether the inclusion of 'high' levels of NSP in weaner diets has a negative influence on carcass composition at market weight is unknown. There is still some conjecture as to whether soluble NSPs are viscous-forming in the small intestine and influence nutrient digestibility, although addition of in-feed enzymes in, for example, hull-less barley-based diets would appear to be beneficial. One of the major research interests at present is the potential for some NSP sources to modulate 'gut health'. Research at Murdoch University has concentrated on interactions between diet and some specific enteric diseases such as PWC, PIS and gastric ulcers. In all cases the type of diet fed, in particular the type of NSP included, influences the expression of each disease. Nevertheless, modulation of the enteric flora to enhance 'gut health' by the use of specific NSPs and (or) the use of technologies such as enzymes and liquid feeding will require detailed studies on bacterial population changes in the gut in response to these strategies. This needs to occur in association with a detailed understanding of the influences of NSP on gut structure and function.

Acknowledgements

The Pig Research and Development Corporation of Australia and the Pig Industry Compensation Fund of Western Australia supported studies reported in this review that were conducted in our laboratory. We also wish to acknowledge the contributions of our colleagues J.M. Accioly, Z. Durmic, B.P. Mullan, N.D. Phillips and I.D. Robertson in the work described herein.

References

Accioly, J.M., Durmic, Z., McDonald, D.E., Oxberry, S.L., Pethick, D.W., Mullan, B.P. and Hampson, D.J. (1998) Dietary effects on the presence of ulcers and urease-producing organisms in the stomach of weaner pigs. In: *Proceedings of the 15th International Pig Veterinary Society Congress*, Birmingham, UK, 3, p. 242.

Agricultural Research Council (1967) *The Nutrient Requirements of Farm Livestock, No. 3: Pigs*. Agricultural Research Council, London.

Åman, P. and Graham, H. (1989) Analysis of total and insoluble mixed-linked (1→3), (1→4)-β-D-glucans in barley grain and other cereals. *Journal of Agricultural and Food Chemistry* 35, 704–712.

Annison, G. (1992) Commercial enzyme supplementation of wheat-based diets raises ileal glycanase activities and improves AME, starch and pentosan digestibility in broiler chickens. *Animal Feed Science and Technology* 38, 105–121.

Annison, G. (1993) The role of wheat non-starch polysaccharides in broiler nutrition. *Australian Journal of Agricultural Research* 44, 405–422.

Annison, G., Moughan, P.J. and Thomas, D.V. (1995) Nutritive activity of soluble rice bran arabinoxylans in broiler diets. *British Poultry Science* 36, 479–488.

Annison, G., Hughes, R.J. and Choct, M. (1996) Effects of enzyme supplementation on the nutritive value of dehulled *lupines*. *British Poultry Science* 37, 157–172.

Aumaître, A., Peiniau, J. and Madec, F. (1995) Digestive adaptation after weaning and nutritional consequences in the piglet. *Pig News and Information* 16, 73N–79N.

Austin, S.C., Wiseman, J. and Chesson, A. (1999) Influence of non-starch polysaccharides structure on the metabolisable energy of U.K. wheat fed to poultry. *Journal of Cereal Science* 29, 77–88.

Baas, T.C. and Thacker, P.A. (1996) Impact of gastric pH on dietary enzyme activity in swine fed β-glucanase supplemented diets. *Canadian Journal of Animal Science* 76, 245–252.

Bach Knudsen, K.E. (1997) Carbohydrate and lignin contents of plant materials used in animal feeding. *Animal Feed Science and Technology* 67, 319–338.

Bach Knudsen, K.E. and Canibe, N. (2000) Breakdown of plant carbohydrates in the digestive tract of pigs fed on wheat- and oat-based rolls. *Journal of the Science of Food and Agriculture* 80, 1253–1261.

Baidoo, S.K. and Liu, Y.G. (1998) Hull-less barley for swine: ileal and faecal digestibility of proximate nutrients, amino acids and non-starch polysaccharides. *Journal of the Science of Food and Agriculture* 76, 397–403.

Baidoo, S.K., Liu, Y.G. and Yungblut, D. (1998) Effect of microbial enzyme supplementation on energy, amino acid digestibility and performance of pigs fed hull-less barley based diets. *Canadian Journal of Animal Science* 78, 625–631.

Ball, R.O. and Aherne, F.X. (1987) Effect of diet complexity and feed restriction on the incidence and severity of diarrhea in early-weaned pigs. *Canadian Journal of Animal Science* 62, 907–913.

van Barneveld, R.J., Baker, J., Szarvas, S.R. and Choct, M. (1995) Effect of lupin non-starch polysaccharides (NSP) on nutrient digestion and microbial activity in growing pigs. *Proceedings of the Nutrition Society of Australia* 19, 43.

Bedford, M.R. and Classen, H.L. (1992) Reduction of intestinal viscosity through manipulation of dietary rye and pentosanase concentration is effected through changes in the carbohydrate composition of the intestinal aqueous phase and results in improved growth rate and food conversion efficiency of broiler chickens. *Journal of Nutrition* 122, 560–569.

Bedford, M.R., Classen, H.L. and Campbell, G.L. (1991) The effect of pelleting, salt and pentosanase on the viscosity of intestinal contents and the performance of broilers fed rye. *Poultry Science* 70, 1571–1577.

Bedford, M.R., Patience, J.F., Classen, H.L. and Inborr, J. (1992) The effect of dietary enzyme supplementation of rye and barley-based diets on digestion and subsequent performance in weanling pig. *Canadian Journal of Animal Science* 72, 97–105.

Bell, J.M. (1993) Factors affecting the nutritional value of canola meal: a review. *Canadian Journal of Animal Science* 73, 679–697.

Bertschinger, H.U., Eggenberger, E., Jucker, H. and Pfirter, H.P. (1978) Evaluation of low nutrient, high fibre diets for the prevention of porcine *Escherichia coli* enterotoxaemia. *Veterinary Microbiology* 3, 281–290.

Bolduan, G., Jung, H., Schnable, E. and Schneider, R. (1988) Recent advances in the nutrition of weaner piglets. *Pig News and Information* 9, 381–385.

Brunsgaard, G. (1998) Effects of cereal type and feed particle size on morphological characteristics, epithelial cell proliferation, and lectin binding patterns in the large intestine of pigs. *Journal of Animal Science* 76, 2787–2798.

Campbell, G.L. and Bedford, M.R. (1992) Enzyme applications for monogastric feeds: a review. *Canadian Journal of Animal Science* 72, 449–446.

Cherbut, C., Albina, E., Champ, M., Coublier, J.L. and Lecannu, G. (1990) Action of guar gums on the viscosity of digestive contents and on the gastrointestinal motor function in pigs. *Digestion* 46, 205–213.

Choct, M. (1997) Non-starch polysaccharides: chemical structures and nutritional significance. *Feed Milling International*, June, pp. 13–19.

Choct, M. and Annison, G. (1990) Anti-nutritive activity of wheat pentosans in poultry diets. *British Poultry Science* 31, 809–819.

Choct, M. and Annison, G. (1992a) The inhibition of nutrient digestion by wheat pentosans. *British Journal of Nutrition* 67, 123–132.

Choct, M. and Annison, G. (1992b) Anti-nutritive effect of wheat pentosans in broiler chickens: role of viscosity and gut microflora. *British Poultry Science* 33, 821–834.

Classen, H.L. and Bedford, M.R. (1991) The use of enzyme to improve the nutritive value of poultry feeds. In: Haresign, W. and Cole, D.J.A. (eds) *Recent Advances in Animal Nutrition – 1991*. Butterworth, London, pp. 95–116.

Coles, G.D., Hartunian-Sowa, S.M., Jamieson, P.D., Hay, A.J., Atwell, W.A. and Fulcher, R.G. (1997) Environmentally-induced variation in starch and non-starch polysaccharide content in wheat. *Journal of Cereal Science* 26, 47–54.

Dréau, D. and Lallès, J.P. (1999) Contribution to the study of gut hypersensitivity reactions to soybean proteins in preruminant calves and early-weaned piglets. *Livestock Production Science* 60, 209–218.

Dunaif, G. and Schneeman, B.O. (1981) The effect of dietary fibre on human pancreatic enzyme activity *in vitro*. *American Journal of Clinical Nutrition* 34, 1034–1035.

Durmic, Z., Pethick, D.W., Pluske, J.R. and Hampson, D.J. (1998) Changes in bacterial populations in the colon of pigs fed different sources of dietary fibre, and the development of swine dysentery after experimental infection. *Journal of Applied Microbiology* 85, 574–582.

Edwards, S.A., Taylor, A.G. and Harland, J.I. (1991) The inclusion of sugar beet pulp in diets for early weaned piglets. *Animal Production* 52, 599–600 (Abstract).

Ellis, P.R., Roberts, F.G., Low, A.G. and Morgan, L.M. (1995) The effect of high-molecular-weight guar gum on net apparent glucose absorption and net apparent insulin and gastric inhibitory polypeptide production in the growing pig: relationship to rheological changes in jejunal digesta. *British Journal of Nutrition* 74, 539–556.

Englyst, H. (1989) Classification measurement of plant polysaccharides. *Animal Feed Science and Technology* 23, 27–42.

Fadel, J.G., Newman, C.W., Newman, R.K. and Graham, H. (1988) Effects of extrusion cooking of barley on ileal and fecal digestibilities of dietary components in pigs. *Canadian Journal of Animal Science* 68, 891–897.

Fengler, A.I. and Marquardt, R.R. (1988) Water soluble pentosans from rye. II. Effects of rate of dialysis on the retention of nutrients by the chick. *Cereal Chemistry* 65, 298–302.

Flourie, B., Vidon, N., Florent, C.H. and Bernier, J.J. (1984) Effect of pectin on jejunal glucose absorption and unstirred layer thickness in normal men. *Gut* 25, 936–941.

Gdala, J. (1998) Composition, properties, and nutritive value of dietary fibre of legume seeds. A review. *Journal of Animal and Feed Sciences* 7, 131–149.

Gdala, J., Jansman, A.J.M., Buraczewska, L., Huisman, J. and van Leeuwin, P. (1997) The influence of α-galactosidase supplementation on the ileal digestibility of lupin seed carbohydrates and dietary protein in young pigs. *Animal Feed Science and Technology* 67, 115–125.

Gee, J., Lee-Finglas, W., Wertley, G. and Johnson, I. (1996) Fermentable carbohydrates elevate plasma enteroglucagon but high viscosity is also necessary to stimulate small bowel mucosal cell proliferation in rats. *Journal of Nutrition* 126, 373–379.

Gibson, G.R., Beatty, E.R., Wang, X. and Cummins, J.H. (1995) Selective stimulation of *Bifidobacteria* in the human colon by oligofructose and inulin. *Gastroenterology* 108, 975–982.

Gill, B.P., Mellange, J. and Rooke, J.A. (2000) Growth performance and apparent nutrient digestibility in weaned piglets offered wheat-, barley- or sugarbeet pulp-based diets supplemented with food enzymes. *Animal Science* 70, 107–118.

Graham, H., Hesselman, K. and Åman, P. (1986a) The influence of wheat bran and sugarbeet pulp on the digestibility of dietary components in a cereal-based diet. *Journal of Nutrition* 116, 242–251.

Graham, H., Hesselman, K., Jonsson, E. and Åman, P. (1986b) Influence of β-glucanase supplementation on digestion of a barley-based diet in the pig gastrointestinal tract. *Nutrition Reports International* 34, 1089–1094.

Graham, H., Fadel, J.G., Newman, C.W. and Newman, P.K. (1989) Effect of pelleting and β-glucanase supplementation on the ileal and fecal digestibility of a barley-based diet in the pig. *Journal of Animal Science* 67, 1293–1298.

Hampson, D.J. and Trott, D.J. (1995) Intestinal spirochaete infections of pigs: an overview with an Australian perspective. In: Hennessy, D.P. and Cranwell, P.D. (eds) *Manipulating Pig Production V.* Australasian Pig Science Association, Werribee, Australia, pp. 139–169.

Hampson, D.J., Robertson I.D., La, T., Oxberry, S.L. and Pethick, D.W. (2000) Influences of diet and vaccination on colonisation of pigs with the intestinal spirochaete *Brachyspira (Serpulina) pilosicoli. Veterinary Microbiology* 73, 75–84.

Hampson, D.J., Pluske, J.R. and Pethick, D.W. (2001) Dietary control of enteric diseases. In: *Proceedings of the 8th Symposium on Digestive Physiology in Pigs.* Swedish University of Agricultural Sciences, Uppsala, Sweden.

Ikegami, S., Tsuchihashi, F., Harada, H., Tsuchihashi, N., Nishide, E. and Innami, S. (1990) Effect of viscous indigestible polysaccharides on pancreatic biliary secretion and digestive organs in rats. *Journal of Nutrition* 120, 253–256.

Jensen, B.B. (1999) Impact of feed composition and feed processing on the gastrointestinal ecosystem in pigs. In: Jansman, A.J.M. and Huisman, J. (eds) *Nutrition and Gastrointestinal physiology – Today and Tomorrow.* TNO Nutrition and Food Research Institute, Wageningen, The Netherlands, pp. 43–56.

Jensen, M.S., Bach Knudsen, K.E., Inborr, J. and Jakobsen, K. (1998) Effect of β-glucanase supplementation on pancreatic enzyme activity and nutrient digestibility in piglets fed diets based on hulled and hull-less barley varieties. *Animal Feed Science and Technology* 72, 329–345.

Jeroch, H. and Dänicke, S. (1995) Barley in poultry feeding: a review. *World's Poultry Science Journal* 51, 271–291.

Johnson, I.T. and Gee, J.M. (1981) Effect of gel-forming gums on the intestinal unstirred layer and sugar transport *in vitro. Gut* 25, 398–403.

Just, A., Fernández, J.A. and Jørgensen, H. (1983) The net energy value of diets for growth in pigs in relation to the fermentative processes in the digestive tract and the site of absorption of the nutrients. *Livestock Production Science* 10, 171–186.

Krakowa, S., Eaton, K.A., Rings, D.M. and Argenzio, R.A. (1998) Production of gastroesophageal erosions and ulcers (GEU) in gnotobiotic swine monoinfected with fermentative commensal bacteria and fed a high-carbohydrate diet. *Veterinary Pathology* 35, 274–282.

Langhout, D.J., Schutte, J.B., de Jong, J., Sloetjes, H., Verstegen, M.W.A. and Tamminga, S. (2000) Effect of viscosity on digestion of nutrients in conventional and germ-free chicks. *British Journal of Nutrition* 83, 533–540.

Larsen, F.M., Moughan, P.J. and Wilson, M.N. (1993) Dietary fibre viscosity and endogenous protein excretion at the terminal ileum of growing rats. *Journal of Nutrition* 124, 833–841.

Larsen, F.M., Wilson, M.N. and Moughan, P.J. (1994) Dietary fibre viscosity and amino acid digestibility, proteolytic digestive enzyme activity and digestive organ weight in growing rats. *Journal of Nutrition* 124, 833–841.

Leterme, P., Froidmont, E., Rossi, F. and Théwis, A. (1998) The high water holding capacity of pea inner fibers affects the ileal flow of endogenous amino acids in pigs. *Journal of Agriculture and Food Chemistry* 46, 1927–1934.

Lewis, B.A. (1993) Fiber chemistry: an historical perspective. In: *Proceedings of the 1993 Cornell Nutrition Conference for Feed Manufacturers.* Department of Animal Science and Division of Nutritional Sciences, New York State College of Agriculture and Life Sciences. Cornell University, Ithaca, New York, pp. 1–8.

Lewis, F.J., McEvoy, J. and McCracken, K.J. (1998) Lack of relationship between wheat *in vitro* viscosity and digestibility parameters for pigs. In: *Proceedings of the British Society of Animal Science*, p. 32.

Lewis, F.J., Schulze, H., McEvoy, J. and McCracken, K.J. (1999) The effects of wheat variety on diet digestibility and on performance of pigs from 10–20 kg. In: *Proceedings of the British Society of Animal Science*, p. 166.

Li, S., Sauer, W.C., Huang, S.X. and Gabert, V.M. (1996) Effect of β-glucanase supplementation to hull-less barley- or wheat-soybean meal diets on the digestibilities of energy, protein, β-glucans, and amino acids in young pigs. *Journal of Animal Science* 74, 1649–1656.

Lizardo, R., Peiniau, J. and Aumaître, A. (1997) The influence of dietary non-starch polysaccharides and protein source on the activities of pancreatic and intestinal enzymes of the weaned piglet. In: Laplace, J.-P., Février, C. and Barbeau, A. (eds) *Proceedings of the VIIth International Symposium on Digestive Physiology in Pigs.* EAAP Publication No. 88. INRA, Saint Malo, France, pp. 630–633.

Longland, A.C., Carruthers, J. and Low, A.G. (1994) The ability of piglets 4 to 8 weeks old to digest and perform on diets containing two contrasting sources of non-starch polysaccharides. *Animal Production* 58, 405–410.

Longstaff, M. and McNab, J.M. (1986) Influence of site and variety on starch, hemicellulose and cellulose composition of wheats and their digestibilities by adult cockerels. *British Poultry Science* 27, 435–461.

Low, A.G. (1989) Secretory response of the pig gut to non-starch polysaccharides. *Animal Feed Science and Technology* 23, 55–65.

Mariscal-Landín, G., Sève, B., Colléaux, Y. and Lebreton, Y. (1995) Endogenous amino nitrogen, collected from pigs with end-to-end ileorectal anastomosis, is affected by the method of estimation and altered by dietary fiber. *Journal of Nutrition* 125, 136–146.

Mathews, C.J., MacLeod, R.J., Zheng, S.X., Hanrahan, J.W., Bennett, H.P. and Hamilton, J.R. (1999) Characterization of the inhibitory effect of boiled rice on intestinal chloride secretion in guinea pig crypt cells. *Gastroenterology* 116, 1342–1347.

May, T., Mackie, R.I., Fahey, G.C., Cremin, J.C. and Garleb, K.A. (1994) Effect of fiber source on short-chain fatty acid production and on the growth and toxin production by *Clostridium difficile*. *Scandinavian Journal of Gastroenterology* 19, 916–922.

McDonald, D.E., Pluske, J.R., van Barneveld, R.J., Pethick, D.W., Mullan, B.P. and Hampson, D.J. (1998) The effects of non-starch polysaccharides and resistant starch on weaner pig performance and digestive tract development. *Proceedings of the Nutrition Society of Australia* 22, 100.

McDonald, D.E., Pethick, D.W., Pluske, J.R. and Hampson, D.J. (1999) Adverse effects of soluble non-starch polysaccharide (guar gum) on piglet growth and experimental colibacillosis immediately after weaning. *Research in Veterinary Science* 67, 245–250.

McDonald, D.E., Pethick, D.W., Mullan, B.P. and Hampson, D.J. (2000a) Increased intestinal viscosity depresses carcass growth and encourages intestinal proliferation of *Escherichia coli* in weaner pigs. In: *Proceedings of the International Pig Veterinary Society Congress*. Melbourne, Australia, p. 21.

McDonald, D.E., Pethick, D.W. and Hampson, D.J. (2000b) Interactions between increased intestinal viscosity, post-weaning colibacillosis and intestinal spirochaetosis in weaner pigs. In: *Proceedings of the International Pig Veterinary Society Congress*, Melbourne, Australia, p. 39.

McDonald, D.E., Pethick, D.W., Mullan, B.P., Pluske, J.R. and Hampson, D.J. (2001) Soluble non-starch polysaccharides from pearl barley exacerbate experimental post-weaning colibacillosis. In: *Proceedings of the 8th Symposium on Digestive Physiology in Pigs*. Swedish University of Agricultural Sciences, Uppsala.

Moore, W.E.C., Moore, L.V.H., Cato, E.P., Wilkins, T.D. and Kornegay, E.T. (1987) Effect of high-fiber and high-oil diets on the fecal flora of swine. *Applied and Environmental Microbiology* 53, 1638–1644.

Morris, V.J. (1986) Gelation of polysaccharides. In: Mitchell, J.R. and Ledward, D.A. (eds) *Functional Properties of Food Macromolecules*. Elsevier Applied Science, London, pp. 121–170.

Mosenthin, R., Hambrecht, E. and Sauer, W.C. (1999) Utilisation of different fibres in piglet feeds. In: Garnsworthy, P.C. and Wiseman, J. (eds) *Recent Advances in Animal Nutrition – 1999*. Nottingham University Press, Loughborough, UK, pp. 227–256.

NRC (1998) *Nutrient Requirements of Swine*. 10th revised edn. National Research Council, Washington, DC.

Ochiai, S., Mori, K. and Adachi, Y. (1997) Unification of the genera *Serpulina* and *Brachyspira*, and proposals of *Brachyspira hyodysenteriae* comb. nov, *Brachyspira innocens* comb. nov and *Brachyspira pilosicoli* comb. nov. *Microbiology and Immunology* 41, 445–452.

Partridge, G. (1998) Enzyme's role in health. *Feed Mix* 6(1), 23–25.

Phillips, N.D. (1999) Molecular detection and identification of gastric bacteria in pigs. Honours thesis, Murdoch University, Western Australia.

Pluske, J.R., Siba, P.M., Pethick, D.W., Durmic, Z., Mullan, B.P. and Hampson, D.J. (1996) The incidence of swine dysentery in pigs can be reduced by feeding diets that limit the amount of fermentable substrate entering the large intestine. *Journal of Nutrition* 126, 2920–2933.

Pluske, J.R., Pethick, D.W. and Mullan, B.P. (1998) Differential effects of feeding fermentable carbohydrate to growing pigs on performance, gut size, and slaughter characteristics. *Animal Science* 67, 147–156.

Pluske, J.R., Pethick, D.W., Durmic, Z., Hampson, D.J. and Mullan, B.P. (1999) Non-starch polysaccharides in pig diets and their influence on intestinal microflora, digestive physiology and enteric disease. In: Garnsworthy, P.C. and Wiseman, J. (eds) *Recent Advances in Animal Nutrition – 1999*. Nottingham University Press, Loughborough, UK, pp. 189–226.

Prohaszka, L. and Baron, F. (1980) The predisposing role of high protein supplies in enteropathogenic *Escherichia coli* infections of weaned pigs. *Zentralblatt für Veterinarmedizin B.* 27, 222–232.

Rainbird, A.L., Low, A.G. and Zebrowska, T. (1984) Effect of guar gum on glucose and water absorption from isolated loops of jejunum in conscious growing pigs. *British Journal of Nutrition* 52, 489–498.

Reid, C.-A. and Hillman, K. (1999) The effects of retrogradation and amylose/amylopectin ratio of starches on carbohydrate fermentation and microbial populations in the porcine colon. *Animal Science* 68, 503–510.

Satchithanandam, S., Vargofcak-Apker, M., Colvert, R.S., Leeds, A.R. and Cassidy, M.M. (1990) Alteration of gastro intestinal mucin by fibre feeding in rat. *Journal of Nutrition* 120, 1179–1184.

Schulze, H., van Leeuwen, P., Verstegen, M.W.A. and van den Berg, J.W.O. (1995) Dietary level and source of neutral detergent fiber and ileal endogenous nitrogen flow in pigs. *Journal of Animal Science* 73, 441–448.

Selvendran, R.R., Stevens, B.J.H. and Du Pont, M.S. (1987) Dietary fibre: chemistry, analysis, and properties. *Advances in Food Research* 31, 117–209.

Smithson, K.W., Millar, D.B., Jacobs, L.R. and Gary, G.M. (1981) Intestinal diffusion barrier: unstirred water layer of membrane surface mucous coat? *Science* 214, 1214–1244.

Smits, C.H.M. and Annison, G. (1996) Non-starch polysaccharides in broiler nutrition – towards a physiologically valid approach to their determination. *World's Poultry Science* 52, 203–221.

Southon, S., Livesey, G., Gee, J.M. and Johnson, I.T. (1985) Differences in international protein synthesis and cellular proliferation in well-nourished rats consuming conventional laboratory diets. *British Journal of Nutrition* 53, 87–95.

Theander, O., Westerlund, E., Åman, P. and Graham, H. (1989) Plant cell walls and monogastric diets. *Animal Feed Science and Technology* 23, 205–225.

Trezona, M., Mullan, B.P., Pluske, J.R., Hanbury, C.D. and Siddique, K.H.M. (2000) Evaluation of Lathyrus (*Lathyrus cicera*) as an ingredient in diets for weaner pigs. *Proceedings of the Nutrition Society of Australia* 24, 119.

Trott, D.J., Stanton, T.B., Jensen, N.S., Duhamel, G.E., Johnson, J.L. and Hampson, D.J. (1996) *Serpulina pilosicoli* sp. nov., the agent of porcine intestinal spirochetosis. *International Journal of Systematic Bacteriology* 46, 206–215.

Varel, V.H. and Yen, J.T. (1997) Microbial perspective on fiber utilization by swine. *Journal of Animal Science* 75, 2715–2722.

Vervaeke, I.J., Graham, H., Dierick, N.A., Demeyer, D.I. and Decuypere, J.A. (1991) Chemical analysis of cell wall and energy digestibility in growing pigs. *Animal Feed Science and Technology* 32, 55–61.

Wiggins, H.S. (1984) Nutritional value of sugars and related compounds undigested in the small gut. *Proceedings of the Nutrition Society* 43, 69–75.

Wilson, F.A. and Dietschy, J.M. (1974) The intestinal unstirred layer: its surface area and effect on active transport kinetics. *Biochimica et Biophysica Acta* 363, 112–126.

Wilson, F.A., Sallee, V.L. and Dietschy, J.M. (1971) Unstirred water layers in intestine: rates of determinant of fatty acid absorption from micellar solutions. *Science* 174, 1031–1033.

Yen, J.T., Nienaber, J.A., Hill, D.A. and Pekas, J.C. (1989) Oxygen consumption by portal vein-drained organs and by the whole animal in conscious growing swine. *Proceedings of the Society for Experimental Biology and Medicine* 190, 393–398.

Younosjai, M.K., Adedoyin, M. and Ranshaw, J. (1978) Dietary components and gastrointestinal growth in rats. *Journal of Nutrition* 108, 341–350.

Zijlstra, R.T., de Lange, C.F.M. and Patience, J.F. (1999) Nutritional value of wheat for growing pigs: chemical composition and digestible energy content. *Canadian Journal of Animal Science* 79, 187–194.

6

Individual Feed Intake of Group-housed Weaned Pigs and Health Status

E.M.A.M. Bruininx[1], C.M.C. van der Peet-Schwering[1] and J.W. Schrama[2]

[1]*Research Institute for Animal Husbandry, PO Box 2176, 8203 AD, Lelystad, The Netherlands;* [2]*Wageningen University and Research Centre, Department of Animal Science, PO Box 338, 6700 AH, Wageningen, The Netherlands*

Introduction

Despite considerable research in the past, post-weaning problems such as the well-known growth depression and the occurrence of diarrhoea are still widespread on modern pig units. Most of the post-weaning problems are part of the post-weaning syndrome (PWS) that includes post-weaning diarrhoea, oedema disease and endotoxin shock. Vellenga (1987) estimated the extra annual costs of these problems within Dutch pig husbandry to be about 11 million euro. In 2000 these annual costs were estimated to be about 7 million euro, due to improvements in housing conditions, management, nutrition and the use of anti-microbial growth promoting agents. However, with more critical use of antimicrobial growth-promoting agents at present and a possible ban in the future, the costs associated with PWS may again increase. Further research is warranted on factors that maintain the health status of the weaned pig.

Nutrition is an important factor in the pathogenesis of PWS. This chapter considers the nutrition and health status of the weaned pig in terms of post-weaning food intake and of food intake behaviour in relation to gut health.

Weaning and Food Intake

It is generally accepted that weaning is a stressful event (e.g. Worsaae and Schmidt, 1980). At weaning, piglets are removed from the dam and allocated to another pen.

After weaning, the piglets are provided with solid food in a way that is completely different from suckling milk. The piglet may respond to these stress factors by ceasing to eat (Bark *et al.*, 1986). Le Dividich and Herpin (1994) observed a period of underfeeding immediately after weaning and concluded that the metabolizable energy requirement for maintenance (ME_m) was not met until day 5 after weaning. These findings were subsequently confirmed in several studies (van Diemen *et al.*, 1995; Gentry *et al.*, 1997; Moon *et al.*, 1997; Sijben *et al.*, 1997). Gross energy intake (which is approximately equal to food intake) was strongly reduced after weaning. The mean gross energy intake during weeks 1 and 2 after weaning were, respectively, 41% and 82% of the average food intake during weeks 4, 5 and 6 after weaning.

The studies cited above also showed a tendency to an increase in energy requirements for maintenance during the first week after weaning (Fig. 6.1), with an average daily ME_m requirement of 461 kJ $kg^{-0.75}$ during week 1 after weaning and 418 kJ $kg^{-0.75}$ averaged over the subsequent 5 weeks. Thus, production was hampered by the reallocation of energy towards maintenance processes, as well as by reduced food intake.

In addition to the variation in food intake observed among groups of weaned pigs, Brooks (1999) showed a great variation in food intake traits within groups. Even though the majority of pigs started eating within 5 h after weaning, some pigs took a very long time (up to 54 h) before they started eating.

During the first days after weaning, piglets depend strongly on body energy reserves, due to low food intake in combination with the increased ME_m requirement. Apart from low food intake, there seems to be a large variation in food intake among individual animals. Quantitative information on individual food intake of weanling pigs housed under practical conditions is scarce.

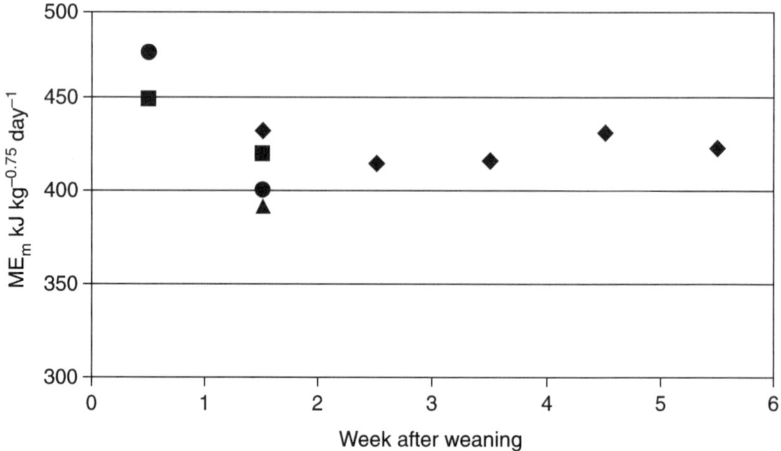

Fig. 6.1. Alterations of metabolizable energy requirement for maintenance (ME_m) of piglets weaned at 4 weeks of age. (◆ van Diemen *et al.*, 1995; ■ Gentry *et al.*, 1997; ▲ Moon *et al.*, 1997; ● Sijben *et al.*, 1997.)

Importance of Early Food Intake

Weaning of pigs is associated with profound changes in gut morphology, such as villus atrophy, crypt hypertrophy and increased epithelial cell mitosis (Nabuurs, 1991; van Beers-Schreurs, 1996). Furthermore, it has been shown that digestive enzymes are present in insufficient quantities and quality for the changes in dietary composition that occur (Lindemann *et al.*, 1986; Makkink, 1993). These changes in intestinal morphology and function are associated with post-weaning digestion problems and malabsorption (Kenworthy, 1976; Hampson and Kidder, 1986; Nabuurs, 1991).

Various factors affecting these changes in intestinal morphology and function have been observed and reported. Deprez *et al.* (1987) and Cera *et al.* (1988) showed that decreased villus height resulted from feeding a dry pelleted diet after weaning. More recently it has become clear that food intake in particular is an important factor. McCracken *et al.* (1995), van Beers-Schreurs (1996) and Pluske (1996a,b) showed the interdependence between voluntary food intake and mucosal architecture. Table 6.1 gives the effect of post-weaning energy intake level on villus height, expressed as a percentage of villus height at weaning.

The provision of milk at the same level of energy intake as from the *ad libitum* provision of a dry diet resulted in similar gut damage. However, when piglets were given free access to cow milk (2.5 times maintenance) pre- and post-weaning villus heights were similar (Pluske, 1996b). This suggests that when food/energy intake is maintained after weaning, the typical villus atrophy at 4–5 days after weaning can be avoided. Makkink (1993) also reported that the typical decline in enzyme

Table 6.1. The relative effect of early post-weaning energy intake on average villus height in the small intestine of pigs.

Reference	Energy source	Energy intake		Age (days)		Villus height (%)
		MJ day^{-1}	GE, ME	Weaning	Slaughter	
Pluske *et al.*						
(1996a)	Starter diet	5.7	GE	28	33	−30
	Ewe milk	7.4	GE	28	33	−2
(1996b)	Cow milk	2.3	GE	29	34	−27
	Starter diet	5.1	GE	29	34	−18
	Cow milk	5.2	GE	29	34	−4
	Cow milk	8.9	GE	29	34	+ 11
(1996c)	Cow milk	5.5	GE	28	33	−4.8
Kelly *et al.*						
(1991)	Starter diet	2.9	GE	14	20	−55
van Beers-Schreurs						
(1996)	Starter diet	0.53	ME BW$^{-0.75}$	28	32	−40
	Sow milk	0.48	ME BW$^{-0.75}$	28	32	−35
	Sow milk	1.4	ME BW$^{-0.75}$	28	32	−11

GE, gross energy; ME, metabolizable energy.

activity just after weaning is most likely related to low food intake after weaning and not the process of weaning *per se*.

Weaning of pigs is often associated with the occurrence of a damaged gut architecture and a decline in digestive enzymes, probably due to a decreased supply of nutrients or energy. Prevention of the reduction in food/energy intake is crucial for maintaining both gut integrity and digestive capacity. This is important for realizing the optimum performance and health status of the weaned pig.

Measuring Individual Food Intake

Based on research regarding individual creep-food intake of suckling pigs (Pajor *et al.*, 1991) and of group-housed growing/finishing pigs (de Haer, 1992; Ramaekers *et al.*, 1996) it appears that there is a very large variation in individual food intake among group-housed weaned pigs. Available data on individual food intake of weaned pigs are based on research with individually housed animals (Makkink, 1993; Pluske *et al.*, 1996a,b). De Haer (1992) showed differences in food intake characteristics between group-housed and individually housed growing/finishing pigs. Group-housed pigs had fewer meals per day than pigs that were housed individually, but the meals of group-housed pigs were larger than those of individually housed pigs. Social interactions among group-housed pigs are considered to be the main cause of these differences.

There are significant social interactions among group-housed weaned pigs in commercial practice. Because food intake is an important determinant of performance and health, a feeding system has been developed as a research tool for monitoring the individual food intake of group-housed weaned pigs. These feeding stations for weaned pigs (IVOG® feeding stations; Insentec B.V., Marknesse, The Netherlands) were adapted from feeding stations for growing/finishing pigs. Briefly, the IVOG® feeding stations for weaned pigs consist of feeders placed on load cells recording to the nearest 10 g, within a range of 0–50 kg. The single-space feeder in a feeding station has a feeding bowl of 220 × 200 × 290 mm. There is a small adjustable fence at 20 cm in front of the feeder which allows competition among pigs and facilitates separate visits. To identify individual animals, each pig has an electronic ear transponder that is read by two antennae in the feeding stations. The feeding stations record feeder weight and time at the beginning and end of each visit, together with the electronic identification number of the pigs.

A feeding station that is suitable for measuring individual food intake data may affect the average performance of group-housed weaned pigs. De Haer (1992) showed that the functioning of feeding stations is not error free. The use of feeding stations for weaned pigs was validated by Bruininx *et al.* (2001a). This validation included an accuracy test, in which the food consumption recorded by the feeding stations was compared with manual weighing. The average food intake per pen computed from individual IVOG® records agreed reasonably well (100.7% over a 34-day period) with food intake per pen computed from manual food weighing. Also, the average performance of weaned pigs fed by feeding stations was similar to that of pigs fed by commercial single-space dry feeders (Table 6.2).

Table 6.2. Growth performance of weaned pigs fed by either commercial single-space dry feeders or IVOG® feeding stations (each pen contained 10 or 11 pigs) (Bruininx *et al.*, 2001a).

			Phase 1 (days 0–13)			Total (days 0–34)		
	No. pens	Weaning weight (kg)	ADFI (kg)	ADLWG (g)	Gain: food ($kg\ kg^{-1}$)	ADFI (kg)	ADLWG (g)	Gain: food ($kg\ kg^{-1}$)
Commercial feeder	11	7.9	0.20[a]	139	0.67	0.52	329	0.64
IVOG® feeding station	9	7.9	0.18[b]	121	0.67	0.50	322	0.64
SEM		0.04	0.006	7.9	0.025	0.009	6.9	0.007

[ab] Means within a column with different superscripts are significantly different ($P < 0.05$.
ADFI = average daily feed intake.
ADLWG = average daily liveweight gain.

During the first 13 days after weaning, average daily food intake (ADFI) of the pigs fed by commercial single-space feeders was higher ($P < 0.05$) than that of pigs fed by the feeding stations. It is probable that pigs needed a longer period to get used to the feeding stations than to the commercial single-space feeders. In the remaining period no effects on food intake were found, whereas growth and food efficiency were similar during the entire period. Because the IVOG® feeding station had no systematic effect on performance, it was considered to be a suitable tool to monitor individual food intake of group-housed weaned pigs.

Individual Food Intake Characteristics of Group-housed Weaned Pigs

There is still a lack of good information on the individual food intake of group-housed weaned pigs. A study was therefore initiated to examine the relationship between health status and individual food/energy intake of the group-housed weaned pig. In the first part of this study, the variation in individual food intake characteristics were addressed. Makkink (1993) and Brooks (1999) emphasized the importance of food intake immediately after weaning for the prevention of reduction of villus length and enzyme activity. Therefore, based on IVOG® data, Bruininx *et al.* (2001b) constructed curves to describe the relationships between the interval (hours) between weaning and first food intake (latency time) for both barrows and gilts (Fig. 6.2).

It can be concluded from Fig. 6.2 that there is considerable variation in latency time among individual pigs. Within the first 4 h after weaning, approximately 50% of both sexes had started to eat, whereas it took about 50 h before 95% of all pigs started to eat. During the experiment, there was only artificial illumination of the nursery room. Lights were on from 0700 h to 1900 h. A striking result was that

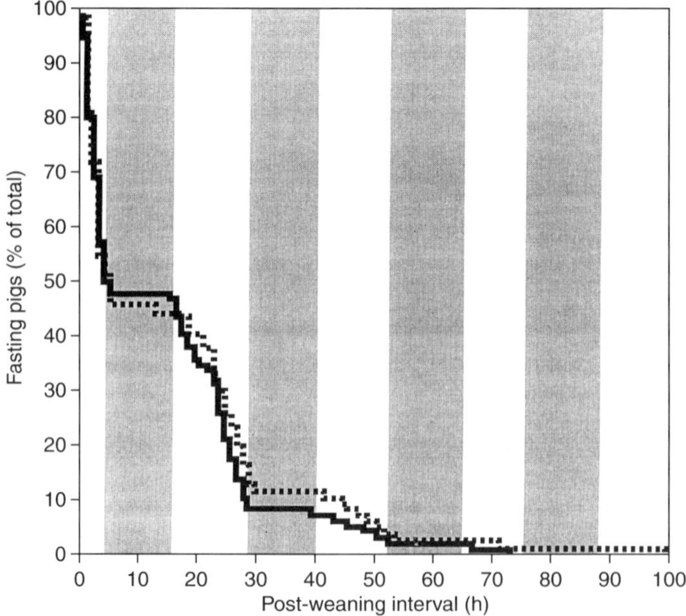

Fig. 6.2. Percentage of weaned pigs that had not eaten after weaning as a function of post-weaning interval. Curves are given for gilts (▬▬) and barrows (▪▪▪▪▪). The dark periods are indicated by shaded bars (Bruininx *et al.*, 2001b).

during the first two night periods the number of animals that had started to eat did not increase. This can be deducted from the almost horizontal lines in Fig. 6.2 during the night periods (indicated by shaded bars). In practice, weaned pigs are sometimes housed in darkness to prevent fighting during the first few days after weaning, but this may have some detrimental effects on the early food intake. Although this experiment was not designed to assess effects of photoperiod on early feeding behaviour, these results suggest that a light–dark cycle may be used as a tool for modulating early feeding behaviour of weaned pigs.

It is likely that the weaning weight of pigs may significantly affect the variation in food intake characteristics. Therefore, Bruininx *et al.* (2001b) studied some individually measured food intake characteristics as affected by weaning weight (Table 6.3). Averaged over the total experimental period, heavy pigs had the highest ADFI and the lowest food intake during the first 24 h after a pig started eating (= initial food intake) (Table 6.3). It may be that the light pigs were more active in their search for food, whereas heavy pigs were more passive. It could be argued that heavy pigs eat later and consequently have a lower early food intake because they have larger energy stores. On the other hand, Brouns and Edwards (1994) suggested that heavy (dominant) sows spend more time defending their food than eating it.

In summary, there are considerable variations in food intake characteristics among individual weaned pigs housed in groups. Moreover, these food intake characteristics seem to be affected by external factors (photoperiod) as well as animal factors (bodyweight).

Table 6.3. Individual performance and food intake (FI) characteristics of weaned pigs (days 0–34) as affected by bodyweight class at weaning (based on Bruininx *et al.*, 2001b).

						Daily visits				
Bodyweight class	No. of pens	Weaning weight (kg)	ADFI (g day^{-1})	ADLWG (g day^{-1})	Gain: food	Total no.	No. with FI	Time per visit (s)	FI per visit (g)	Initial FI (g $kg^{-0.75}$ per day)
Light	65	6.7^{a}	452^{a}	298^{a}	0.67	34.4^{a}	12.2	365^{ab}	42^{a}	27.9^{a}
Middle	61	7.9^{b}	486^{a}	320^{ab}	0.66	29.2^{b}	12.9	338^{a}	41^{a}	18.9^{b}
Heavy	60	9.3^{c}	535^{b}	345^{b}	0.65	26.7^{b}	12.0	383^{b}	50^{b}	17.7^{b}
SEM		0.05	13.0	9.1	0.008	1.53	0.42	13.4	2.0	2.60

^{ab} Means within a column with different superscripts are significantly different ($P < 0.05$).
ADFI = average daily feed intake.
ADLWG = average daily liveweight gain.
SEM = pooled standard error of the mean.
Initial food intake = amount of food consumed during the 24 h following the first visit to the food station with food intake.

Conclusions

During the first few days after weaning, piglets depend strongly upon body energy reserves. A continuous supply of nutrients (energy) seems to be very important to maintain villus architecture as well as digestive capacity. For an accurate assessment of the relationships between food intake and post-weaning health, quantitative information on individual food intake of pigs housed under practical conditions is needed. Feeding stations for weaned pigs are a valuable tool to monitor the individual food intake of group-housed weanling pigs. Data collected from feeding stations on individual food intake characteristics of group-housed weaned pigs show a considerable between-animal variation in early food intake characteristics. Given the concept that food intake and food intake characteristics are related to post-weaning health problems (e.g. diarrhoea, growth depression) these data suggest that early feeding behaviour of weaned pigs can be modulated (e.g. by light patterns). The precise nature of the relationships between food intake characteristics of weaned pigs and post-weaning health remains to be established.

References

Bark, L.J., Crenshaw, T.D. and Leibbrandt, V.D. (1986) The effect of meal intervals and weaning on feed intake of early weaned pigs. *Journal of Animal Science* 62, 1233–1239.

van Beers-Schreurs, H.M.G. (1996) The changes in the function of the large intestine of weaned pigs. PhD dissertation, Faculty of Veterinary Medicine, Utrecht University, The Netherlands.

Brooks, P.H. (1999) Strategies and methods for the allocation of food and water in the post-weaning period. Paper 5.4 presented to the 50th Meeting of the European Association of Animal Production, Zurich, August 22–26.

Brouns, F. and Edwards, S.A. (1994) Social rank and feeding behaviour of group housed sows fed competitively or ad libitum. *Applied Animal Behaviour Science* 39, 225–235.

Bruininx, E.M.A.M., van der Peet-Schwering, C.M.C., Schrama, J.W., Vesseur, P.C., Everts, H. and Beynen, A.C. (2001a) The IVOG® feeding station: a tool for monitoring the individual feed intake of group-housed weanling pigs. *Journal of Animal Physiology and Animal Nutrition* 85, 81–87.

Bruininx, E.M.A.M., van der Peet-Schwering, C.M.C., Schrama, J.W., Vereijken, P.F.G., Vesseur, P.C., Everts, H., den Hartog, L.A. and Beynen, A.C. (2001b) Individually measured feed intake characteristics and growth performance of group-housed weanling pigs: effects of sex, initial body weight, and body weight distribution within groups. *Journal of Animal Science* 79, 301–308.

Cera, E., Charlier, G. and Houvenaghel, A. (1988) Effect of age, weaning and post weaning diet on small intestinal growth and jejunal morphology in young swine. *Journal of Animal Science* 66, 574–584.

Deprez, P., Deroose, P., van den Hende, C., Muylle, E. and Oyaert (1987) Liquid versus dry feeding in weaned pigs: the influence on small intestinal morphology. *Journal of Veterinary Medicine* 34, 254–259.

van Diemen, P.M., Schrama, J.W., van der Hel, W., Verstegen, M.W.A. and Noordhuizen, J.P.T.M. (1995) Effects of atrophic rhinitis and climatic environment on the performance of pigs. *Livestock Production Science* 43, 275.

Easter, R.A. (1988) Acidification of pig diets for pigs. In: Haresign, W. and Cole, D.J.A. (eds) *Recent Advances in Animal Nutrition.* Butterworths, London, pp. 61–72.

Gentry, J.L., Swinkels, J.W.G.M., Lindemann, M.D. and Schrama, J.W. (1997) Effect of hemoglobin and immunisation status on energy metabolism of weanling pigs. *Journal of Animal Science* 75, 1032–1040.

de Haer, L.C.M. (1992) Relevance of eating pattern for selection of growing pigs. PhD dissertation, Research Institute for Animal Production (IVO-DLO) 'Schoonoord', Wageningen Agricultural University, The Netherlands.

Hampson, D.J. and Kidder, D.E. (1986) Influence of creep feeding and weaning on brush border enzyme activities in the piglet small intestine. *Research in Veterinary Science* 40, 24–31.

Kelly, D., Smyth, J.A. and McCracken, K.J. (1991) Digestive development in the early-weaned pig. II. Effect of level of food intake on digestive enzyme activity during the immediate post-weaning period. *British Journal of Nutrition* 65, 181–188.

Kenworthy, R. (1976) Observations on the effects of weaning in the young pig. Clinical and histopathological studies of intestinal function and morphology. *Research in Veterinary Science* 2(1), 69–75.

Le Dividich, J. and Herpin, P. (1994) Effects of climatic conditions on the performance, metabolism and health status of weaned pigs: a review. *Livestock Production Science* 38, 79–90.

Lindemann, M.D., Cornelius, S.G., El Kandelgy, S.M., Moser, R.M. and Pettigrew, J.E. (1986) Effect of age, weaning and diet on digestive enzyme levels in the piglet. *Journal of Animal Science* 69, 1298–1307.

Makkink, C.A. (1993) Of piglets, dietary proteins, and pancreatic proteases. PhD dissertation, Department of Animal Nutrition, Wageningen Agricultural University, The Netherlands.

McCracken, B.A., Gaskins, H.R., Ruwe-Kaiser, P.J., Klasing, K.C. and Jewell, J.E. (1995) Diet-dependent and diet-independent metabolic responses underlie growth stasis of pigs at weaning. *Journal of Nutrition* 125, 2838–2845.

Moon, H.K., Han, I.K., Parmentier, H.K. and Schrama, J.W. (1997) Effects of a cell mediated immune response on energy metabolism in weanling piglets. In: McCracken, K., Unsworth, E.F. and Wylie, A.R.G. (eds) *Proceedings 14th Symposium on Energy Metabolism of Farm Animals.* CAB International, Wallingford, UK, pp. 143–146.

Nabuurs, M.J.A. (1991) Etiologic and pathogenic studies on post weaning diarrhoea. PhD dissertation, State University, Utrecht, The Netherlands.

Odle, J., Zijlstra, S.M. and Donovan, S.M. (1996) Intestinal effect of milkborne growth factor in neonates of agricultural importance. *Journal of Animal Science* 74, 2509–2522.

Pajor, E.A., Fraser, D. and Kramer, D.L. (1991) Consumption of solid food by suckling pigs: individual variation and relation to weight gain. *Applied Animal Behaviour Science* 32, 139–155.

Pluske, J.R., Williams, I.H. and Aherne, H.X. (1996a) Maintenance of villous height and crypt depth in piglets by providing continuous nutrition after weaning. *Animal Science* 62, 131–144.

Pluske, J.R., Williams, I.H. and Aherne, H.X. (1996b) Villous height and crypt depth in piglets in response to increases in the intake of cows' milk after weaning. *Animal Science* 62, 145–158.

Pluske, J.R., Thompson, M.J., Atwood, C.S., Bird, P.H., Williams, I.H. and Hartmann, P.E. (1996c) Maintenance of villous height and crypt depth, and enhancement of disaccharide digestion and monosaccharide absorption, in piglets fed on cows' whole milk after weaning. *British Journal of Nutrition* 76, 409–422.

Ramaekers, P.J.L. (1996) Control of individual daily growth in group-housed pigs using feeding stations. PhD dissertation, Department of Animal Nutrition, Wageningen Agricultural University, The Netherlands.

Sijben, J.W.C., van Vugt, P.N.A., Swinkels, J.W.G.M., Parmentier, H.K. and Schrama, J.W. (1997) Energy metabolism of immunised weanling pigs is not affected by dietary nucleotides. *Journal of Animal Physiology and Animal Nutrition* 79, 153–161.

Vellenga, L. (1987) Het slingerziekte-syndroom bij varkens. Werkgroep Slingerziekte/speendiarree. *Dier en arts* 249.

Worsaae, H. and Schmidt, M. (1980) Plasma cortisol and behaviour in early weaned piglets. *Acta Veterinaria Scandinavica* 21, 640–657.

7

The Weaner Pig – Enzymes and Biotechnology for the Future

G. Partridge

Finnfeeds International Ltd, PO Box 777, Marlborough, Wiltshire SN8 1XN, UK

Introduction

The 1990s saw a steady rise in the use of certain biotechnological products as feed additives, and in research on new possibilities. Bonneau and Laarveld (1999) reviewed this whole area and its potential impact on farm animal nutrition, physiology and health and readers are referred to that paper for a comprehensive review of the subject.

Amongst the biotechnological additives, feed enzymes are seen perhaps as having made most progress and impact in the past decade, following the extensive and increasing use of crystalline amino acids, many of which are also produced from industrial fermentation. Bonneau and Laarveld (1999) identified the use of enzymes in a number of applications for improving feed characteristics, such as: removal of specific anti-nutritional factors; improving digestibility of nutrients; aiding the digestibility of non-starch polysaccharides; supplementing endogenous enzymes for improved digestion; reducing effluent losses.

The weaner pig has always provided a unique set of nutritional challenges that need to be overcome in order to maximize the young animal's growth potential. Some of the principal factors that limit post-weaning growth rate have been outlined by others (Partridge and Gill, 1993; Thacker, 1999a) and include: inadequate levels of endogenous enzyme secretion; reduced absorptive capacity in the gut; inadequate acid secretion; and inadequate food and water intake. Many of these factors are clearly interrelated and all, potentially, can be influenced to varying degrees by the application of feed enzyme technology.

Young pig nutrition is also being increasingly influenced by current trends in agricultural production, specifically to find potential alternatives to therapeutic and sub-therapeutic antibiotics at all stages of the production cycle. There is increasing research interest towards a better understanding of the role that certain feed additives, or combinations of them, can play in manipulating the gut microflora in a positive way against a background of reduced use of antibiotics.

This chapter reviews the role for feed enzymes in the nutrition of the weaner pig. It focuses on the main enzyme activities for which there is a growing body of published data: carbohydrases (with emphasis on xylanases and β-glucanases), phytases and proteases. It also examines current views on the modes of action of these enzymes and how they can offer particular benefits in the young, newly weaned animal. The interaction between feed enzymes and the gut microflora is receiving increasing attention in current research, using novel microbiological techniques, and this topic is considered in the light of the above constraints on antibiotic usage.

Carbohydrases

Carbohydrase enzymes for application in diets for the pig after weaning are targeted primarily against the non-starch polysaccharide (NSP) and starch components. Maize, wheat and barley, the main grain components of post-weaning pig feeds globally, all contain significant cell wall material, but they differ in both their level and composition (Table 7.1).

Maize- and sorghum-based diets

NSP levels in maize and sorghum (where used) are similar (9–10%; Dierick and Decuypere, 1994), with insoluble arabinoxylans comprising over 40% of this total. Standard milling procedures (e.g. roller or hammer milling through screens of varying diameter) are known to result in intact packages of cell wall material, with enclosed nutrients, entering the stomach and small intestine. As the cell wall structure of these particular grains is predominantly insoluble fibre and unresponsive to endogenous enzymes, varying quantities of 'packaged' material have been observed at the end of the small intestine, entering the hindgut (M.R. Bedford, Finnfeeds, 2000, personal communication). The observation that the post-weaned pig, in particular, appears to respond positively in growth rate and feed utilization to progressively fine grinding of maize and sorghum (Healy *et al.*, 1994) is, most likely, a reflection of reductions in nutrient 'packaging' by this cell wall material. Although

Table 7.1. The starch and fibre composition of the key grains used in weaner feeds globally (g kg^{-1}) (from various sources).

	Maize	Sorghum	Wheat	Barley
Starch	640	620	610	560
Total fibre	100	110	110	180
Crude fibre	25	25	25	45
β-Glucans	–	7	7	40
Arabinoxylans	44	45	58	65
Cellulose	20	21	25	39
Lignin	5	6	7	20

fine grinding of grains offers an undoubted opportunity to improve productive performance in young pigs, through improvements in nutrient digestibility, it also has to be considered alongside the economic costs of such processing. This led to recommendations on particle size reduction for grains in diets for weaned pigs (Healy *et al.*, 1994) that basically represent a compromise between these two conflicting requirements.

Exogenous enzymes of particular relevance to the fibre components of maize and sorghum are xylanase (targeting the insoluble arabinoxylan) and, to a lesser extent, cellulase. Relatively few published studies have reported the successful effects of exogenous enzymes on 'simple' maize/soya or sorghum/soya diets for weaned pigs. Li *et al.* (1996b) showed no response with maize/soya diets to the addition of β-glucanase, which was presumably chosen for this study for its 'cellulase' activity. Li *et al.* (1999), working with a mixed enzyme product (amylases; β-glucanase; pectinase, cellulase, proteases) showed non-significant improvements in growth rate and feed utilization in maize/soya diets in the presence or absence of an acidifier. In grower/finisher pigs fed similar diets, Lindemann *et al.* (1997) also found numerical improvements (non-significant) in growth and feed:gain ratio in pigs offered diets containing an enzyme complex (protease, cellulase, pentosanase, alpha-galactosidase, amylase), whereas Kim *et al.* (1998) showed no such effects in a sorghum/soya diet supplemented with cellulase. In contrast, Evangelista *et al.* (1998) found a significant improvement in growth rate in pigs (28–85 kg) offered maize-based diets supplemented with an enzyme product containing amylase, β-glucanase, xylanase and pectinase activities. Interpretation of responses is, of course, more difficult with any mixed enzyme product, even if the activity levels are quantified, which is the exception rather than rule in a number of published studies.

Starch digestibility in maize-based diets, with and without exogenous enzyme addition, appears not to have been studied systematically. Variations in amylose:amylopectin ratio, for example, might be expected to influence the degree of digestibility of maize starch. Topping *et al.* (1997) fed pigs on diets containing either standard maize starch or a high amylose material and observed large differences in the amounts of starch reaching the caecum (Fig. 7.1). This situation has implications not only for starch availability and digestibility in the small intestine, but also with respect to substrate-related changes in the microflora of the hindgut. There is much recent interest in the use of high amylose maize starch as an ingredient in human foods, to stimulate potentially beneficial butyrate fermentation in the hindgut. Certain exogenous amylases, with different properties to endogenous amylase, could have a role in improving starch digestibility in the small intestine of the young pig and thereby influence microbial numbers and/or composition. This opportunity appears to have been little studied, as judged by published literature to date.

Barley- and wheat-based diets

The effect of exogenous enzymes on diets for weaned pigs based on barley, wheat or mixtures of these grains has received far greater attention than maize or sorghum as

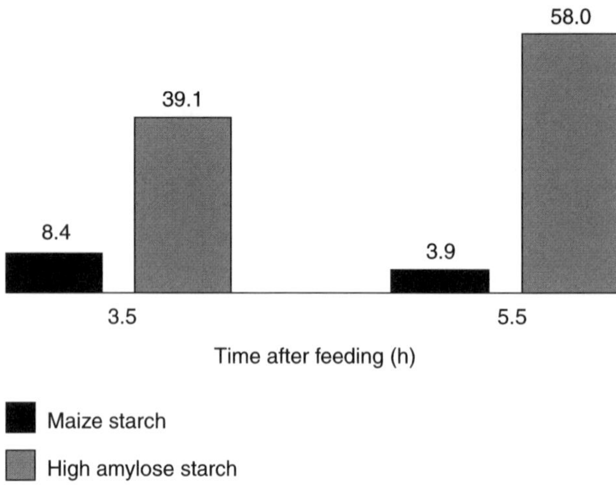

Fig. 7.1. Concentration of starch (mg g^{-1} dry matter) in caecal digesta of pigs fed diets containing maize starch or high amylose starch (mean of two pigs per treatment) (Topping *et al.*, 1997).

substrates in the scientific literature. This is unsurprising, given the history of feed enzyme application, which has been dominated by the use of, predominantly, xylanase and β-glucanase activities into diets for poultry and pigs containing these 'viscous' grains in Europe, Canada and Australia over the period 1990–2000. Hulled barley contains around 180 g NSPs kg^{-1}, with β-glucans and arabinoxylans constituting the majority of this cell wall material (Table 7.1). The concentration of soluble β-glucans is highest in the endosperm of the grain. Insoluble fibre predominates in the outer (aleurone) layer, with the surface of the aleurone cell walls containing high levels of xylan, surrounding a core of mixed-linked β-glucans (Autio *et al.*, 1996). For this reason optimal responses in weaner pig diets based on, or containing, barley are only likely to be seen when both xylanase and β-glucanase activities are supplied in sufficient quantities (van Lunen and Schulze, 1996; Ramaswamy *et al.*, 1996).

The last few years has seen increasing interest from certain markets, particularly Canada, in the opportunities for the use of hull-less barley as a feed ingredient in diets for weaned pigs. Hull-less barley differs from conventional barley grain in that the hull is less firmly attached to the kernel and therefore becomes detached during threshing. Despite, consequently, having lower insoluble fibre levels and higher levels of protein, its feeding value for piglets and poultry has often failed to live up to its theoretical potential (Baidoo *et al.*, 1998b; Thacker, 1999b). The reason appears to be that removal of the hull effectively concentrates the soluble fibre fraction in the grain endosperm, typically giving β-glucan levels of 4.5–7% vs. 3.5–4.5% in hull-less and hulled barley, respectively. These soluble, high molecular weight fibres are a problem for the pig in the small intestine: they can interfere with effective nutrient digestion and absorption, slow digesta transit rate and encourage microbial proliferation in the foregut. Feed processing (e.g. conditioning/pelleting/expansion) can exacerbate this situation by further solubilizing more fibre (Fig. 7.2). Feed processing can therefore be seen as a double-edged sword in these situations – positive in terms of its effects on feed hygiene and starch gelatinization (depending on moisture content during processing and the temperatures achieved) and negative in terms of some of these effects on dietary fibre, and their potential consequences.

Not surprisingly, exogenous enzymes have been found to be particularly beneficial in barley-based diets for young pigs (Table 7.2). Several studies used β-glucanase activities alone; although positive effects were almost invariably seen, it is questionable whether they were maximized based on the structure of the barley aleurone layer, as described above.

The predominant NSPs present in the endosperm and aleurone layers of wheat are arabinoxylans, comprising 50–60% of the total NSP content (Table 7.1). In contrast to barley and oats, β-glucan is a minor component. The arabinoxylans found in the thin cell walls within the endosperm are water-soluble to varying degrees, depending on variety, growing conditions, etc. In contrast, those in the thick cell walls of the aleurone layer are predominantly insoluble where they encircle, and complex with, potentially useful nutrients. Xylanases are therefore the main type of exogenous enzyme used for wheat-based diets and a number of trials in recent years have looked at their effects in diets for young pigs (Table 7.3). Overall, the consistency of response in published studies appears to be more variable than that of the barley-based trials (Table 7.2) but, equally, few studies have attempted to superimpose some predetermined aspect of wheat quality before appraising enzyme response. In the few recent studies where this has been attempted (Choct *et al.*, 1999; Partridge *et al.*, 1999) the ability of xylanase to raise significantly the productive performance of a 'poor' wheat to that of a 'good' wheat was clearly demonstrated, irrespective of the age of pigs studied (Table 7.4).

The relative importance of soluble arabinoxylans to the response to xylanase in the young pig is thrown into question by a number of studies where rye was the predominant grain (Thacker *et al.*, 1991, 1992a; Bedford *et al.*, 1992; Thacker and Baas, 1996). Dusel *et al.* (1997b) showed that the use of a mixed grain diet based

Cont'd on p. 137

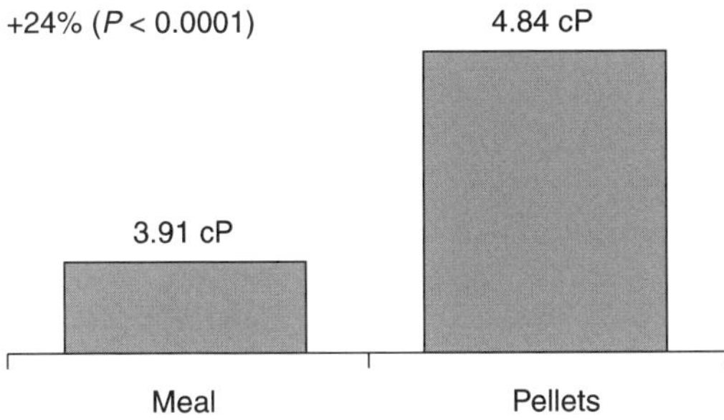

Fig. 7.2. The effect of conditioning and pelleting on solubilization of fibre and resultant viscosity value. Measured by 'Avicheck'™ (Finnfeeds), an estimation of viscosity in the water-soluble phase, following a simulated predigestion step *in vitro*. Eleven paired pig feed samples, wheat- and barley-based mixtures (cP = centipoise).

Table 7.2. Some published studies (1990–2000) on the effects of enzyme preparations on barley-based diets for weaner pigs (note that where digestibility measurements were made, data are presented preferentially at the ileal level).

References	Basal diet, main ingredients	Pelleted (P) or meal-mash (M)	Pig weight (kg)	Enzymes added		Responses (*$P < 0.05$)		
				Enzyme	Inclusion rate (t^{-1})	Area of response	Control	Plus enzymes
Bohme (1990)	Barley, wheat, SBM	P	11–25	Cellulase, β-glucanase, α-amylase, glucoamylase	1 kg	Growth performance (g day^{-1})	381	440
						Feed:gain	1.88	1.70*
Bedford *et al.* (1992)	Hull-less barley, SBM	M	12–18	β-Glucanase, (1086 U g^{-1})	2 kg	Weight gain (kg)	5.24	6.14*
						Feed:gain	1.75	1.55 ($P = 0.10$)
Thacker *et al.* (1992b)	Hull-less barley, SBM (C1 = no acid; C2 = + fumaric acid)	P (< 60°C)	8–18	β-Glucanase (750 U g^{-1})	2.5 kg	Growth performance (g day^{-1})		
						C1	254	274
						C2	275	275
						Feed:gain		
						C1	2.20	2.03
						C2	1.99	1.91
						Faecal digestibility of protein (%)		
						C1	78.4	77.8
						C2	77.6	81.7
						Faecal digestibility of energy (%)		
						C1	82.6	81.9
						C2	81.0	82.3

Inborr (1994)	Hulled or hull-less barley, SBM	M	9.5–14	β-Glucanase 89–95 mU g^{-1} feed	2.5 kg	Growth performance (g day^{-1})		
						Hull-less	223	229
						Feed:gain	1.60	1.52
						Hulled	200	208
						Feed:gain	1.71	1.65
						Effect of enzyme on growth ($P = 0.07$)		
						Effect of enzyme on feed:gain ($P = 0.06$)		
Power *et al.* (1996)	Hulled barley, wheat, wheat middlings	P (~70°C)	6–15	β-Glucanase (0, 500, 1000, 2000 ppm)		Growth performance and digestibility at optimum inclusion level (2000 ppm)		
						(g day^{-1})	201	227*
						Feed:gain	2.11	1.95*
						Faecal digestibility of protein (%)	77.4	80.3
Li *et al.* (1996b)	Hull-less barley, SBM	M	6–11	β-Glucanase (1000 U g^{-1})	2.5 kg (this highest inclusion gave max. response)	Faecal digestibility of protein (%) (significant linear effects of dose rate were seen)	81.6	88.5*
						Faecal digestibility of energy (%) (significant linear effects of dose rate were seen)	85.2	89.5*

Continued

Table 7.2. *Continued*

References	Basal diet, main ingredients	Pelleted (P) or meal-mash (M)	Pig weight (kg)	Enzymes added: Enzyme	Enzymes added: Inclusion rate (t^{-1})	Responses (*$P < 0.05$): Area of response	Control	Plus enzymes
Li *et al.* (1996a)	Hull-less barley, SBM	M	7–11	β-Glucanase (1000 U g^{-1})	2 kg	Ileal digestibility of protein (%)	65.2	73.5*
						Ileal digestibility of energy (%)	64.9	71.1*
Baidoo *et al.* (1998b)	Hull-less barley (varieties 'Condor' and 'Buck'), canola meal	M	14 (ileal cannulated)	Cellulase; β-glucanase; xylanase (11 U g^{-1}; 27 U g^{-1}; 43 U g^{-1})	100 g	Ileal digestibility of protein (%)	57.6	61.8*
						Effects were seen on both varieties, hence these are overall values		
						Ileal digestibility of energy (%)	57.1	63.3*
						Effects were seen on both varieties, hence these are overall values		
Baidoo *et al.* (1998b)	Hull-less barley, canola meal, SBM	M or P (80°C)	9–60	Cellulase; β-glucanase; xylanase (11 U g^{-1}; 27 U g^{-1}; 43 U g^{-1})	100 g	Growth performance (g day^{-1})	721	782*
						Feed:gain	2.01	1.84*
						Positive effects to enzyme addition		

						were seen in both mash and pelleted feed, hence overall values are shown		
Baidoo, *et al.* (1998a)	Hull-less barley (Falcon)	M	10–14 (ileal cannulated)	E1: Xylanase E2: β-glucanase E3: Xylanase; β-glucanase; protease		Ileal digestibility of protein (%)	59.6	
						E1		60.1
						E2		66.7*
						E3		67.2*
						Ileal digestibility of energy (%)	62.1	
						E1		64.6
						E2		69.8*
						E3		71.2*
Baidoo *et al.* (1998a)	Hull-less barley (Buck, Condor or Falcon), canola meal	M	14 (ileal cannulated)	Xylanase; β-glucanase; protease		Ileal digestibility of protein (%)		
						Buck	60.3	64.5
						Condor	54.8	60.7
						Falcon	67.6	67.2
						Ileal digestibility of energy (%)		
						Buck	57.7	67.0
						Condor	56.2	66.3
						Falcon	60.6	61.3
Gill *et al.* (2000)	Hulled barley, SBM	P (75°C)	8–18	β-glucanase; xylanase; amylase	1 kg	Growth performance (g day^{-1})	329	350
						Feed:gain	1.60	1.53
						DE (MJ kg^{-1})	14.86	15.08
						Faecal digestibility of protein (%)	76.8	75.1

SBM = soybean meal.
* $P < 0.05$.

Table 7.3. Some published studies (1990–2000) on the effects of enzyme preparations on wheat-, rye- or triticale-based diets for weaner pigs (where digestibility measurements were made, data are presented preferentially at the ileal level).

References	Basal diet, main ingredients	Pelleted (P) or meal-mash (M)	Pig weight (kg)	Enzymes added: Enzyme	Enzymes added: Inclusion rate (t^{-1})	Responses: Area of response	Responses: Control	Responses: Plus enzymes
Bohme (1990)	Wheat, barley, maize SBM	P	11–25	Cellulase; β-glucanase; α-amylase; glucoamylase	1 kg	Growth performance ($g\ day^{-1}$)	411	458
						Feed:gain	1.81	1.56*
Bedford *et al.* (1992)	Rye, SBM	M	11–15	Xylanase (850 U g^{-1})	2 kg	Weight gain (kg)	4.24	4.24
						Feed:gain	1.64	1.57
McLean *et al.* (1992)	Wheat, SBM		Weaner pigs	E1: Amylase; xylanase; cellulase; pectinase	700 g	Ileal digestibility of protein (%)	74.0	
						E1		76.6
						E2		73.1
				E2: β-glucanase; pectinase; cellulase; hemicellulase	1 kg	Ileal digestibility of energy (%)	70.8	
						E1		73.5*
						E2		72.9
Inborr *et al.* (1993)	Wheat:barley (50:50), SBM	M	~6–10	E1: β-glucanase # 1; xylanase; amylase (35; 590; 3300 U g^{-1})	1 kg	Growth performance ($g\ day^{-1}$)	204	
						Feed:gain	1.38	
				E2: β-glucanase # 2; xylanase; amylase (41; 740; 3300 U g^{-1})	950 g	Feed:gain		1.38
						E1		223
						E2		205
								1.39
						Ileal digestibility of protein (%)	76.8	
						E1		78.4

						E2		77.9
						Ileal digestibility of starch (%)	94.9	
						E1		97.7*
						E2		96.8
Officer (1995)	Wheat, fishmeal, meat meal, SBM	M	~6.5– 14	E1: protease; lipase; β-glucanase; amylase; cellulase	2 kg	Growth performance (after weaning) (g day^{-1})		
				E2: protease; lipase; β-glucanase; amylase; cellulase; hemicellulase; pentosanase	2 kg	E1	309	319
						Feed:gain	1.54	1.40 ($P < 0.10$)
						E2	299	282
						Feed:gain	1.76	1.82
						E3	361	332
						Feed:gain	1.56	1.49 ($P < 0.10$)
				E3: amylase; xylanase; β-glucanase; pectinase	2 kg			
Campbell *et al.* (1995)	Wheat, lupin (W/L) or wheat, SBM (W/S)	P (70°C)	6–13	Xylanases; pentosanases; β-glucanases	400 g	Growth performance		
						W/L	355	388
						Feed:gain	1.21	1.15
						W/S	291	353
						Feed:gain	1.29	1.23
						Overall effects of enzyme on growth rate ($P < 0.05$) and feed:gain ($P = 0.08$)		

Continued

Table 7.3. *Continued*

References	Basal diet, main ingredients	Pelleted (P) or meal-mash (M)	Pig weight (kg)	Enzymes added: Enzyme	Enzymes added: Inclusion rate (t^{-1})	Responses: Area of response	Responses: Control	Responses: Plus enzymes
Schulze *et al.* (1996)	Wheat, SBM	P	8–22	E1: Xylanase # 1 (5000 U g^{-1})	1 kg	Growth performance (g day^{-1})	304	
						Feed:gain	1.69	
				E2: Xylanase # 1/ protease (5000/500 U g^{-1})	1 kg	E1		394*
						Feed:gain		1.49
				E3: Xylanase # 2 (5000 U g^{-1})	1 kg	E2		385*
						Feed:gain		1.49*
						E3		402*
						Feed:gain		1.45*
Li *et al.* (1996a)	Wheat, SBM	M	7–11	β-glucanase (1000 U g^{-1})	2 kg	Ileal digestibility of protein (%)	68.8	75.9
						Ileal digestibility of energy (%)	66.9	71.2
Li *et al.* (1996b)	Wheat, SBM (W) or rye, SBM (R)	M	6–11	β-glucanase, 1000 U g^{-1} (A dose–response study – the numerically max. responses are shown)	2 kg (W) 1 kg (R)	Faecal digestibility of protein (%)		
						W	85.1	89.1
						R	87.0	89.3
						Faecal digestibility of energy (%)		
						W	86.8	88.4
						R	87.2	88.1
Dusel *et al.* (1997b) and Jeroch *et al.* (1999)	Wheat (high extract viscosity), rye barley, SBM	P	10–25	E1: Xylanase (5000 U g^{-1}) or	1 kg	Growth performance (g day^{-1})	354	
				E2: Xylanase	1 kg	Feed:gain	1.92	

				(5000 U g^{-1}) Protease (500 U g^{-1})		E*		462*
						Feed:gain		1.66*
						E*		480*
						Feed:gain		1.65*
						DE (measured MJ kg^{-1} dry matter)	13.87	15.08*
Partridge *et al.* (1998)	Wheat:barley (50:50), SBM – C1	P	8–30	Xylanase; β-glucanase; amylase; pectinase (4000; 150; 1000; 25 U g^{-1})	1 kg	Growth performance (g day^{-1})		
	Wheat:corn (50:50), wheat pollard, SBM – C2	P				C1	575	581
						Feed:gain	1.67	1.59*
						C2	512	557*
						Feed:gain	1.69	1.62*
Jeroch *et al.* (1999)	Wheat (86% of diet), low (L) or high (H) extract viscosity	P	10–25	Xylanase; (5000 U g^{-1})	1 kg	Growth performance (g day^{-1})		
						L	376	383
						Feed:gain	1.96	1.93
						H	398	400
						Feed:gain	1.90	1.80
						DE (measured, MJ kg^{-1} dry matter)		
						L	16.23	16.29
						H	15.91	16.07
Choct *et al.* (1999)	High (H), medium (M) and low (L) feed intake wheats – pre-determined		7–16	Xylanase; β-glucanase; cellulase	120 g	Growth performance (g day^{-1})		
						L	230	466*
						Feed:gain	1.38	1.23
						M	425	445
						Feed:gain	1.27	1.20
						H	460	479
						Feed:gain	1.14	1.20

Continued

Table 7.3. *Continued*

References	Basal diet, main ingredients	Pelleted (P) or meal-mash (M)	Pig weight (kg)	Enzymes added		Responses		
				Enzyme	Inclusion rate (t^{-1})	Area of response	Control	Plus enzymes
Gill *et al.* (2000)	Wheat/SBM (W) Wheat/beetpulp/ SBM (WSBP)	P (75°C)	8–18	Xylanase; amylase; pectinase (W) β-glucanase; amylase; pectinase (WSBP)	1 kg	Growth performance (g day^{-1}) (W)	338	360
						Feed:gain	1.53	1.48
						(WSBP)	345	363
						Feed:gain	1.56	1.48
						Faecal digestibility of protein (%) (WSBP)	72.1	73.9
						DE (MJ kg^{-1}) (WSBP)	14.10	14.33

SBM = soybean meal.
* $P < 0.05$.

Table 7.4. Effect of xylanase addition to wheats pre-selected for productive performance (determined in prior growth studies with weaner pigs, 28–60 kg) (Partridge *et al.*, 1999).

Wheat performance	−/+ Xylanase	Finish weight (kg)	Daily gain (g)	Daily feed intake (g)	Feed:gain
High	−	61.9	960[a]	1772[a]	1.84
Medium	−	60.2	918[ab]	1622[ab]	1.80
Medium	+	61.2	945[a]	1710[ab]	1.81
Low	−	58.3	878[b]	1576[b]	1.76
Low	+	61.5	952[a]	1728[a]	1.81
SEM		0.651(NS)	12.4*	33*	0.021(NS)

*$P < 0.05$.
NS, non-significant.
[ab] Means with different superscripts in the same column are significantly different ($P < 0.05$).

Table 7.5. Effect of enzyme addition to a wheat-based diet with a high extract viscosity, piglets 10–25 kg (Dusel *et al.*,1997b, and personal communication).

Wheat extract viscosity*	Daily gain (g)	Daily feed (g)	Feed:gain	Viscosity in the small intestine (cP)
Low (positive control)	388[ab]	717	1.85[b]	3.3[a]
High	354[b]	727	1.92[b]	6.4[b]
High + xylanase	462[a]	766	1.66[a]	3.9[a]
High + xylanase + protease	480[a]	794	1.65[a]	3.1[a]

*Low = 1.3 mPa; high = 3.3 mPa.
[ab] Means with different superscripts in the same column are significantly different ($P < 0.05$).

on a combination of wheat (high extract viscosity), rye and barley gave responses to enzyme addition that went beyond the apparent effects on viscosity reduction (Table 7.5). As in the studies of Choct *et al.* (1999) and Partridge *et al.* (1999), addition of the xylanase or xylanase/protease combination in these diets released an apparent limitation on voluntary food intake in the animal such that the growth response seen was a cumulative effect of both increased nutrient availability and intake. Other, more subtle changes in the microbial flora may also have influenced the response to the enzyme, a factor that will be examined later in this chapter. Pigs in the xylanase/protease-supplemented group appeared to show decreased bacterial proliferation in the small intestine (Fig. 7.3) when compared with both control diets, suggesting that some degree of nutrient 'sparing' may also have contributed to the productive response seen. These studies illustrate the difficulty in interpreting responses, or apparent lack of them, to exogenous enzymes in the pig unless many parameters are examined synchronously. The 'microbial environment' in which the enzyme is used will almost certainly influence the response seen, which has implications for trials based on 'low-challenge' research vs. more commercial situations.

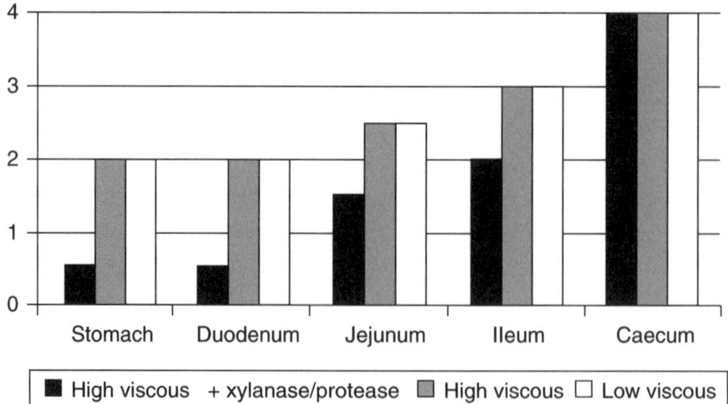

Fig. 7.3. Effect on microbial populations in digesta when piglets are fed diets based on high viscous wheat −/+ xylanase/protease, or low viscous wheat. Microscopical analysis of digesta (VTT Biotechnology and Food Research, Finland); samples from the studies of Dusel *et al.* (1997a). *Y*-axis is microbial loading in digesta; 0 = few microbes present; 4 = high levels of microbes present.

Phytases

Of the total phosphorus found in common ingredients used in pig and poultry feeds, around 60–80% is in the form of phytates, which are salts of phytic acid. The bioavailability of phosphorus in cereal grains, their by-products and legumes is very low (e.g. 10–30% in maize/soybean diets) because both pigs and poultry have a very limited capability to utilize phytate phosphorus (Kornegay, 1996). Phytic acid also has chelating potential in the gut, forming a wide variety of insoluble salts with di- and trivalent cations at neutral pH. Manganese, iron and magnesium can all be complexed, but zinc and copper have the strongest binding affinity. This binding potentially renders these minerals less available for intestinal absorption. Under neutral conditions, the carboxyl groups of some amino acids may also bind to phytate through a divalent or trivalent mineral. Phytate–protein or phytate–mineral–protein complexes may therefore reduce, to some degree, the utilization of dietary protein.

The phosphate groups of phytate must be hydrolysed before the phosphorus can be utilized by pigs or poultry and the supplementation of feeds with microbially synthesized phytase offers this opportunity. The last decade has seen a big investment in research on phytase, and latterly in its application in diets for different classes of stock when it is economically interesting relative to inorganic phosphorus, or desirable for environmental reasons.

Benefits of phytase supplementation have been found in a number of studies in the weaned pig (e.g. Beers and Jongbloed, 1992; Mroz *et al.*, 1994; Yi *et al.*, 1996; Qian *et al.*, 1996; Cadogan *et al.*, 1999). In the latter study superior feed:gain ratios were seen in phytase-supplemented feeds, where calcium and available phosphorus levels were reduced by 0.12% relative to the positive control diet (Table 7.6). The opportunity that phytase supplementation offers for reductions in acid-binding capacity, due to reduced calcium levels, is also considered a bonus in formulations for young pigs with limited acid-secreting capability.

Table 7.6. The effect of phytase addition and lysine level on the performance of male pigs over 21 days after weaning (Cadogan *et al.*, 1999).

Phytase level (FTU kg^{-1})	Available lysine (g MJ^{-1} DE)	Daily gain (g)	Daily feed intake (g)	Feed:gain
0	0.90	361	427	1.18^{ab}
550	0.90	437	489	1.13^{b}
0	0.85	373	436	1.17^{ab}
550	0.85	359	398	1.11^{b}
0	0.80	353	437	1.24^{a}
550	0.80	393	446	1.14^{b}
P value (phytase effect)		0.072	0.592	0.009

[ab] $P < 0.05$, no significant effects of available lysine level, or interaction (available lysine × phytase).
Control : 0.89% calcium, 0.49% available phosphorus.
+ Phytase : 0.77% calcium; 0.37% available phosphorus.

Proteases

Maximizing the digestibility of raw materials is a primary concern in the formulation of diets for young pigs immediately after weaning, especially in segregated early-weaning systems, where animals will have a particularly immature digestive system. Protein sources in the most effective diets for weaned pigs have historically been highly processed and relatively expensive raw materials, such as low temperature-processed fishmeal, milk products and plasma products, coupled with limited amounts of certain isolated vegetable proteins. Although these diets undoubtedly work successfully, there comes a stage in most production systems at which the young animal has to adapt to simpler diets heavily based on vegetable protein, usually soybeans, in extracted or full fat form.

Concerns about the extensive use of soya in young pig diets are mainly fuelled by the residual anti-nutrients it can contain (e.g. trypsin inhibitors, lectins), which are known to stimulate excessive endogenous nitrogen losses (Jansman *et al.*, 1994), as well as its tendency to induce damaging, transient hypersensitivity reactions in the immature gut (Li *et al.*, 1991; Friesen *et al.*, 1993). The main storage proteins in the bean (glycinin and β-conglycinin) have been particularly implicated in this respect. Tokach *et al.* (1994) outlined the practical dilemma, under USA production conditions, whereby feeding all-milk protein diets after weaning merely seemed to postpone this damaging hypersensitivity response giving no net growth advantage and a clear economic disadvantage over a 5-week period after weaning (Table 7.7). There seems, therefore, to be a physiological benefit in introducing the young animal to soybean protein as soon as possible but without inducing these problems, which seem to be further influenced by the degree of immunological 'stress' that the young pigs are under, such as health status and presence or absence of an antibiotic in the diet (Yen and Pond, 1987; Friesen *et al.*, 1993).

Table 7.7. The effect of post-weaning exposure to a soybean-based diet on starter pig performance over a 5-week period (Tokach *et al.*, 1994). All pigs received the same phase II diet from days 14 to 35.

Parameters	Milk-based diet days 0–14 after weaning	Soybean-based diet days 0–14 after weaning	P
Days 0–14			
Daily gain (g)	299	218	< 0.01
Daily feed intake (g)	295	277	< 0.05
Food conversion ratio	0.99	1.27	< 0.01
Days 14–35			
Daily gain (g)	404	503	< 0.01
Daily feed intake (g)	739	807	< 0.01
Food conversion ratio	1.83	1.60	< 0.01
Days 0–35			
Daily gain (g)	363	390	< 0.05
Daily feed intake (g)	562	594	< 0.05
Food conversion ratio	1.54	1.51	NS

Milk-based diet: maize 44%; skimmed milk 20%; whey powder 20%; casein 7%; soya oil 6%.
Soybean-based diet: maize 25%; soybean meal 40%; lactose 24%; soya oil 6%.
Phase II diet (days 14–35): maize 56%; soybean meal 23%; whey powder 10%; fishmeal 4%; soya oil 4%.

Further processing of soya products, particularly moist extrusion, appears to be one means of achieving improvements (Friesen *et al.*, 1992) but a potentially more cost-effective approach is the use of appropriate exogenous enzymes, particularly proteases, which have been investigated in recent years.

Caine *et al.* (1998) described a series of *in vitro* studies with a subtilisin protease, which showed the potential of the enzyme to solubilize protein in toasted soybean meal and, at the same time, reduce levels of residual soybean trypsin inhibitors (Table 7.8). Similar effects were described for a range of proteases, with differing pH characteristics, examined by Huo *et al.* (1993) and Beal *et al.* (1998a).

In growth trials, it certainly appears that some of the benefits seen *in vitro* can be translated into improved productive performance, which could be of particular value to the post-weaned pig. Rooke *et al.* (1996) studied the use of soybean meal, pretreated with protease, in diets for weaned pigs and showed clear benefits in growth performance in the critical first 7 days after weaning (Fig. 7.4). Similarly, Beal *et al.* (1998b), working with both raw and autoclaved soybeans in wet-fed diets for growing pigs (32 kg start weight), also saw improvements in performance following protease addition (Fig. 7.5). It was clear that the proteases studied were unable to give raw soya the feeding value of conventionally toasted material, but equally the enzyme's ability to hydrolyse the soybean storage protein into lower molecular weight units (Beal *et al.*, 1998c), coupled with its effects on the proteina-

Table 7.8. Effect of protease addition on total soluble matter, soluble crude protein and soybean trypsin inhibitor level (SBTI) under the optimum temperature (50°C), concentration 1 mg g^{-1} soybean meal and pH conditions (pH 4.5) defined by Caine *et al.* (1998) (*in vitro* studies with subtilisin protease).

	Control (no enzyme addition)	+ Protease
Soluble matter (g kg^{-1})	212.4[a]	356.0[b]
Soluble crude protein (g kg^{-1})	90.5[a]	318.7[b]
SBTI (mg kg^{-1})	3.55[a]	3.08[b]

[a,b] Means with different superscripts in the same row are significantly different ($P < 0.05$).

ceous anti-nutrients (trypsin inhibitors, lectins), offered interesting possibilities for the future. Added-value soya-based products using enzyme technology are now already available to the feed industry, offering good opportunities for more cost-effective piglet diet formulation. Equally, with the increasing interest in wet feeding systems for different weight classes of pig in some areas of the world, a process involving enzymatic pre-digestion of certain components prior to diet mixing also becomes feasible.

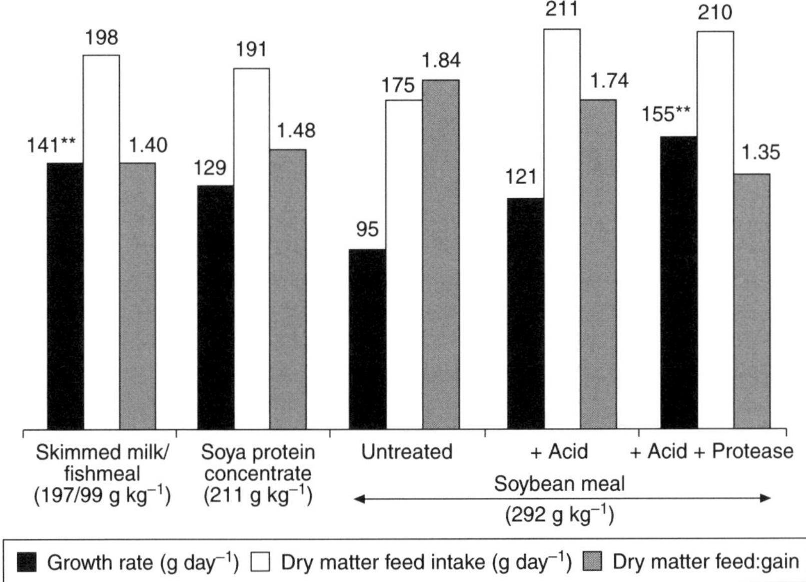

Fig. 7.4. The effect of pre-treatment of soybean meal with acid −/+ protease on piglet performance in the first 7 days after weaning (Rooke *et al.*, 1996). Soybean meal treated for 3 h at 50°C, pH 4.5 and at 20% dry matter; product neutralized and then dried (65°C) before inclusion into a maize diet + amino acids + vitamin/mineral, lysine 1.2%, digestible energy 14.5 MJ kg^{-1} (3465 kcal). $^{**}P < 0.01$.

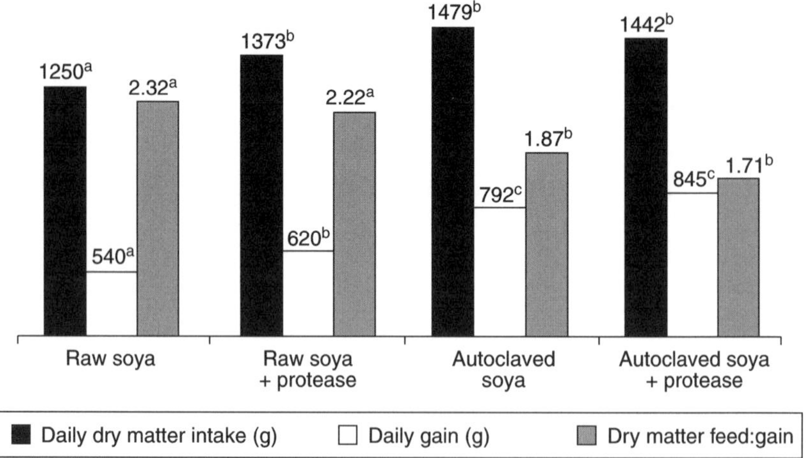

Fig. 7.5. The effect of protease pretreatment of raw or autoclaved full-fat soybeans on grower pig performance (32 kg start weight) (Beal *et al.*, 1998b).
[a–c] within parameters, means with different superscripts are significantly different ($P < 0.05$).

Modes of Action of Exogenous Enzymes and Their Particular Relevance to the Nutrition of the Weaned Pig

Low voluntary food intake in the post-weaning period places a well-recognized constraint on weaner performance (Partridge and Gill, 1993; Thacker, 1999a). Poor growth after weaning is clearly a multifactorial problem, but feed enzyme addition can potentially alleviate some of the potential disadvantages experienced by the young animal through a combination of positive effects on nutrient digestibility and certain aspects of digestive physiology.

It is well known that soluble, high molecular weight NSPs from viscous grains (e.g. barley, wheat, maize, triticale) interfere with nutrient digestion and absorption, with consequences for microbial proliferation in the gut (Apajalahti and Bedford, 1998; Bedford and Schulze, 1998). Compared with poultry, for example, the naturally higher water content of pig digesta (~10 percentage units higher than poultry on similar diets) leads to dilution effects that will negate, to some degree, this viscosity problem in the pig's gut (Fig. 7.6) (Danicke *et al.*, 1999). This is not to say that viscosity effects in pigs are irrelevant to carbohydrase responses in viscous grain diets, but rather that they are likely to be of a lower order of importance relative to the responses seen on similar diets in broilers. Reductions in digesta viscosity have been seen in many trials in pigs offered viscous cereal-based diets (Fig. 7.7) but it is clear that starting viscosity levels are always considerably lower than in equivalent trials in poultry. The influence of soluble fibre in pigs seem to be more subtle than just direct effects on nutrient digestibility. Rather, it could be mediated indirectly through effects on the gut microflora (e.g. Fig. 7.3), with implications for nutrient sparing in the small intestine, if microbial loading is reduced in the presence of feed enzymes.

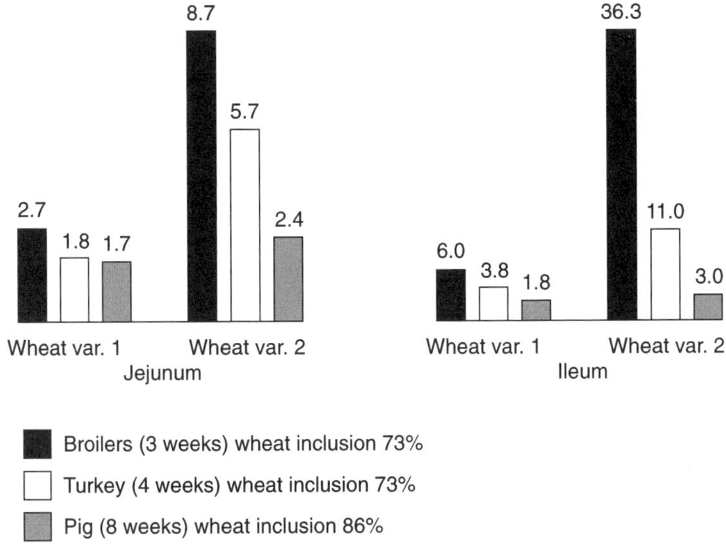

Fig. 7.6. Effect of wheat variety on intestinal viscosity (mPa) in broilers, turkeys and piglets (adapted from Danicke *et al.*, 1999). Ileum = small intestine (lower two-thirds for the pig); wheat var. 1 = 'Ibis' (11 g soluble arabinoxylan kg^{-1} dry matter); wheat var. 2 = 'Alidos' (17 g soluble arabinoxylan kg^{-1} dry matter).

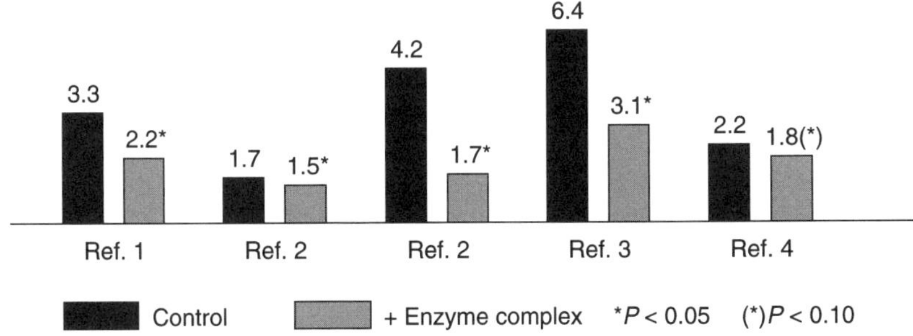

Fig. 7.7. Effects of enzymes on viscosity in the small intestine of piglets. Ref. 1, Inborr (1994); Ref. 2, Sudendey and Kamphues (1995) wheat/piglet and barley/grower; Ref. 3, Dusel *et al.* (1997a); Ref. 4, Partridge, *et al.* (1998).

Other physiological effects have been found following the addition of exogenous enzymes to diets for young pigs. Sudendey and Kamphues (1995) (Fig. 7.8) described studies in which viscosity reduction following enzyme addition in both wheat- and barley-based diets was associated with significant increases in digesta dry matter flow rate from the stomach down into the small intestine. Ellis *et al.* (1996) noted that an increase in gastric viscosity in the pig can lead to a failure in the 'sieving mechanism' whereby larger digesta particles normally fall to the bottom of the stomach – a feature seemingly unique to the pig. In a viscous environment, these large particles stay suspended and are thereby more likely to exit the pylorus before

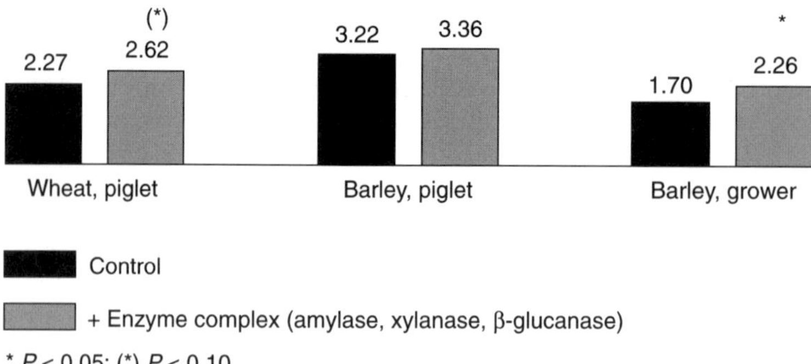

Fig. 7.8. Effect of enzyme addition to wheat- or barley-based diets on dry matter outflow rate from the stomach (Sudendey and Kamphues, 1995). Values are dry matter (g) leaving the stomach per hour per kilogram body weight.

being exposed to gastric secretions or particle size reduction, which may have negative implications for subsequent digestion. Further down the gastrointestinal tract, viscosity is also known to disturb peristalsis, creating laminar rather than disturbed flow, and also leads to disrupted pancreatic secretion via its effects on gastric inhibitory polypeptide. Viscous NSPs may also physically coat starch granules, further reducing the rate of digestion (Ellis *et al.*, 1996).

In successful pig trials using enzyme supplementation, a stimulation of voluntary food intake is frequently an important contributor to the observed improvements in daily liveweight gain (Haberer *et al.*, 1998). The fact that gastric emptying rate and distension of the stomach are two of a number of factors involved in satiety signals in the pig (Forbes, 1995) is unlikely to be coincidental. It is proposed that some of the positive responses to exogenous enzymes on voluntary food intake could be due to various influences on digesta flow rate as well as to improvements in nutrient availability and digestibility in the small intestine that appear to have an influence on food intake. For example, Whittemore (1993) proposed the following relationship between food intake and diet digestibility in the pig:

Limit to voluntary food intake (kg day^{-1}) $\simeq$ [0.013 $\times$ liveweight (kg)]/(1 $-$ digestibility coefficient).

Feedback loops involving gut hormones could also be contributors to carbohydrase effects, as outlined by Bedford and Schulze (1998). In this case it was proposed that caecal fermentation, enhanced by breakdown products from xylanase addition (fermentable oligomers), could influence enteroglucagon concentrations, which in turn influence stomach motility by depressing gastrin concentrations. These many possibilities suggest that exogenous enzymes could be potent influencers of gastrointestinal physiology of the young pig through a variety of different mechanisms, but these aspects have been little studied to date.

The water-holding capacity (WHC) of feed raw materials has also been shown to influence voluntary food intake in young pigs (Kyriazakis and Emmans, 1995), with feed intake falling as WHC in the diet rises. As carbohydrase enzymes have

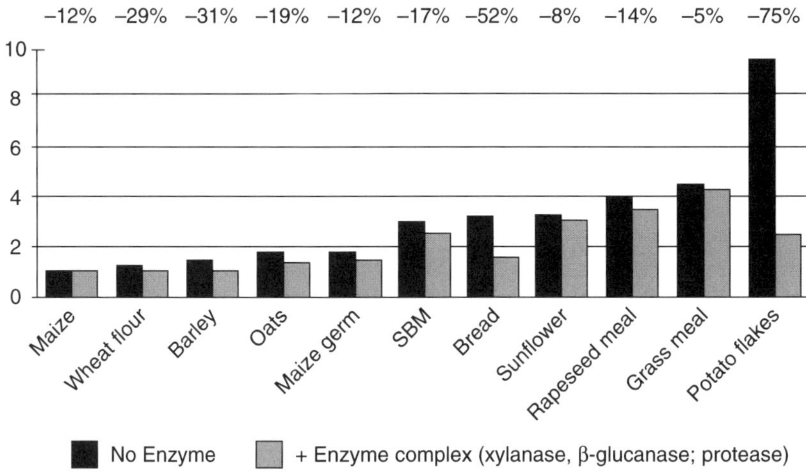

Fig. 7.9. Effect of an enzyme complex on water-holding capacity (g g^{-1}) of raw materials. Sources: Finnfeeds (personal communication, 1995).

been found to reduce the WHC of feedstuffs *in vitro* (Fig. 7.9), it follows that some of the benefits of enzyme addition on voluntary food intake in piglets may arise from influences on these physical factors in the gut. These benefits would be in addition to the potential benefits of enzyme use arising from both water and water-soluble nutrient release from dietary fibre residues in the gut.

Exogenous Enzymes – Effects on the Gut Microflora and Implications for Weaner Pig Nutrition and Health

The important interactions between diet composition, particularly the fibre components, and gut microflora changes during both chronic and acute disease challenge have been illustrated by Pluske *et al.* (1996, 1998). In these studies, soluble fermentable fibre and resistant starch from various feed raw materials were found to be particularly provocative to a specific disease condition, swine dysentery, induced by the anaerobic spirochaetal bacterium *Serpulina hyodysenteriae*. Although, to date, attempts to influence this condition by the strategic use of carbohydrase enzymes have given equivocal results (Durmic *et al.*, 1997), this concept is receiving increasing research interest as attention has focused on prophylaxis rather than routine medication in modern pig production systems.

Undigested soluble fibre has been identified as one of the key negative influences on the non-specific colitis syndrome affecting, particularly, pigs in the 15–40 kg weight range offered pelleted wheat-based rations (Taylor, 1989). Addition of an appropriate xylanase to these pelleted diets had a beneficial effect on the syndrome, negating the need to feed meal diets and/or involve costly reformulation on affected units (Hazzledine and Partridge, 1996).

Other studies have reported positive interactions between the use of feed enzymes and the incidence of diet-induced diarrhoea, particularly in the young pig after weaning (Inborr and Ogle, 1988; Flourou-Paneri *et al.*, 1998; Kantas *et al.*, 1998). These, together with reports of potential synergies between enzymes and therapeutic and sub-therapeutic addition of antibiotics (Flourou-Paneri *et al.*, 1998; Kantas *et al.*, 1998; Gollnisch *et al.*, 1999) offer interesting opportunities for future production methods where strategic rather than routine antibiotic use will become the norm.

Recent advances in methodologies to appraise the gut microflora have enabled quantification of the changes that exogenous enzymes can illicit. These methods overcome the limitations of conventional bacterial plating techniques, allowing changes and disturbances in bacterial populations to be studied in the presence or absence of various feed additives. Apajalahti and Bedford (1998), for example, described the use of DNA-based techniques to identify changes in the populations of bacterial genera in the guts of broilers fed enzymes. The addition of xylanase to wheat-based diets altered the relative abundance of bacterial DNA collected from the caecal flora (Table 7.9), with an increase in abundance of *Peptostreptococcus, Bacteroides, Proprionibacterium, Eubacterium* and *Bifidobacterium* and decreased abundance of *Clostridium, Enterobacteriacaea* and *Campylobacter*. The production of volatile fatty acids (VFAs) in the caecum was also changed, with a relative rise in proprionate production and stimulation of total VFAs produced. Huge shifts in bacterial groups were also reported, with changes in wheat:rye ratio in the broiler (Table 7.9), with clear implications for bird performance and the likelihood of disease incidence. Total microbial numbers in the small intestine of broilers fed wheat-based diets were also reduced by addition of xylanase compared with control animals, though to a lesser extent than the addition of an antimicrobial (avoparcin).

There are clear parallels in this work with the increasing interest in non-digestible oligosaccharides (NDOs) as potential 'prebiotics' (functional foods) for use in human foods and animal feeds (Mul and Perry, 1994; Grizard and Barthomeuf, 1999; van Loo *et al.*, 1999; Mosenthin *et al.*, 1999). Xylo-oligosaccharides are one such group of potentially beneficial NDOs and it is of interest in the studies of Apajalahti and Bedford (1998) that addition of xylanase resulted in increased concentrations of short-chain xylo-oligomers in the ileum of broilers fed wheat-based diets (Table 7.9). Whether these breakdown products of xylanase addition influence the microbial population in a beneficial way remains an active area of current research and hints at potential opportunities for *in situ* prebiosis, if it can be adequately characterized and manipulated.

Future Perspectives

Some of the nutritional problems faced by the young pig in the post-weaning period have proved to be positively influenced by the use of an increasingly effective range of specific exogenous enzymes over the past decade. The most cost-effective use of feed enzymes will always demand a good understanding of the anti-nutritional factors that influence raw material variability and (of equal importance

Table 7.9. The effect of xylanase addition to wheat- (and rye)-based broiler feeds on microflora changes in the caecum and xylo-oligomer production in the ileum. Data from Apajalahti and Bedford (1998).

(a) Relative abundance (%) of bacterial DNA collected from the caeca of broiler chickens fed wheat-based diets with or without 2500 U xylanase kg^{-1} feed.

Bacterial genus	Control	+ Xylanase
Lactobacillus and *Enterococcus*	26	26
Clostridium	14	11
Escherichia and *Salmonella*	11	6
Peptostreptococcus	10	11
Bacteroides	10	16
Proprionibacterium	10	11
Campylobacter	9	6
Eubacterium	8	10
Bifidobacterium	2	4

(b) Influence of the proportion of wheat:rye in broiler diets on the relative abundance of bacterial DNA (%) in the caeca of broiler chickens.

	Proportion wheat:rye (%)			
Bacterial genus	100:0	67:33	33:67	0:100
Enterococcus	17	36	39	43
Lactobacillus	49	35	16	6
Escherichia	8	7	27	25
Salmonella	3	2	4	8
Clostridium	12	13	8	10
Campylobacter	10	7	7	8

(c) Effect of the addition of xylanase to wheat-based broiler diets on the concentration (g kg^{-1} digesta) of xylo-oligomers of varying degrees of polymerization in the ileum.

	Degree of polymerization				
	< 10	< 100	< 500	< 2,000	< 10,000
Control	0.1	0.6	1.8	2.1	1.9
+ Xylanase	0.5	2.1	4.2	2.2	2.4

commercially) their rapid estimation at realistic cost. This remains an area of active research, with considerable scope for improvements in animal performance.

Changes in feeding systems for young pigs, specifically the potential use of wet-feeding (where applicable), also offer a good opportunity for enzyme solutions and, ultimately, more cost-effective production.

Finally, the changing nature of animal production with an increasing emphasis on prophylactic nutrition focuses even more attention on understanding the complex interrelationships between the animal, its microflora and the feed it receives. New methodologies in this area offer much scope for improvements in nutritional welfare, particularly for the young newly weaned animal.

References

Apajalahti, J. and Bedford, M.R. (1998) Nutrition effects on the microflora of the GI tract. In: *Proceedings of the 19th Western Nutrition Conference*, Saskatoon, Canada, pp. 60–68.

Autio, K., Mannonen, L., Pietila, K., Kosinken, M., Siika-aho, M. and Linko, M. (1996) Incubation of barley kernel sections with purified cell wall degrading enzymes. *Journal of the Institute of Brewing* 102, 427–432.

Baidoo, S.K., Liu, Y.G. and Grandhi, R.R. (1998a) Exogenous enzymes and hull-less barley utilisation by young pigs. In: *1998 Feed grain quality conference,* 8–10 November, Edmonton, Alberta, Canada, pp. 67–71.

Baidoo, S.K., Liu, Y.G. and Yungblut, D. (1998b) Effect of microbial enzyme supplementation on energy, amino acid digestibility and performance of pigs fed hull-less barley based diets. *Canadian Journal of Animal Science* 78, 625–631.

Beal, J.D., Brooks, P.H. and Schulze, H. (1998a) The effect of pre-treatment with different proteases on the *in vitro* digestibility of nitrogen in raw soy bean and four different full fat soya bean meals. In: van Arendonk, J.A.M. (ed.) *Abstracts of the 49th Meeting of EAAP, Warsaw, Poland*, PN2.7, p. 264.

Beal, J.D., Brooks, P.H. and Schulze, H. (1998b) The effect of the addition of a protease enzyme to raw or autoclaved soya bean on the growth performance of liquid fed grower/finisher pigs. *Proceedings of the British Society of Animal Science*, p. 161.

Beal, J.D., Brooks, P.H. and Schulze, H. (1998c) The hydrolysation of protein in raw and autoclaved soya bean meals by a microbial protease. *Proceedings of the British Society of Animal Science*, p. 167.

Bedford, M.R. and Schulze, H. (1998) Exogenous enzymes in pigs and poultry. *Nutrition Research Reviews* 11, 91–114.

Bedford, M.R., Patience, J.F., Classen, H.L. and Inborr, J. (1992) The effect of dietary enzyme supplementation of rye- and barley-based diets on digestion and subsequent performance in weanling pigs. *Canadian Journal of Animal Science* 72, 97–105.

Beers, S. and Jongbloed, A.W. (1992) Effect of supplementary *Aspergillus niger* phytase in diets for piglets on their performance and apparent digestibility of phosphorus. *Animal Production* 55, 425–430.

Böhme, H. (1990) Untersuchungen zur Wirksamkeit von Enzymzusatzen in der Ferkelaufzucht. *Landbauforschung volkenrode* 3, 213–217.

Bonneau, M. and Laarveld, B. (1999) Biotechnology in animal nutrition, physiology and health. *Livestock Production Science* 59, 223–241.

Cadogan, D.J., Selle, P.H. and Campbell, R.G. (1999) The effects of phytase supplementation and available lysine content of weaner diets on pig growth performance. *Recent Advances in Animal Nutrition in Australia* 12, 10A.

Caine, W.R., Verstegen, M.W.A., Sauer, W.C., Tamminga, S. and Schulze, H. (1998) Effect of protease treatment of soybean meal on content of total soluble matter and crude protein and level of soybean trypsin inhibitors. *Animal Feed Science and Technology* 71, 177–183.

Campbell, R.G., Whitaker, A., Hastrup, T. and Rasmussen, P.B. (1995) Interrelationship between protein source and Biofeed Plus on the performance of weaned pigs from 23 to

42 days of age. In: Hennessy, D.P. and Cranwell, P.D. (eds) *Proceedings of the 5th Conference of the Australasian Pig Science Association*, Canberra, p.194.

Choct, M., Cadogan, D.J., Campbell, R.G. and Kershaw, S. (1999) Enzymes can eliminate the difference in the nutritive value of wheats for pigs. In: Cranwell, P.D. (ed.) *Manipulating Pig Production*, Proceedings of the 7th Conference of the Australasian Pig Science Association, Adelaide, p. 39.

Danicke, S., Dusel, G., Jeroch, H. and Kluge, H. (1999) Factors affecting efficiency of NSP-degrading enzymes in rations for pigs and poultry. *Agribiological Research* 52, 1–24.

Dierick, N.A. and Decuypere, J.A. (1994) Enzymes and growth in pigs. In: Cole, M., Wiseman, J. and Varley, M.A. (eds) *Principles of Pig Science*. Nottingham University Press, Loughborough, UK, pp. 169–195.

Durmic, Z., Pethick, D.W., Mullan, B.P., Schulze, H. and Hampson, D.J. (1997) The effects of extrusion and enzyme addition in wheat based diets on fermentation in the large intestine and expression of swine dysentery. In: *Manipulating Pig Production*, Proceedings of the 6th Biennial Conference of the Australasian Pig Science Association, Canberra, 7–10 December 1997, p.180.

Dusel, G., Kluge, H., Simon, O., Jeroch, H. and Schulze, H. (1997a) [Investigation on the effect of nsp-degrading enzymes in wheat-based diets on intestinal viscosity and nutrient digestibility of piglets]. *Proceedings of the Society for Nutrition and Physiology* 6 (in German).

Dusel, G., Schulze, H., Kluge, H., Simon, O. and Jeroch, H. (1997b) The effect of wheat quality measured by extract viscosity and dietary addition of feed enzymes on performance of young pigs. *Journal of Animal Science* 75 (Supplement 1), 200.

Ellis, P.R., Rayment, P. and Wang, Q. (1996) A physio-chemical perspective of plant polysaccharides in relation to glucose absorption, insulin secretion and the entero-insular axis. *Proceedings of the Nutrition Society* 55, 881–898.

Evangelista, J.N.B., Sales Bastos, F.J. and Amarante, V. (1998) [Use of enzymes in maize–soya diets for pigs]. *Anaporc* 177, 89–97 (in Spanish).

Florou-Paneri, P., Kantas, D., Alexopoulos, C., Tsinas, A.C., Vassilopoulos, V. and Kyriakis, S.C. (1998) A comparative study of the effect of a dietary multi-enzyme system and/or virginiamycin on weaned piglet performance. *Proceedings of the 15th IPVS Congress*, Birmingham, UK, p. 35.

Forbes, J.M. (1995) *Voluntary Food Intake and Diet Selection in Farm Animals*. CAB International, Wallingford, UK.

Friesen, K.G., Nelssen, J.L., Behnke, K.C. and Goodband, R.D. (1992) Making soy-based baby pig feeds: effect of extrusion parameters. *Feed International*, September, pp. 50–55.

Friesen, K.G., Goodband, R.D., Nelssen, J.L., Blecha, F., Reddy, P.G. and Kats, L.J. (1993) The effect of pre- and post-weaning exposure to soybean meal on growth performance and on the immune response in the early-weaned pig. *Journal of Animal Science* 71, 2089–2098.

Gill, B.P., Mellange, J. and Rooke, J.A. (2000) Growth performance and apparent nutrient digestibility in weaned piglets offered wheat-, barley- or sugarbeet pulp-based diets supplemented with food enzymes. *Animal Science* 70, 107–118.

Gollnisch, K., Vahjen, W., Simon, O. and Schulz, E. (1999) [Influence of an antimicrobial (avilamycin) and an enzymatic (xylanase) feed additive alone or in combination on pathogenic microorganisms in the intestine of pigs (*E. coli, C. perfringens*)]. *Landbauforschung Volkenrode* 193, 337–342 (in German).

Grizard, D. and Barthomeuf, C. (1999) Non-digestible oligosaccharides used as prebiotic agents: mode of production and beneficial effects on animal and human health. *Reproduction Nutrition Development* 39, 563–588.

Haberer, B., Schulz, E. and Flachowsky, G. (1998) Effects of β-glucanase and xylanase supplementation in pigs fed a diet rich in nonstarch polysaccharides: disappearance and disappearance rate of nutrients including the nonstarch polysaccharides in stomach and small intestine. *Journal of Animal Physiology and Animal Nutrition* 78, 95–103.

Hazzledine, M. and Partridge, G.G. (1996) Enzymes in animal feeds – application technology and effectiveness. *Proceedings of the 12th Annual Carolina Swine Nutrition Conference*, pp. 12–33.

Healy, B.J., Hancock, J.D., Kennedy, G.A., Bramel-Cox, P.J., Behnke, K.C. and Hines, R.H. (1994) Optimum particle size of corn and hard and soft sorghum for nursery pigs. *Journal of Animal Science* 72, 2227–2236.

Huo, G.C., Fowler, V.R., Inborr, J. and Bedford, M. (1993) The use of enzymes to denature antinutritive factors in soybean. In: van der Poel, A.F.B., Huisman, J. and Saini, H.S. (eds) *Recent Advances in Research on Antinutritional Factors in Legume Seeds*, Wageningen Pers, Wageningen, The Netherlands, pp. 517–521.

Inborr, J. (1994) Supplementation of pig starter diets with carbohydrate-degrading enzymes – stability, activity and mode of action. *Agricultural Science in Finland*, vol. 3 (Supplement no. 2).

Inborr, J. and Ogle, R.B. (1988) Effect of enzyme treatment of piglet feeds on performance and post weaning diarrhoea. *Swedish Journal of Agricultural Research* 18, 129–133.

Inborr, J., Schmitz, M. and Ahrens, F. (1993) Effect of adding fibre and starch degrading enzymes to a barley/wheat based diet on performance and nutrient digestibility in different segments of the small intestine of early weaned pigs. *Animal Feed Science and Technology* 44, 113–127.

Jansman, A.J.M., Schulze, H., van Leeuwen, P. and Verstegen, M.W.A. (1994) Effects of protease inhibitors and lectins from soya on the true digestibility and endogenous excretion of crude protein in piglets. In: *VIth International Symposium on Digestive Physiology*, Bad Doberan, Germany, pp. 322–324.

Jeroch, H., Dusel, D., Kluge, H. and Nonn, H. (1999) The effectiveness of microbial xylanase in piglet rations based on wheat or wheat, rye and barley. *Landbauforschung Volkenrode* 193, 223–228.

Kantas, D., Florou-Paneri, P., Vassilopoulos, V. and Kyriakis, S.C. (1998) The effect of a dietary multi-enzyme system on piglet post-weaning performance. *Proceedings of the 15th IPVS Congress*, Birmingham, UK, p 36.

Kim, I.H., Hancock, J.D., Hines, R.H. and Kim, C.S. (1998) Effects of cellulase enzymes and bacterial feed additives on the nutritional value of sorghum grain for finishing pigs. *American Journal of Animal Science* 11(5), 538–544.

Kornegay, E.T. (1996) Using microbial phytase to improve the bioavailability of phosphorus, calcium, zinc and amino acids in swine and poultry diets. In: *Proceedings of BASF Symposium*, Pacific Northwest Animal Nutrition Conference, Seattle, Washington, pp. 68–106.

Kyriazakis, I. and Emmans, G.C. (1995) The voluntary feed intake of pigs given feeds based on wheat bran, dried citrus pulp and grass meal, in relation to measurements of feed bulk. *British Journal of Nutrition* 73, 191–207.

Li, D.F., Nelssen, J.L., Reddy, P.G., Blecha, F., Klemm, R.D. and Goodband, R.D. (1991) Interrelationship between hypersensitivity to soybean proteins and growth performance in early weaned pigs. *Journal of Animal Science* 69, 4062–4069.

Li, S., Sauer, W.C., Huang, S.X. and Gabert, V.M. (1996a) Effect of β-glucanase supplementation to hull-less barley or wheat-soybean meal diets on the digestibilities of energy, protein, β-glucans and amino acids in young pigs. *Journal of Animal Science* 74, 1649–1656.

Li, S., Sauer, W.C., Mosenthin, R. and Kerr, B. (1996b) Effect of β-glucanase supplementation of cereal-based diets for starter pigs on the apparent digestibilities of dry matter, crude protein and energy. *Animal Feed Science and Technology* 59, 223–231.

Li, D., Liu, S.D., Qiao, S.Y., Yi, G.F., Liang, C. and Thacker, P. (1999) Effect of feeding organic acid with or without enzyme on intestinal microflora, intestinal enzyme activity and performance of weaned pigs. *Asian-Australian Journal of Animal Science* 12, 411–416.

Lindemann, M.D., Gentry, J.L., Monegue, H.J., Cromwell, G.L. and Jacques, K.A. (1997) Determination of the contribution of an enzyme combination (Vegpro) to performance in grower-finisher pigs. *Proceedings of the 6th Conference of the Australasian Pig Science Association*, Canberra, p. 247.

van Loo, J., Cummings, J., Delzenne, N., Englyst, H., Franck, A., Hopkins, M., Kok, N., Macfarlane, G., Newton, D., Quigley, M., Roberfroid, M., van Vliet, T. and van den Heuvel, E. (1999) Functional food properties of non-digestible oligosaccharides: a consensus report from the ENDO project (DGXII AIRII-CT94-1095). *British Journal of Nutrition* 81, 121–132.

van Lunen, T.A. and Schulze, H. (1996) Evaluation of enzyme treatment of barley/rapeseed meal diets on digestibility and availability of nutrients for pigs from 40 to 90kg live weight. In: *Proceedings of the 14th International Pig Veterinary Congress*, Bologna, Italy, 7–10 July, p. 466.

McLean, D., McEvoy, J. and McCracken, K.J. (1992) Effects of processing and feed enzymes on nutrient digestibility in diets for weaned pigs. *Proceedings of the Nutrition Society* 52, 211A.

Mosenthin, R., Hambrecht, E. and Sauer, W.C. (1999) Utilisation of different fibres in piglet feeds. In: Garnsworthy, P.C. and Wiseman, J. (eds) *Recent Advances in Animal Nutrition.* Nottingham University Press, Loughborough, UK, pp. 227–256.

Mroz, Z., Jongbloed, A.W. and Kemme, P.A. (1994) The influence of graded calcium supply on microbial phytase efficacy in starter diets for pigs. *Proceedings of the 45th EAAP Meeting*, Edinburgh, 29pp.

Mul, A.J. and Perry, F.G. (1994) The role of fructo-oligosaccharides in animal nutrition. In: Garnsworthy, P.C. and Cole, D.J.A. (eds) *Recent Advances in Animal Nutrition.* Nottingham University Press, Loughborough, UK, pp. 57–79.

Officer, D.I. (1995) Effect of multi-enzyme supplements on the growth performance of piglets during the pre- and post-weaning periods. *Animal Feed Science and Technology* 56, 55–65.

Partridge, G.G. and Gill, B.P. (1993) New approaches with pig weaner diets. In: Garnsworthy, P.C. and Cole, D.J.A. (eds) *Recent Advances in Animal Nutrition.* Nottingham University Press, Loughborough, UK, pp. 221–248.

Partridge, G.G., de la Fuente, J.M., Flores, A., Sanz, M. and Tan, M.Y. (1998) The use of a multi-enzyme product in mixed grain diets for young pigs. *Proceedings of the 8th World Conference on Animal Production*, Seoul, Korea, pp. 116–117.

Partridge, G.G., Simmins, P.H. and Cadogan, D.J. (1999) Influence of xylanase addition to diets containing wheat co-products and nutritionally defined wheat on growing pig performance. In: Cranwell, P.D. (ed.) *Proceedings of the 7th Conference of the Australasian Pig Science Association*, Adelaide, p. 37.

Pluske, J.R., Siba, P.M., Pethick, D.W., Durmic, Z., Mullan, B.P. and Hampson, D.J. (1996) The incidence of swine dysentery in pigs can be reduced by feeding diets that limit the amount of fermentable substrate entering the large intestine. *Journal of Nutrition* 126, 2920–2933.

Pluske, J.R., Durmic, Z., Pethick, D.W., Mullan, B.P. and Hampson, D.J. (1998)

Confirmation of the role of rapidly fermentable carbohydrates in the expression of swine dysentery in pigs after experimental infection. *Journal of Nutrition* 128, 1737–1744.

Power, R., de Koning, W. and Fremaut, D. (1996) The effect of β-glucanase supplementation on production parameters in piglets receiving barley-based diets. *Animal Science* 62, 662–663.

Qian, H., Kornegay, E.T. and Conner, D.E. (1996) Adverse effects of wide calcium:phosphorus ratios on supplemental phytase efficacy for weanling pigs fed two dietary phosphorus levels. *Journal of Animal Science* 74, 1288–1297.

Ramaswamy, C.M., van Lunen, T.A. and Schulze, H. (1996) Effect of xylanase and β-glucanase on nutrient digestibility in pigs fed hull-less or hulled barley based diets. *Proceedings of the Eastern Nutrition Conference*, Canada, 15–17 May.

Rooke, J.A., Fraser, H., Shanks, M. and Morgan, A. (1996) The potential for improving soya bean meal in diets for weaned pigs by protease treatment. *Proceedings of the British Society of Animal Science*, Paper No. 136.

Schulze, H., Partridge, G.G. and Gill, B.P. (1996) The use of different xylanase sources and a protease in wheat-based diets for weaner pigs. *Animal Production* 62, 663.

Sudendey, C. and Kamphues, J. (1995) [Effect of enzyme addition (amylase, xylanase and β-glucanase) on digestive processes in the intestinal tract of force fed weaner pigs]. *Proceedings of the Society for Nutrition and Physiology* 3, 88A (in German).

Taylor, D.J. (1989) *Pig Diseases*, 5th edn. Burlington Press, Cambridge, UK, pp. 269–271.

Thacker, P.A. (1999a) Nutritional requirements of early weaned pigs: a review. *Pig News and Information* 20, 13N–24N.

Thacker, P.A. (1999b) Effect of micronisation on the performance of growing/finishing pigs fed diets based on hulled and hull-less barley. *Animal Feed Science and Technology* 79, 29–41.

Thacker, P.A. and Baas, T.C. (1996) Effects of gastric pH on the activity of exogeneous pentosanase and the effect of pentosanase supplementation of the diet on the performance of growing-finishing pigs. *Animal Feed Science and Technology* 63, 187–200.

Thacker, P.A., Campbell, G.L. and GrootWassink, J. (1991) The effect of enzyme supplementation on the nutritive value of rye-based diets for swine. *Canadian Journal of Animal Science* 71, 489–496.

Thacker, P.A., Campbell, G.L. and GrootWassink, J.W.D. (1992a) Effect of salinomycin and enzyme supplementation on nutrient digestibility and the performance of pigs fed barley- or rye-based diets. *Canadian Journal of Animal Science* 72, 117–125.

Thacker, P.A., Campbell, G.L. and GrootWassink, J.W.D. (1992b) The effect of organic acids and enzyme supplementation on the performance of pigs fed barley-based diets. *Canadian Journal of Animal Science* 72, 395–402.

Tokach, M.D., Goodband, R.D. and Nelssen, J.L. (1994) Recent developments in nutrition for the early-weaned pig. *The Compendium*, March issue, pp. 408–419.

Topping, D.L., Gooden, J.M., Brown, I.L., Biebrick, D.A., McGrath, L., Trimble, R.P., Choct, M. and Ilman, R.J. (1997) A high amylose (amylomaize) starch raises proximal large bowel starch and increases colon length in pigs. *Journal of Nutrition* 127, 615–622.

Whittemore, C.T. (1993) *The Science and Practice of Pig Production*. Longman, Harlow, UK, p. 378.

Yen, J.T. and Pond, W.G. (1987) Effect of dietary supplementation with vitamin C or carbadox on weanling pigs subjected to crowding stress. *Journal of Animal Science* 64, 1672.

Yi, Z., Kornegay, E.T., Ravindran, V., Lindemann, M.D. and Wilson, J.H. (1996) Effectiveness of 'Natuphos' phytase in improving the bioavailabilities of phosphorus and other nutrients in soybean meal-based semipurified diets for young pigs. *Journal of Animal Science* 74, 1601–1611.

Liquid feeding for the Young Piglet 8

P.H. Brooks, C.A Moran, J.D. Beal, V. Demeckova and A. Campbell

Seale-Hayne Faculty, University of Plymouth, Newton Abbot, Devon TQ12 6NQ, UK

Introduction

The newly weaned pig poses the biggest management challenge on many pig production units. Even piglets that have been growing at 300 g per day while suckling the sow often grow at half that rate, or less, in the week following weaning. Many individual weaned piglets make no growth or have negative growth immediately after weaning. This not only represents an immediate loss in productivity but also predisposes the pig to a future loss of growth potential. Pigs that are disadvantaged at this stage of their growth perform less well throughout the growing period. The reasons for this poor performance are multifactorial. Food and water intake, housing and environment and the social setting all influence post-weaning performance. The effects of some of these individual factors are well documented but the interactions between them have been less well researched.

Weaning – a Process or an Event?

In the wild pig, weaning comprises a gradual transition from a liquid diet, exclusively based on sow milk (*c.* 200 g dry matter kg^{-1}), to a solid diet, over a period of several weeks. The wild pig is an opportunistic feeder and is primarily herbivorous (Baber and Coblenz, 1987). According to Spitz (1986), 90% of the diet of the wild pig consists of vegetable matter, of which 50% comprises seeds and fruits. The remainder of this diet is made up of ground-dwelling insects, molluscs and earthworms. Consequently, when the wild piglet starts taking solid food, that food will have a dry matter (DM) content of between 15 and 30%.

The situation on commercial pig units is very different. The domestic piglet is typically weaned abruptly, at between 14 and 35 days of age (depending on the coun-

(eds M.A. Varley and J. Wiseman)

try). It is expected to make an instant transition from a liquid diet, of around 20% DM, to a compound diet with around 85% DM, usually presented in pelleted form. In nature, the gradual transition from a milk diet through a mixed diet to a diet devoid of milk provides the stimulus, and allows time, for the enzyme systems of the immature gastrointestinal tract (GIT) to develop (Kidder and Manners, 1978; Gestin *et al.*, 1997). It also provides time for changes in the ecology of the GIT as the microbial population is influenced by and adapts to the new feed inputs (Buddington, 1998).

The long transitional period found in nature also has behavioural implications. When suckling the sow, piglets have been acclimatized to having both hunger and thirst satisfied by the sow's milk. The stimulus provided by the sow also has two important effects on their behaviour. The sow calling her piglets to feed at intervals of 40–50 min conditions them to suckle at regular intervals. It also programmes the litter to feed as a group (Brooks and Burke, 1998). When weaning occurs as a gradual process, piglets have the opportunity to learn by experience about food and water, without their nutrient intake (from the sow) being interrupted. In addition, they develop individual foraging behaviour. On commercial pig units the provision of creep food and water attempts to provide a similar opportunity. However, because piglets are weaned at a young age on commercial pig units, the creep feed intake by individual piglets may be limited and very variable. Metz and Gonyou (1990) reported that piglets weaned at 2 weeks of age consumed only 7 g food per day in the 2 days before weaning, whereas piglets weaned at 4 weeks of age consumed 127 g. Within litters, creep food consumption may also vary dramatically, with some piglets eating up to ten times as much as their littermates (Pajor *et al.*, 1991). Large variations in intake were also observed when liquid milk-replacer was offered to suckling piglets (Azain *et al.*, 1996). Water intake may also vary considerably in the pre-weaning period and may be affected by the type of drinker provided. Newborn pigs have been shown to learn to drink much more quickly from a bowl than from a nipple drinker (Phillips and Fraser, 1991).

Following abrupt weaning, piglets have to learn to distinguish between the physiological drives of hunger and thirst. They also have to learn how to satisfy these drives by consuming water and solid food. If piglets have had little experience of consuming either food or water before weaning, this can result in a discontinuity in feed and water intake that seriously compromises their future performance.

The Interaction Between Water and Feed Intake After Weaning

Voluntary or involuntary deprivation of water, or food, after weaning has serious consequences for the piglet. Gill (1989) showed that it could take more than a week for the weaned piglet to restore its daily fluid intake to the equivalent of that on the day before weaning. Piglets experiencing such reduced fluid intakes can become seriously dehydrated. The resulting disturbance of the pig's homeostatic balance has important repercussions on its physiology, as does any subsequent rapid rehydration that occurs when the pig finally starts drinking. It is important to encourage piglets to maintain fluid intake after weaning. They may find some difficulty identifying nipple drinkers as a supply of water, but there is little evidence that the provision of dripping water devices encourages water consumption (Ogunbameru *et al.*, 1991). Providing readily available water in a bowl has been shown to encourage water consumption and increase feed intake (English *et al.*, 1981). The management of bowls is critical, as fouling may reduce the palatability and consumption of water (Brooks and Carpenter, 1990).

Water consumption levels for 5 days after weaning appear to have little relationship with presumed physiological need (McLeese *et al.*, 1992). Brooks *et al.* (1984) demonstrated that, in the immediate post-weaning period, piglets consumed water rather than food (Fig. 8.1). Only when pigs learned to recognize food did they develop a more normal water:feed ratio (Table 8.1). It may be that in this early post-weaning period piglets fail to discriminate between water and food and as a consequence consume water to provide gut fill. Pigs that have learned to eat and drink will minimize the water:feed ratio when fed *ad libitum* and will take additional water, presumably to produce feelings of satiety, when their feed is restricted (Yang *et al.*, 1981, 1984). In the weaned pig, water availability affects feed intake and performance (Table 8.2). Restricting the flow rate of nipple waterers significantly affected both water and food intake, with consequent effects on pig performance (Barber *et al.*, 1989).

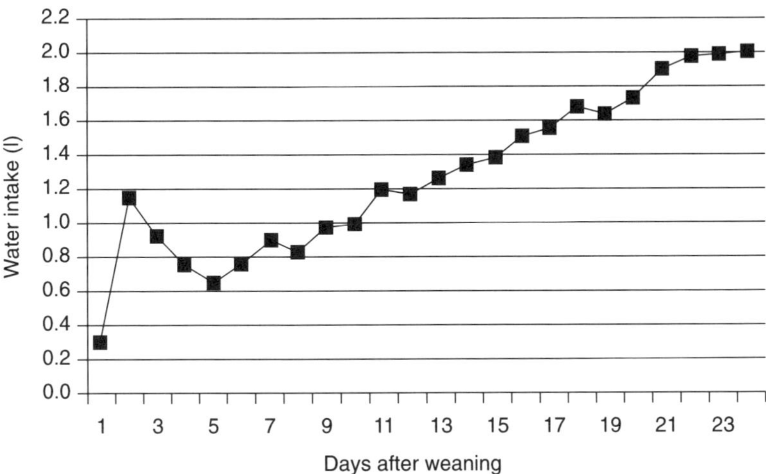

Fig. 8.1. Water intake (l) of pigs following weaning at 21 days of age (after Brooks *et al.*, 1984).

Table 8.1. Ratio of water consumed (g) to food consumed (g) by pigs weaned at 21 ± 1 days of age and fed on two commercial diets (Brooks *et al.*, 1984).

Week	Diet A	Diet B
1	4.3:1	4.0:1
2	3.2:1	3.5:1
3	2.9:1	3.6:1
4	2.8:1	3.7:1

Table 8.2. The effects of water delivery rate on the voluntary food intake and water use of weaned pigs (Barber *et al.*, 1989).

	Water delivery rate (cm^3 min^{-1})				
	175	350	450	700	SE_D
Daily feed intake (g)	303[c]	323[b]	341[a]	347[a]	3.68
Daily gain (g)	210[c]	235[b]	250[a]	247[a]	5.57
Feed conversion ratio	1.48	1.39	1.37	1.42	0.03
Daily water used (l)	0.78[d]	1.04[c]	1.32[b]	1.63[a]	0.01
Time spent drinking (s day^{-1})	268[b]	176[a]	175[a]	139[a]	14.4

[a, b, c, d] Within a row, means with the same superscript are not significantly different ($P > 0.05$)
SE_D, standard error of difference between means.

The Significance of Maintaining Continuity of Food Intake After Weaning

Pig producers tend to consider lack of food intake only in terms of lost growth rate resulting from a failure to provide the necessary nutrients for growth. However, failure to eat has more profound and long-lasting effects on performance.

The time interval between weaning and the pig taking its first feed or first drink has been studied by Brooks and Brice (1999, unpublished data). The results (Fig. 8.2) show two important factors. Firstly, there was great variation between individual pigs. Secondly, although a majority of pigs had taken their first meal within 3 min of weaning, some pigs took a very long time (up to 54 h) before eating their first meal, even though their pen-mates had already found and were consuming feed.

This observation has important implications for practical management and for experimental design. Managers of pig units need to recognize that just because some pigs are observed to be using the feed and water provided in the pen, this does not mean that all are. Thus, management strategies must aim to ensure that feed and water acquisition by all pigs in the group takes place as soon after weaning as possible. Because experiments usually consider the mean performance of groups of pigs, apparent treatment differences may result either from a change in the behaviour of some pigs in the group or from a change in the biological response of the whole population.

A retrospective analysis of data collected in a number of studies at the University of Plymouth demonstrated a highly significant ($P < 0.001$) effect of DM feed intake in the first week after weaning on the weight of pigs at 28 days after weaning (Geary and Brooks, 1998). The results of this analysis suggest that each 50 g day^{-1} increase in DM feed intake in the week following weaning increased the 28-day post-weaning weight by 870 g.

DM feed intake in the week after weaning accounted for as much variation in the 28-day post-weaning weight as any combination of weaning weight, weaning

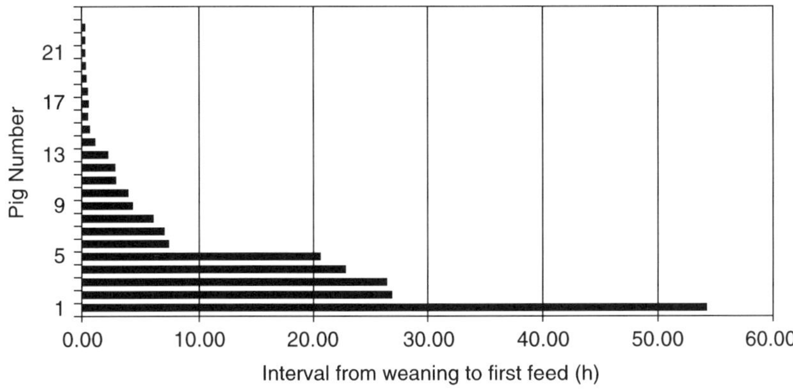

Fig. 8.2. Interval from weaning (24 days) to first feed (h) (Brooks and Brice, unpublished data).

age, sex and dietary treatment. Data from our own unit confirms reports from commercial farms that pigs with a low weight for age at weaning can sometimes grow faster than their heavier littermates (Brooks, 1998), though the increase was not sufficient to make up the difference in initial weight. Behavioural observations suggested that this was related to latency to the first feed after weaning. Smaller pigs were more likely to have consumed creep feed prior to weaning and hence take more readily to solid food following weaning. The heavier pigs at weaning were often those that had had a good teat and had little experience of solid food; these pigs took longer to establish a pattern of consuming solid food after weaning.

It is easy to forget that the epithelial lining of the gut is the most rapidly developing tissue in the body and that many of the nutrients required for gut growth are absorbed direct from the gut lumen. The work of Pluske *et al.* (1996, 1997) showed that a continuous supply of nutrients is essential in order to maintain the villus architecture of the gut after weaning.

Previous studies suggested that villus height was greater when pigs were fed a liquid diet rather than a dry diet (Fig. 8.3) (Deprez *et al.*, 1987). It was assumed that the physical form of the food was responsible for this effect. However, the studies undertaken by Pluske *et al.* (1996) demonstrated that the change in villus height was not a function of diet form, but of nutrient intake (Fig. 8.4). In that study, feeding liquid diets at a maintenance energy level still resulted in a reduction in villus height 5 days after weaning. However, when pigs were fed an allowance equivalent to three times the requirement for maintenance, villus height was actually greater than that immediately before weaning. A recent study, using a larger data set (Ward and Moran, unpublished data), has again demonstrated the positive relationship between DM intake and villus height (Fig. 8.5).

There is good evidence, from rats, that transient starvation can result in atrophy of the villi (Steiner *et al.*, 1968; Rudo *et al.*, 1976). A reduction in villus height reduces absorption and allows more nutrients to escape digestion and enter the lower gut. This promotes the development of an inappropriate gut microflora that

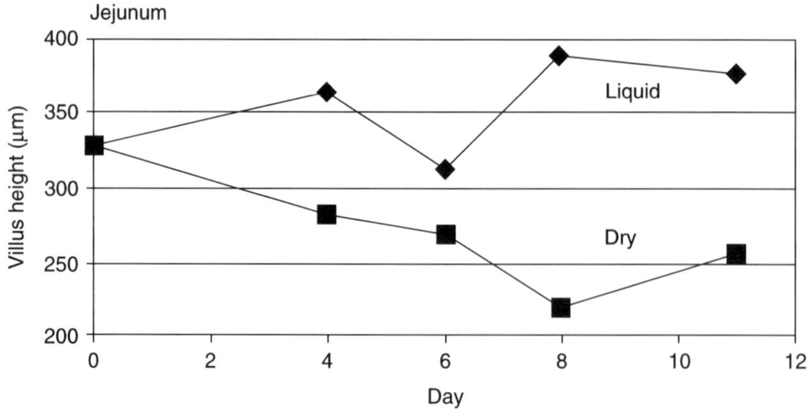

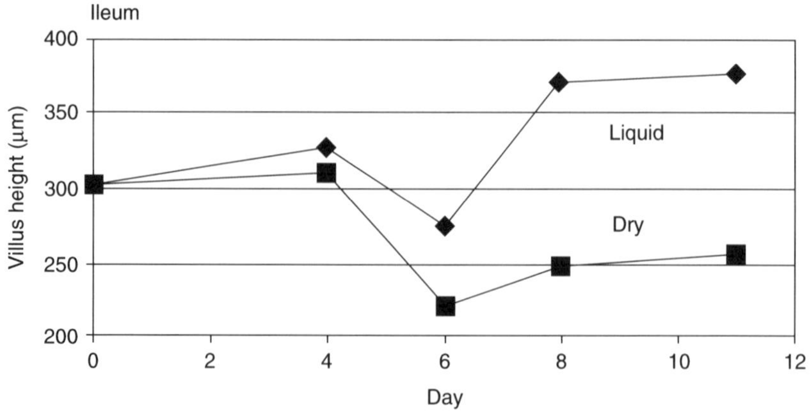

Fig. 8.3. Effect on villus height of feeding weaned pigs liquid or dry feed (after Deprez *et al.*, 1987).

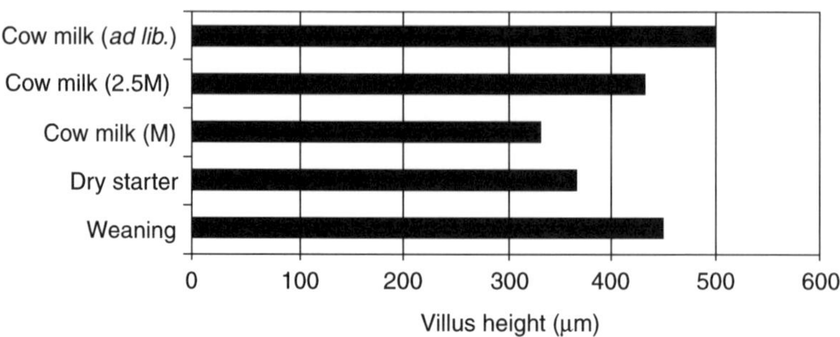

Fig. 8.4. Effect of form and quantity of feed on villus height 5 days after weaning. M = maintenance, 2.5M = 2.5 × maintenance (after Pluske *et al.*, 1996).

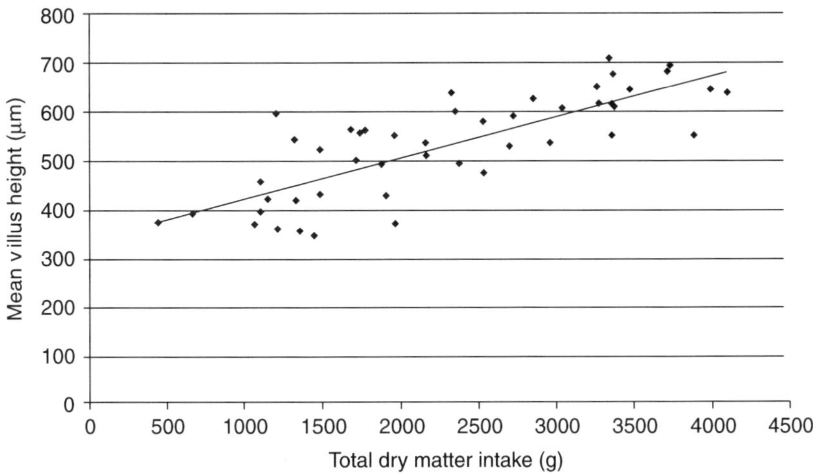

Fig. 8.5. Relationship between total dry matter feed intake (days 5–13 after weaning) and villus height in piglets fed liquid diets for 14 days after weaning.

in turn can result in enteric disease. The practical implication of these findings is that every effort must be made to maintain continuity of food and water intake following weaning.

The Case for Liquid Feeding

In view of the above, it might be anticipated that post-weaning performance would be improved by offering a liquid diet after weaning. Liquid feeding could be seen as having potential advantages for the following reasons:

- It could provide a diet with a DM concentration more like that of sow milk and more like the solid food that the pig would encounter in the wild, which consequently might encourage intake and maintain continuity of nutrient supply.
- It could provide a diet that more closely meets the piglet's need for both nutrients and water.
- It might overcome some of the problems posed by the piglets having to learn to satisfy their drives of hunger and thirst separately.

In order for *ad libitum* liquid feeding to be successful, some major problems have to be overcome:

- The feed has to be provided in a management system that minimizes waste.
- The feeding system must not result in an unacceptable increase in labour.
- The feed must be palatable.
- The microbiological quality of the feed has to be maintained.

Liquid feeding, using milk-based diets, has been successfully used for the artificial rearing of piglets and for pigs weaned at very young ages (Braude and Newport, 1977; Lecce *et al.*, 1979; Armstrong and Clawson, 1980; English *et al.*, 1981; Efird *et al.*, 1982; Ratcliffe *et al.*, 1986; Taverner *et al.*, 1987; Maswaure and Mandisodza, 1995; Dunshea *et al.*, 2000). However, liquid feeding of milk products has often been associated with the development of acute diarrhoea caused by enterotoxigenic bacteria (Lecce, 1986; Dunshea *et al.*, 2000). Liquid-feeding systems can easily become contaminated and the labour involved in cleaning and replenishing feed troughs has generally inhibited commercial producers from using liquid feed.

Until recently, researchers tended to look for complex engineering solutions in order to prevent the feed becoming unwholesome and unpalatable. For example, an automatic wet-feeding system developed for use with weaned pigs (Partridge *et al.*, 1992) was shown to increase significantly both feed intake and daily gain in the first 3 weeks after weaning (Table 8.3), but achieved little commercial uptake.

The development of computerized liquid-feeding systems capable of feeding pigs *ad libitum* has reawakened interest in the possibility of liquid feeding for weaner pigs. These systems rely upon liquid-sensing probes suspended in the trough to determine the fill level in the trough. These are interrogated by the computer at predetermined intervals. If the probes are dry, pneumatic valves are activated to dispense feed into the trough until the feedback by the probes terminates the fill cycle.

The Significance of Fermentation in Liquid-feeding Systems

A major problem with most automated and all *ad libitum* liquid-feeding systems is that uncontrolled fermentation will always occur unless positive action is taken to prevent microbial activity. Smith (1976) showed that lactic acid bacteria (LAB) that

Table 8.3. Intake and growth rate of weaned piglets fed using a conventional dry feed hopper or an automated wet feeding system (the equipment dispensed varying quantities of a compound pelleted diet plus water in a 1:1 ratio, either 6, 8, 12 or 24 times throughout the day) (Partridge *et al.*, 1992).

	Daily feed intake (g)			Daily gain (g)		
Week	Dry hopper	Wet antifeeder	SE_D	Dry hopper	Wet antifeeder	SE_D
1	149	176	14.0**	133	147	15.5
2	327	357	23.6	330	355	14.4*
3	453	518	36.3**	380	434	17.2***
Overall	310	351	16.8*	281	312	12.1**

SE_D, standard error of difference between means.
**P* < 0.05; ***P* < 0.01; ****P* < 0.001.

occur naturally on cereal grains will proliferate in a wet-feed system and reduce the pH (increase the acidity) of the diet. In Smith's study, adding water to the meal at feeding time produced a diet with a pH of 5.8. Soaking the mixture for 24 h resulted in a massive proliferation of LAB, which produced a beneficial increase in the acidity of the diet.

Because pipeline liquid-feeding systems are not normally sterilized between feeds, they are generally microbiologically active and the feed system acts as a microbiological fermenter. A Danish survey revealed that it took 3–5 days for the lactobacillus levels to elevate and stabilize in pipeline feeding systems (Hansen and Mortensen, 1989). The survey also found that it could be detrimental to sterilize pipeline-feeding systems as this removed the lactobacilli and reduced the acidity of the system. This in turn allowed coliform bacteria to proliferate for a period of 1–5 days until the lactobacilli re-established themselves and once again lowered the pH. The coliform 'bloom' that followed the sterilization of pipeline systems was often associated with outbreaks of diarrhoea. These problems generally disappeared when the acidity increased again in the system.

It should be noted that recent developments in liquid-feeding equipment and feeding practice sometimes produce undesirable effects. There is an increasing trend to feed pigs *ad libitum* on liquid diets. Liquid-feed equipment manufacturers have started to market systems that mix small quantities of feed at regular intervals and deliver this into short troughs that always contain some feed. When components used to construct the diet are sufficiently acid, or have been deliberately acidified in order to eliminate bacteria and yeasts (pH < 2.5), fermentation in the trough is usually inhibited and the feed in the trough retains its palatability (note that in these circumstances the final diet fed to the pig will still be around pH 4). However, when the pH of the diet is above 4.5 an undesirable fermentation can occur and enteropathogens such as coliforms and salmonellas will proliferate.

A more recent approach relies upon controlled fermentation with LAB to maintain the palatability of the diet and to prevent the growth of enteropathogens in the feed. The production and use of fermented liquid feed (FLF) is simple in concept, relying on a natural process that has been used for centuries to preserve human food and in the last century for ensiling herbage for ruminants. Feed is mixed with two to three times its weight of water and allowed to steep. The natural flora carried by the raw materials initiates a fermentation (Fig. 8.6). Initially, the value of the mixture is around pH 6–7 and there is a rapid growth of coliform bacteria. LAB then start to grow and as they produce acid (primarily lactic acid) the pH of the mixture falls. Once it drops below pH 4.0, coliform bacteria are progressively eliminated from the feed and populations are generally below the limits of detection within 4 days at an ambient temperature of *c.* 16°C (Fig. 8.6a). If insufficient lactic acid is produced, the pH is not reduced sufficiently and coliform bacteria persist in the system (Fig. 8.6b). A continuous fermentation can be maintained by retaining about half of the feed each day and adding half fresh feed and water to the next mixture. Yeasts also grow in these continuous fermentation systems. If the conditions are not optimized for the growth of LAB, and in particular if the fermentation temperature is too low, yeasts can become the predominant organisms.

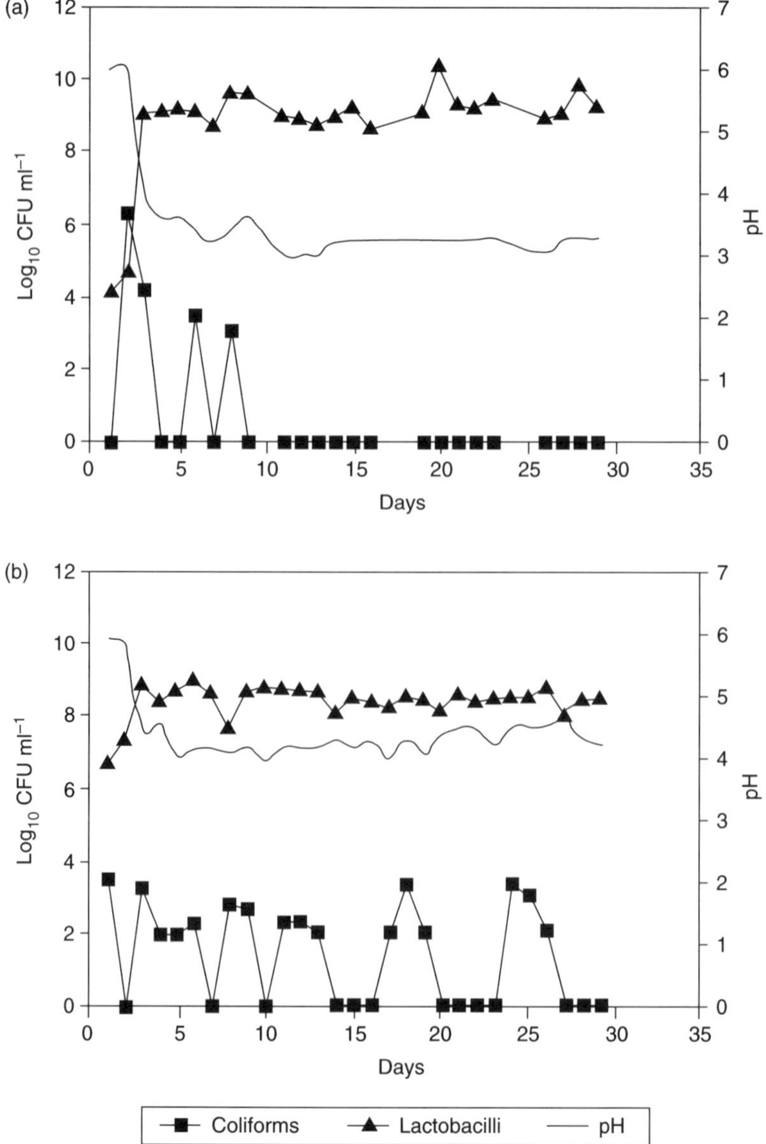

Fig. 8.6. Elimination of coliform bacteria from spontaneous lactic acid bacteria fermentations. CFU, colony forming units.

This can be problematic, as many yeast species generate off-taints in the feed that make it unpalatable. CFU, colony-forming units.

The effects of presenting pigs with their feed in dry, liquid (freshly prepared) or fermented liquid feed have been reviewed by Jensen and Mikkelsen (1998). The results of this review are summarized in Table 8.4. In weaner pigs, daily liveweight

Table 8.4. Improvement (%) in growth rate and food conversion ratio in experiments in which the performance of pigs fed dry feed (DF), liquid feed (LF) or fermented liquid feed (FLF) was compared (Jensen and Mikkelsen, 1998).

		Improved daily weight gain		Improved food conversion ratio	
	No. of trials	Mean ± SD	Range	Mean ± SD	Range
LF vs. DF	10	12.3 ± 9.4	–7.5–34.2	–4.1 ± 11.8	–32.6–10.1
FLF vs. DF	4	22.3 ± 13.2	9.2–43.8	–10.9 ± 19.7	–44.3–5.8
FLF vs. LF	3	13.4 ± 7.1	5.7–22.9	–1.4 ± 2.4	–4.8–0.6

SD, standard deviation.

gain was improved by an average of 12.3% when the feed was presented in liquid form compared with dry, and by a further 13.4% if the liquid feed was fermented. However, the feed conversion ratio (FCR) was generally somewhat poorer on liquid feed and FLF compared with dry feed, probably as a result of increased feed wastage. In these studies fermentation occurred spontaneously, initiated by the native microflora present on feed ingredients.

Spontaneous fermentation was used in the early studies at the University of Plymouth (Russell *et al.*, 1996). In these studies, pigs fed FLF had significantly higher feed intakes than pigs fed dry pelleted feed (DPF), particularly in the first 3 weeks following weaning (Table 8.5). High quality commercial dry diets were used, which had very high digestibility and palatability and included lactose (provided from milk by-products) and/or glucose. The fermented diets were well accepted by weaners and there was no evidence of post-weaning scours. Growth rate was also improved by feeding FLF rather than DPF (Table 8.6).

It was notable in this study that the improved growth rate of pigs fed FLF rather than DPF appeared to decline in later weeks. Subsequent reanalysis of this and other studies indicated that this was because diets formulated on the basis of anticipated DM intake as dry pelleted feed provided excess protein to piglets on FLF, which had feed intakes up to 20% greater than anticipated. Under these circumstances feed utilization was reduced, water consumption (and hence effluent

Table 8.5. Dry matter intake of piglets fed dry pelleted feed (DPF) or fermented liquid feed (FLF) (Russell *et al.*, 1996).

Weeks after weaning	Trial 1			Trial 2		
	DPF	FLF	SE_D	DPF	FLF	SE_D
1	130	416	11***	199	271	9*
2	354	741	19***	418	560	16**
3	636	1068	16***	686	819	12***
4	889	120	35	877	954	9**
Overall	443	807	11***	545	654	10***

SE_D, standard error of difference between means.
*$P < 0.05$; ** $P < 0.01$; *** $P < 0.001$.

Table 8.6. Growth rate (g day^{-1}) of piglets fed dry pelleted feed (DPF) or fermented liquid feed (FLF) (Russell *et al.*, 1996).

Weeks after weaning	Trial 1			Trial 2		
	DPF	FLF	SE_D	DPF	FLF	SE_D
1	162	123	23**	140	178	21
2	264	426	37***	340	425	18**
3	529	635	38**	511	602	21***
4	674	630	44	594	610	22
Overall	343	428	21***	397	454	14***

** $P < 0.01$; *** $P < 0.001$.
SE_D, standard error of difference between means.

production) was increased and ultimately DM intake was affected. This implies that when constructing diets that are going to be fed as FLF, care must be taken to formulate diets on the basis of nutrient intake per pig per day, using a realistic estimate of DM intake.

Geary *et al.* (1996) subsequently investigated the effect of the diet's DM concentration on performance. Young pigs were capable of high DM intakes even on diets with a very low DM content (Table 8.7). However, in order to avoid excess effluent production it is recommended that the DM content of the diet should not fall below 200 g kg^{-1}. The data in this study suggested that performance might be improved further by feeding a higher DM concentration (> 250 g kg^{-1}) following weaning and reducing the DM concentration (to around 200 g kg^{-1}) by 4 weeks after weaning.

Jensen and Mikkelsen (1998) reported that approximately 3% of the DM and energy content of the feed were lost during the fermentation process (Table 8.8). Some caution must be exercised in interpreting data where energy values are estimated from proximate analysis of the DM and the application of equations developed to predict energy value in dry feed. In liquid feed, some oligosaccharides and sugars present in the diet are turned into volatile fatty acids and alcohol, which are lost when liquid feed is dried prior to proximate analysis being undertaken. Short-chain fatty acids are efficiently used as an energy source by the pig (Mosenthin, 1998) and con-

Table 8.7. Dry matter intake of piglets fed fermented liquid feed with different dry matter concentrations (Geary *et al.*, 1996).

Weeks after weaning	Dry matter concentration of fermented liquid feed (g kg^{-1})				
	255	224	179	149	SE_D
1	204[a]	178[b]	109[ab]	161[a]	19**
2	397[a]	367	380	320[a]	24*
3	600	539	599	578	26
4	700	648	774	746	36
Overall	475	433	466	451	22

[a, b] Within a row, means with the same superscript differ at $P < 0.05$.
SE_D, standard error of difference between means.

Table 8.8. Effect of fermentation temperature on the chemical composition of liquid feed (Jensen and Mikkelsen, 1998).

Treatment	DM(%)	Energy (MJ kg^{-1})	Starch (g kg^{-1})	Total N (g kg^{-1})
Non-fermented	24.1[a]	4.77[a]	97.5	9.8
15°C	23.4[b]	4.62[b]	97.6	9.4
20°C	23.4[b]	4.65[b]	65.9	9.3
25°C	23.2[b]	4.58[b]	96.5	9.4
30°C	23.4[b]	4.63[b]	97.5	9.8
Loss (%)	3.1	3.1		

[a,b] Figures within a column not sharing a common superscript differ significantly ($P < 0.05$).

tribute between 20 and 30% of the pig's energy requirement (Kennelly and Aherne, 1980; Yen *et al.*, 1991). To overcome this problem an energy equation has been developed in The Netherlands (CVB, 1998) that takes into account the contribution of the volatile components in the liquid diets. In addition, changes in the gut architecture produced by feeding FLF (discussed later) may increase the pig's absorptive capacity. Thus, even though fermentation is an energy-demanding process, the positive benefits may cancel out the effect on energy supply to the pig.

Controlling Fermentation

A major problem in the practical application of FLF is the control of fermentation. Jensen and Mikkelsen (1998) demonstrated the importance of temperature in controlling the fermentation and lowering the pH of the feed. Using a 50% residue and 8 h replenishment of the tank, they found that a steady state was reached in 50 h when the tank was maintained at 25°C, whilst a peiod of around 100 h was required when the tank was maintained at 15°C. Studies at the University of Plymouth have shown that the temperature at which the fermentation takes place also affects the rate at which lactic acid is produced and pH is reduced (Fig. 8.7). It is important that water added to the system is at the correct temperature, as water added direct from the tap (at around 5°C) will cold-shock the system. This has three adverse effects: (i) it adversely affects the growth of the LAB; (ii) it allows the yeasts to become dominant; and (iii) it induces the production of 'cold shock' proteins in enteropathogens. These 'cold shock' proteins protect them and allow them to persist in the feed for much longer.

In early studies at Plymouth, we relied upon naturally occurring LAB to ferment the diets. Although this produced satisfactory results, the uncontrolled nature of this system gave cause for concern. In particular, there was concern that pig producers would not be able to exercise the same levels of control on farm that could be achieved in experimental facilities. Therefore, more recent studies have investigated the use of specific LAB inoculants to control fermentation. In one study (Geary *et al.*, 1999) diets were acidified to pH 4.2, either with lactic acid or through fermentation with *Pediococcus acidilactici*. Performance of the pigs did not differ between treatments and was better than that obtained in previous experiments, where uncontrolled LAB fermentation was used (Table 8.9).

Acidification was more effective than fermentation in reducing the initial 'bloom' of coliform bacteria. However, it is worth noting that, even when the diet was acidified with lactic acid, a population of LAB still developed. Over time, as many LAB were present in the lactic acid treatment as in the treatment that had been fermented with *P. acidilactici* (Fig. 8.8).

To produce a controlled fermentation, either the organism used as an inoculant must be capable of outcompeting the native flora on the raw materials, or the native

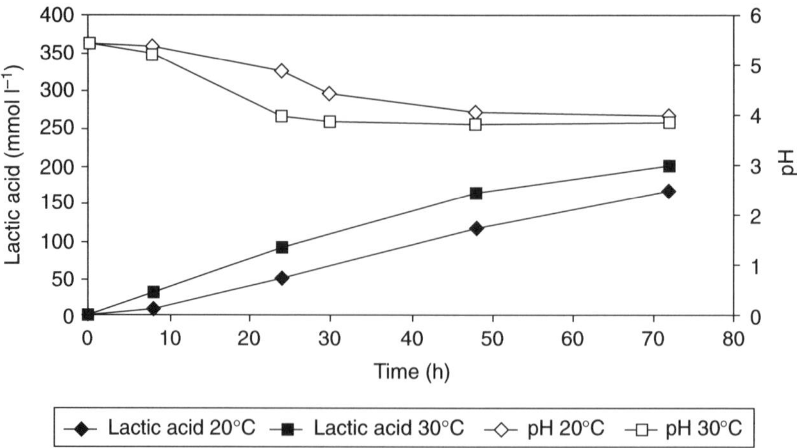

Fig. 8.7. The effect of fermentation temperature on the production of lactic acid and reduction of pH in feed (J.D. Beal, unpublished data).

Table 8.9. Performance of pigs fed liquid feed acidified with either lactic acid (Control) or as a result of fermentation of the diet with *Pediococcus acidilactici* (Geary *et al.*, 1998). None of treatment differences was statistically significant.

	Treatment		
Parameter	Control supplemented with lactic acid	Fermented with *Pedioccocus acidilactici*	SE_D
Dry matter feed intake (g day^{-1})	536	563	71
Daily gain (g day^{-1})	474	496	25
Dry matter feed conversion ratio	1.15	1.11	0.09
Total water intake (ml per pig day^{-1})	2078	2283	252
Average water intake from drinkers (ml per pig day^{-1})	511	638	133
Average effluent production (ml per pig day^{-1})	1118	1359	185

SE_D, standard error of difference between means.

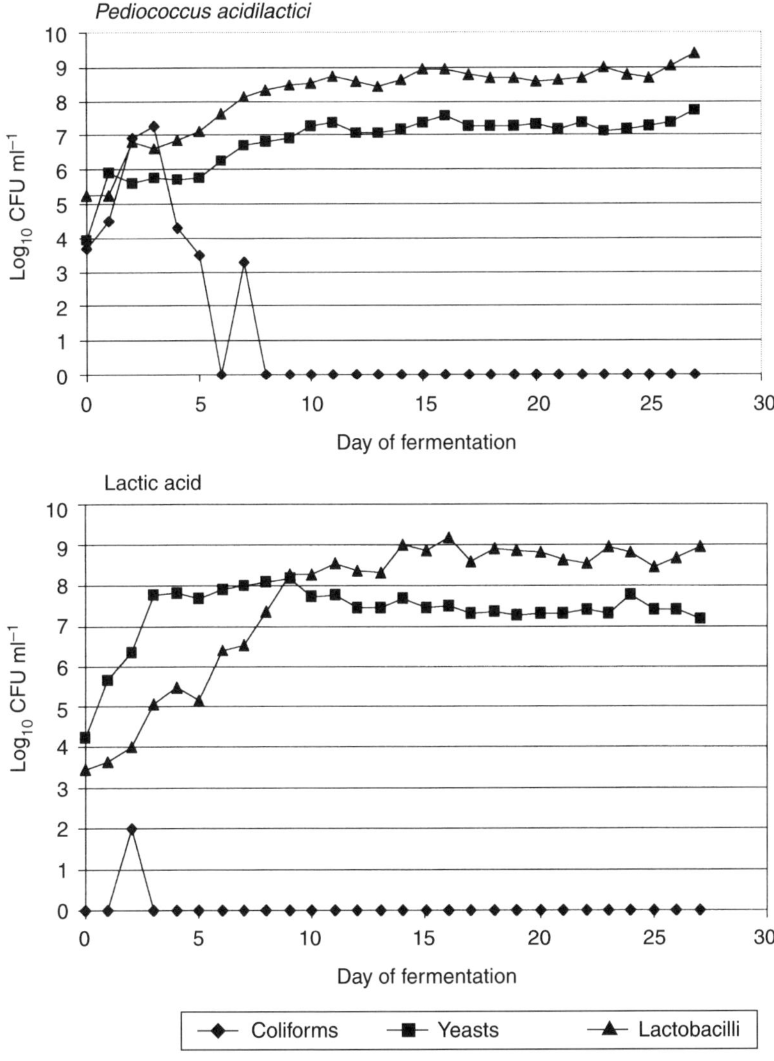

Fig. 8.8. Effect on microflora of acidifying a diet to pH 4 with either lactic acid or by fermenting with *Pediococcus acidilactici* (Geary *et al.*, 1999). CFU, colony-forming units.

flora must be eliminated before inoculation, to remove competition. The data above indicates that to eliminate the native flora by acidification is unlikely to be feasible: the low pH required would make the feed unacceptable to the pig. To date, we have been unable to find any sterilant material that we can add to liquid diets which is both economically viable and maintains the palatability of the feed. However, it may not be necessary to sterilize the feed completely. Demeckova *et al.* (2000) demonstrated that chlorine dioxide is very effective at removing coliform bacteria from liquid feed prior to the addition of a LAB inoculant (Table 8.10). In prefer-

Table 8.10 Effect of chlorine dioxide and time on coliform populations in liquid feed (Demeckova *et al.*, 2000).

Time (h)	Chlorine dioxide concentration (ppm) 0	100	200	300	400	500	Main effect of time
0	3.2[a]	3.2[a1]	2.7[a1]	< 2.0	< 2.0	< 2.0	2.8[1]
3	4.3	3.3[1]	2.5[1]	< 2.0	< 2.0	< 2.0	3.0[1,2]
6	4.9	3.7[1]	2.4[1]	< 2.0	< 2.0	< 2.0	3.3[2]
24	8.2	6.3	4.0	< 2.0	< 2.0	< 2.0	5.1
Main effect of ClO_2	5.1	4.1	2.9	< 2.0	< 2.0	< 2.0	

[a,b,c] Within rows, means with the same superscript are not significantly different ($P > 0.05$).
[1,2,3] Within columns, means with the same superscript are not significantly different ($P > 0.05$).

ence studies, pigs tended to consume more of the feed that had been sanitized with chlorine dioxide, both when fed fresh and following subsequent fermentation. This could imply that the growth of the native flora produces compounds that impair the flavour of the food. When offered liquid feed without choice, there was no difference in the intake of fermented feed that had or had not been sanitized before fermentation (Moran, 2000).

Dry Matter Preference of Weaners

Although weaner pigs are willing and able to consume sufficient DM when fed low-DM diets, this does not mean that it would be their choice. In an unpublished study by Brooks and Wattam, piglets were offered either DPF, or FLF (180 g DM kg^{-1} diet), or both diets simultaneously (Table 8.11). Daily gain in the first 4 weeks following weaning was higher for the pigs fed FLF than the pigs fed DPF, or the pigs fed both FLF and DPF, but the differences were not statistically significant. Over the 4 weeks of the trial, the pigs fed both FLF and DPF took only 7% of their

Table 8.11. Performance of weaner pigs offered fermented liquid feed (FLF), dry pelleted feed (DPF) or both FLF and DPF (Brooks and Wattam, unpublished data).

	Dietary treatment FLF	DPF	FLF and DPF	SEM
Dry matter feed intake (g day^{-1})	399	342	430	23.6
Daily gain (g)	310	262	287	9.37
Dry matter feed conversion ratio	1.28	1.3	1.49	0.06
Total water intake (ml per pig day^{-1})	1820[ab]	889[a]	946[b]	158[**]
Average water intake from drinkers (ml per pig day^{-1})	1087	729	811	57.9

[a,b] Means with the same superscript differ at $P < 0.01$.
SEM, standard error of the mean.

DM as liquid feed and the remainder as dry feed, which equated to an 87% DM diet overall (Brooks and Wattam, unpublished data). Thus, it would appear that, given the choice, pigs of this age would self-select a DM closer to that of dry feed but that this diet would not produce an intake of DM or a growth rate as high as that produced by a liquid diet. It should be noted that the FLF used in this study had a very low DM (180 g kg^{-1}), which had previously been shown to reduce intake compared with higher DM concentrations. This may have created an initial reluctance by the pigs to consume this material.

Diet Form and Behaviour

The behaviour of the pigs in this trial was also studied (Brooks and Murray, unpublished data). It was noticeable that in the last week of the trial there were significant differences in the behaviour of the pigs on the three dietary treatments. Pigs on the liquid diet were noticeably more restful and indulged in less aberrant behaviour (belly-nosing and navel-suckling) than pigs on the other two treatments (Table 8.12).

Studies on the relationship between behaviour and performance of individual piglets showed that there was a significant negative relationship between the amount of time spent belly-nosing and the growth rate. Those pigs involved in belly-nosing other pigs for less than 1.5% of the observation time (Group 1) had significantly higher daily gains than all the other pigs (Group 2) (Table 8.13). Many more of the pigs given fermented liquid feed were in the low belly-nosing group and this might provide part of the explanation for the superior growth rate of weaners fed FLF. It also suggests that liquid feeding may result in a significant improvement in the welfare of weaner pigs. Further studies are in progress to try to determine what factors relating to liquid feeding are involved in these behavioural changes.

Brooks and van Zuylen (1998) considered two other important aspects of managing pigs fed FLF after weaning: (i) the effect of the form of the creep feed offered before weaning (FLF or DPF); and (ii) their weight at weaning: 5.5–7.5 kg (low) and 7.5–9.5 kg (high). There were no significant differences in post-weaning growth rate but low liveweight pigs and pigs fed FLF tended to have higher growth rates (Table 8.14). Examination of the growth rate data on a weekly basis revealed a

Table 8.12. Proportion of time spent on various behaviours by weaner pigs fed fermented liquid feed (FLF), dry pelleted feed (DPF) or both FLF and DPF (Brooks and Murray, unpublished data).

	Dietary treatment		
Behaviour	FLF	DPF	FLF and DPF[a]
Sleeping / resting	0.87	0.84	0.82
Feeding	0.03	0.06	0.06
Other	0.09	0.07	0.09

[a] Pigs offered both FLF and DPF diets simultaneously.

Table 8.13. Effect of belly-nosing for more or less than 1.5% of time on growth rate of weaner pigs (Brooks and Murray, unpublished data).

	Belly-nosing for less than 1.5% of time	Belly nosing for more than 1.5% of time
Proportion of time belly-nosing	0.002*	0.035
Proportion of time sleeping / resting	0.87***	0.82
Daily gain (g)	309*	265
Number of pigs fed fermented liquid feed (FLF)	14	4
Number of pigs fed dry pelleted feed (DPF)	9	9
Number of pigs fed FLF and DPF[a]	7	11
Totals	30	24

[a] Pigs offered both FLF and DPF diets simultaneously.
*$P < 0.05$; ***$P < 0.001$.

Table 8.14. Post-weaning performance of pigs fed fermented liquid feed (FLF) which were weaned at low (L) or high (H) weights and had been offered FLF or dry pelleted feed (DPF) during the suckling period (Brooks *et al.*, 1998).

	Treatments				
	L[a]	H[b]	FLF	DPF	SE_D
Dry matter feed intake (g day^{-1})	427	402	422	406	82
Daily gain (g)	372	353	373	351	31
Dry matter feed conversion ratio	1.15	1.13	1.14	1.14	0.1
Water intake from drinkers (ml day^{-1})	624	658	659	623	212

[a] Pigs in the L group weighed between 5.5 and 7.5 kg at weaning.
[b] Pigs in the H group weighed between 7.5 and 9.5 kg at weaning.
SE_D, standard error of difference between means.

Table 8.15. Post-weaning growth rate (g day^{-1}) of pigs weaned at low (L) or high (H) weights and offered fermented liquid feed (FLF) or dry pelleted feed (DPF) during the suckling period (Brooks *et al.*, 1998).

	Treatments				
	L	H	FLF	DPF	SE_D
Week 1	204[a]	97[a]	156	144	37
Week 2	333	411	378	366	41
Week 3	577	552	585	544	37
Overall	427	402	422	406	82

[a] means with the same superscript differ significantly $P < 0.05$.
SE_D, standard error of difference between means.

Table 8.16. Proportion of total dry matter intake taken as fermented liquid feed (FLF) or dry pelleted feed (DPF) in the third week after weaning[a] (Brooks *et al.*, 1998).

	Weaning weight		Creep feed		
	Low	High	FLF	DPF	SE_D
DPF (g day^{-1})	166	168	167	167	21
FLF (g day^{-1})	468	484	475	477	102
Proportion taken as FLF	0.74	0.74	0.74	0.74	

[a] Pigs were fed FLF for 2 weeks after weaning and then given a choice of FLF and DPF in week 3.
SE_D, standard error of difference between means.

significant difference in growth rate between the two weight groups during the first week after weaning (Table 8.15). Pigs of low weaning weight had a growth rate more than double that of their heavier littermates. However, the higher growth rate was not sufficient for them to catch up with pigs of high weaning weight by 3 weeks after weaning.

It is worth noting that, both in our own unit and in commercial units with which we have worked, no problems have been experienced when the feeding regime has been changed to liquid from dry. The data in Table 8.16 show the proportion of DM taken as DPF or FLF in four different treatment groups (Brooks and van Zuylen, 1998).

Effects of Fermented Liquid Feed on Feed and Gut Microbiology

The increase in acidity resulting from fermentation has a marked effect on the microbial population in the feed and in the piglet's gut. In studies in The Netherlands (van Winsen *et al.*, 1997) it was found that pig feed fermented with *Lactobacillus plantarum* induced a bacteriostatic effect on *S. typhimurium* during the first 2 h following inoculation and a bactericidal effect thereafter (Fig. 8.9). Six hours after inoculation, *Salmonella typhimurium* was not detected in the FLF. In contrast, in non-fermented feed, *S. typhimurium* survived and multiplied during the first 10 h of storage. Studies in our Plymouth laboratory have shown that when *S. typhimurium* DT104:30 and *Pediococcus pentosaceus* are co-inoculated into liquid feed, *P. pentosaceus* rapidly dominates the fermentation and reduces *S. typhimurium* to undetectable levels (Fig. 8.10). The elimination of *S. typhimurium* occurs much more rapidly if the feed is maintained at 30°C than in feed kept at 20°C. FLF is also very effective at eliminating pathogenic *Escherichia coli* (Beal *et al.*, 2000) (Table 8.17).

The weaned pig has a general insufficiency of stomach acid, which is the first line of defence against bacterial invasion (Smith and Jones, 1963; Cranwell *et al.*, 1976). Manipulation of stomach acidity, through lactic acid supplementation of the diet (Thomlinson and Lawrence, 1981), feeding fermented milk (Ratcliffe *et al.*, 1986;

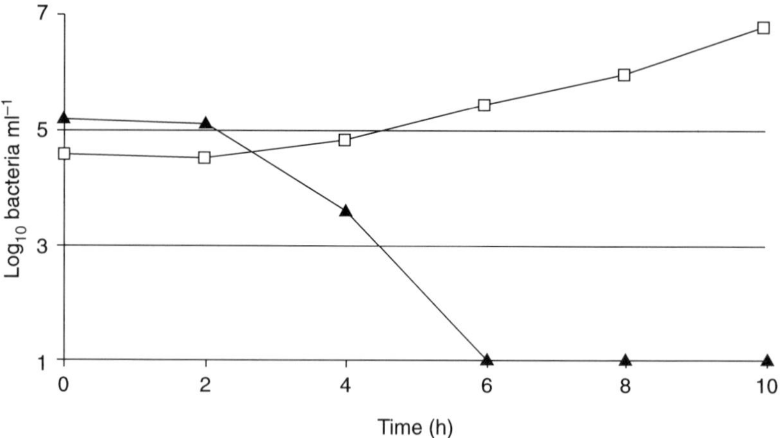

Fig. 8.9. Growth of *Salmonella typhimurium* in liquid and fermented liquid feed (after van Winsen *et al*, 1997).

Table 8.17. Decimal reduction time of six strains of *Escherichia coli* in fermented liquid feed fermented for 48 h, 72 h or 96 h (Beal *et al*., 2000).

	E. coli strain						
Time	K88(99)	K88(100)	K88(101)	O157:H7	K99(185)	K99(230)	SE_D
48 h	25.2[a]	26.1[a]	22.3[b]	12.16	22.0[b]	22.2[b]	0.96
72 h	23.7[a]	23.6[a]	24.2[a]	9.3	16.5[b]	14.6[b]	0.96
96 h	22.9[a]	17.4[b]	24.3[a]	10.3	15.8[b]	14.0[b]	0.96

[a,b] Means with the same superscript within a row are not significantly different, $P > 0.05$.
SE_D, standard error of difference between means.

Table 8.18. Concentration of short-chain fatty acids in the gastrointestinal tract of piglets fed unfermented liquid (control) feed (NLF) or fermented liquid feed (FLF) (after Mikkelson and Jensen, 1997).

	Acetic acid		Lactic acid	
	NLF	FLF	NLF	FLF
Stomach	16.6	16.5	37.3	81.7**
SI 1	10.6	8.8	7.7	26.3**
SI 2	10.2	9.1	13.7	24.5
SI 3	11.7	9.8	38.0	32.7
SI 4	12.8	13.3	47.6	40.2
Caecum	73.8	93.0***	2.3	2.9
Colon 1	61.0	87.3***	0.5	0.0
Colon 2	55.6	72.8**	0.7	0.0
Colon 3	60.1	69.9	1.0	0.0

* $P < 0.05$; ** $P < 0.01$; *** $P < 0.001$.

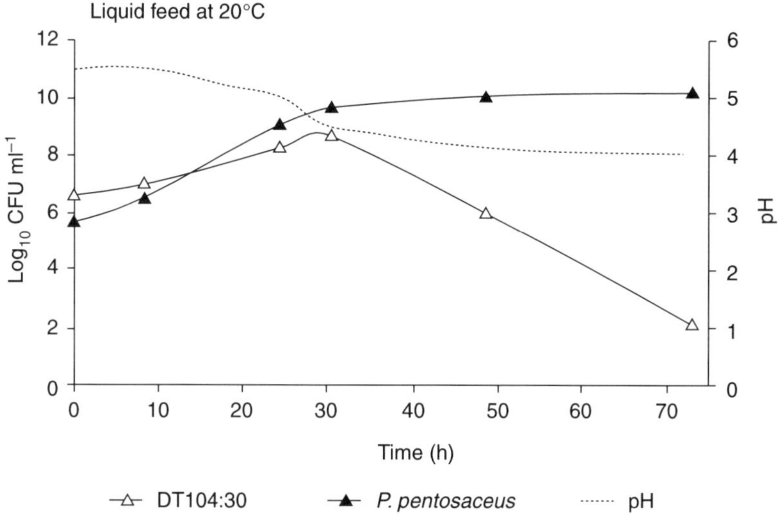

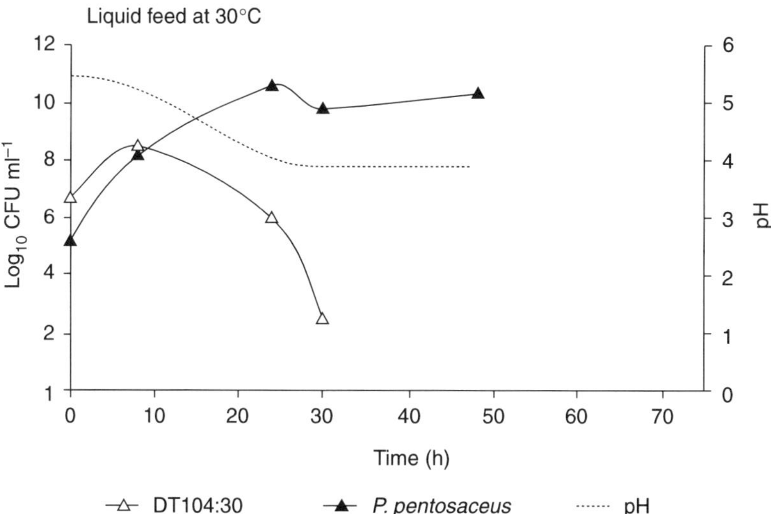

Fig. 8.10. Growth of *Salmonella typhimurium* DT104:30 and *Pediococcus pentosaceus* co-inoculated in feed at time zero and maintained at 20 or 30°C (J.D. Beal, unpublished data).

Dunshea *et al.*, 2000) or through water (Cole *et al.*, 1968), reduced gastric pH and the number of coliforms in the stomach. Similarly, feeding FLF resulted in a significant increase in stomach acidity. Mikkelson and Jensen (1997) found that fermented liquid feed resulted in a significant increase in lactic acid content in the stomach (Table 8.18) and some small but significant changes in other sections of the gut.

Table 8.19. Effect of dietary treatment on the pH of the intestinal contents of piglets 14 days after weaning (Moran and Ward, unpublished data).

	Dietary treatment			
	Suckled	Dry pelleted feed	Liquid feed	Fermented liquid feed
Stomach	2.4^{b}	3.9^{ab}	4.8^{a}	3.9^{ab}
Terminal ileum	5.9^{bc}	6.3^{abc}	6.4^{abc}	6.1^{abc}
Caecum	6.1^{ab}	5.8^{b}	6.0^{ab}	6.0^{ab}
Colon	6.6^{a}	5.9^{b}	6.0^{ab}	6.2^{ab}

^{abc}Means with the same superscript within a row are not significantly different, $P < 0.05$.

Recent studies (Moran and Ward, unpublished data) have shown that feed form can have an effect on the pH of the GIT (Table 8.19) with the lowest pH being found in pigs that continued to suckle their dams. However, pH variations need to be treated with caution, due to the buffering effect of the feed components and the difficulties of sampling (Kidder and Manners, 1978; Bolduan *et al.*, 1988).

Feeding FLF does not appear to produce a significant effect on the number of LAB present throughout the gut but it does dramatically reduce the number of coliforms in the lower small intestine, caecum and colon (Jensen and Mikkelsen, 1998). The ratio of LAB to coliforms in the lower gut of pigs weaned on to liquid diets is very similar to that of pigs that continue to suckle the sow (Fig. 8.11), but if the pigs are weaned on to dry diets there is a significant shift in the ratio towards the coliforms. Conversely, if they are weaned on to FLF the ratio shifts in favour of the LAB.

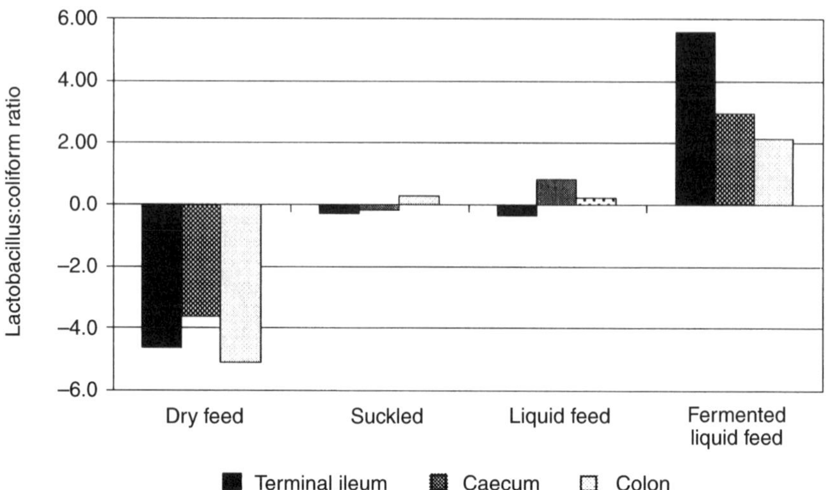

Fig. 8.11. Lactobacillus:coliform ratio (log transformed numbers) in lower gastrointestinal tract as determined by dietary treatment (Moran, 2001).

Conclusions

Providing feed for the weaner pig in a liquid form increases the intake of feed in the immediate post-weaning period and helps to maintain the continuity of nutrient supply. However, there are considerable problems in maintaining the hygienic quality of liquid diets fed *ad libitum*. If steps are not taken to prevent the proliferation of enteropathogens, liquid feeding can contribute to rather than reduce post-weaning diarrhoea. Fermentation of liquid feed can overcome this problem. If fermentation is properly controlled, the resultant feed is well accepted by weaner pigs and has a beneficial effect on gut architecture and microbiology. The challenge now is to refine equipment and develop management practices that will enable the potential of this feeding method to be realized on commercial pig units.

References

Armstrong, W.D. and Clawson, A.J. (1980) Nutrition and management of early weaned pigs: effect of increased nutrient concentrations and (or) supplemental liquid feeding. *Journal of Animal Science* 50(3), 377–383.

Azain, M.J., Tomkins, T., Sowinski, J.S., Arentson, R.A. and Jewell, D.E. (1996) Effect of supplemental pig milk replacer on litter performance: seasonal variation in response. *Journal of Animal Science* 74(9), 2195–2202.

Baber, D.W. and Coblenz, B.E. (1987) Diet, nutrition and conception in feral pigs on Santa Catalina island. *Journal of Wildlife Management* 51(2), 306–317.

Barber, J., Brooks, P.H. and Carpenter, J.L. (1989) The effects of water delivery rate on the voluntary food intake, water use and performance of early-weaned pigs from 3 to 6 weeks of age. In: Forbes, J.M., Varley, M.A. and Lawrence, T.L.J. (eds) *The Voluntary Feed Intake of Pigs.* Occasional publication no. 13, British Society of Animal Production, Edinburgh, pp. 103–104.

Beal, J.D., Moran, C.A., Campbell, A. and Brooks, P.H. (2001) The survival of potentially pathogenic *E. coli* in fermented liquid feed. In: Lindberg, J.E. and Ogle, B. (eds) *Digestive Physiology of Pigs.* 8th Symposium on Digestive Physiology in Pigs, Uppsala, Sweden. CAB International, Wallingford, UK.

Bolduan, G., Jung, H., Schnabel, E. and Schneider, R. (1988) Recent advances in the nutrition of weaner piglets. *Pig News and Information* 9(4), 381–385.

Braude, R. and Newport, M.J. (1977) A note on a comparison of two systems for rearing pigs weaned at 2 days of age, involving either a liquid or a pelleted diet. *Animal Production* 24, 271–274.

Brooks, P.H. (1998) Fermented liquid feed (FLF) for weaned piglets. In: van Arendonk, J.A.M., Ducrocq, V., van der Honing, Y., Madec, F., van der Lende, T., Puller, D., Folch, J., Fernandez, E.W. and Bruns, E.W. (eds) *Book of Abstracts of the 49th Meeting of the European Association of Animal Production. Warsaw 24–29 August.* Wageningen Pers, Wageningen, The Netherlands, p. 262 (abstract).

Brooks, P.H. and Burke, J. (1998) Behaviour of sows and piglets during lactation. In: Verstegen, M., Schrama, J. and Maughan, P. (eds) *Lactation in the Sow.* Wageningen Pers, Wageningen, The Netherlands, pp. 299–336.

Brooks, P.H. and Carpenter, J.L. (1990) The water requirement of growing–finishing pigs –

theoretical and practical considerations. In: Haresign, W. and Cole, D.J.A. (eds) *Recent Advances in Animal Nutrition.* Butterworths, London, pp. 115–136.

Brooks, P.H. and van Zuylen, B. (1998) The effect of feeding a fermented liquid diet to suckling pigs on their pre- and post-weaning performance and the effect of weaning weight on post-weaning performance. In: van Arendonk, J.A.M., Ducrocq, V., van der Honing, Y., Madec, F., van der Lende, T., Puller, D., Folch, J., Fernandez, E.W. and Bruns, E.W. (eds) *Book of Abstracts of the 49th Meeting of the European Association of Animal Production. Warsaw 24–29 August.* Wageningen Pers, Wageningen, The Netherlands, p. 268 (abstract).

Brooks, P.H., Russell, S.J. and Carpenter, J.L. (1984) Water intake of weaned piglets from three to seven weeks old. *The Veterinary Record* 115, 513–515.

Buddington, R.K. (1998) The influences of dietary inputs on the neonatal gastrointestinal tract: managing the development of a complex ecosystem. *Journal of Animal and Feed Sciences* 7 (Supplement 1), 155–165.

Cole, D.J.A., Beal, R.M. and Luscombe, J.R. (1968) The effect on performance and bacterial flora of lactic acid, propionic acid, calcium propionate and calcium acrylate in the drinking water of weaned pigs. *The Veterinary Record* 83, 459–464.

Cranwell, P.D., Noakes, D.E. and Hill, K.J. (1976) Gastric secretion and fermentation in the suckling pig. *British Journal of Nutrition* 36, 71–86.

Demeckova, V., Moran, C.A., Caveney, C., Campbell, A., Kuri, V. and Brooks, P.H. (2001) The effect of fermentation and/or sanitization of liquid diets on the feeding preferences of newly weaned pigs. In: Lindberg, J.E. and Ogle, B. (eds) *Digestive Physiology of Pigs. 8th Symposium on Digestive Physiology in Pigs*, Uppsala, Sweden. Swedish University of Agricultural Sciences. CAB International, Wallingford, UK.

Deprez, P., Deroose, P., Vandenhende, C., Muylle, E. and Oyaert, W. (1987) Liquid versus dry feeding in weaned piglets: the influence on small intestinal morphology. *Journal of Veterinary Medicine B* 34, 254–259.

Dunshea, F.R., Kerton, D.J., Eason, P.J. and King, R.H. (2000) Supplemental fermented milk increases growth performance of early-weaned pigs. *Asian–Australasian Journal of Animal Sciences* 13(4), 511–515.

Efird, R.C., Armstrong, W.D. and Herman, D.L. (1982) The development of digestive capacity in young pigs: effects of weaning regimen and dietary treatment. *Journal of Animal Science* 55(6), 1370–1379.

English, P.R., Anderson, P.M., Davidson, F.M. and Dias, M.F.M. (1981) A study of the value of readily available liquid supplements for early-weaned pigs. *Animal Production* 32, 395.

Geary, T.M. and Brooks, P.H. (1998) The effect of weaning weight and age on the post-weaning growth performance of piglets fed fermented liquid diets. *The Pig Journal* 42, 10–23.

Geary, T.M., Brooks, P.H., Morgan, D.T., Campbell, A. and Russell, P.J. (1996) Performance of weaner pigs fed *ad libitum* with liquid feed at different dry matter concentrations. *Journal of Science of Food and Agriculture* 72, 17–24.

Geary, T.M., Brooks, P.H., Beal, J.D. and Campbell, A. (1999) Effect on weaner pig performance and diet microbiology of feeding a liquid diet acidified to pH 4 with either lactic acid or through fermentation with *Pediococcus acidilactici. Journal of the Science of Food and Agriculture* 79(4), 633–640.

Gestin, M., Le Huerou-Luron, I., Le Drean, G., Peiniau, J., Rome-Philouze, V., Aumaître, A. and Guilloteau, P. (1997) Effect of age and feed intake on pancreatic enzyme activities in piglets. In: Laplace, J.-P., Fevrier, C. and Barbeau, A. (eds) *Digestive Physiology of the Pig.* EAAP Publication no. 88, St Malo, France, pp. 127–130.

Gill, B.P. (1989) Water use by pigs managed under various conditions of housing, feeding, and nutrition. PhD thesis, University of Plymouth, Newton Abbot, UK.

Hansen, I.D. and Mortensen, B. (1989) Pipe-cleaners beware. *Pig International* 19(11), 9–10.

Jensen, B.B. and Mikkelsen, L.L. (1998) Feeding liquid diets to pigs. In: Garnsworthy, P.C. and Wiseman, J. (eds) *Recent Advances in Animal Nutrition 1998.* Nottingham University Press, Thrumpton, Nottingham, UK, pp. 107–126.

Kennelly, J.J. and Aherne, F.X. (1980) Volatile fatty acid production in the hindgut of swine. *Canadian Journal of Animal Science* 60(4), 1056.

Kidder, D.E. and Manners, M.J. (1978) *Digestion in the Pig.* Scientechnica, Bristol, UK, 201 pp.

Lecce, J.G. (1986) Diarrhea: the nemesis of the artificially reared, early weaned piglet and a strategy for defense. *Journal of Animal Science* 63, 1307–1313.

Lecce, J.G., Armstrong, W.D., Crawford, P.C. and Ducharme, G.A. (1979) Nutrition and management of early weaned piglets: liquid vs dry feeding. *Journal of Animal Nutrition* 48(5), 1007–1014.

Maswaure, S.M. and Mandisodza, K.T. (1995) An evaluation of the performance of weaner pigs fed diets incorporating fresh sweet liquid whey. *Animal Feed Science and Technology* 54, 193–201.

McLeese, J.M., Tremblay, M.L., Patience, J.F. and Christison, G.I. (1992) Water intake patterns in the weanling pig: effect of water quality, antibiotics and probiotics. *Animal Production* 54, 135–142.

Metz, J.H.M. and Gonyou, H.W. (1990) Effect of age and housing conditions on the behavioural and haemolytic reaction of piglets to weaning. *Applied Animal Behaviour Science* 27, 299–309.

Mikkelson, L.L. and Jensen, B.B. (1997) Effect of fermented liquid feed (FLF) on growth performance and microbial activity in the gastrointestinal tract of weaned piglets. In: Laplace, J.P., Fevrier, C. and Barbeau, A. (eds) *Digestive Physiology in Pigs.* EAAP Publication no. 88, INRA, Paris, pp. 639–642.

Moran, C.A. (2000) Development and benefits of fermented liquid diets for newly weaned pigs. PhD thesis, University of Plymouth, Newton Abbot, UK.

Mosenthin, R. (1998) Physiology of small and large intestine of swine – review. *Asian-Australasian Journal of Animal Sciences* 11(5), 608–619.

Ogunbameru, B.O., Kornegay, E.T. and Wood, C.M. (1991) A comparison of drip and non–drip nipple waterers used by weanling pigs. *Canadian Journal of Animal Science* 71, 581–583.

Pajor, E.A., Fraser, D. and Kramer, D. (1991) Consumption of solid food by suckling pigs: individual variation and relation to weight gain. *Applied Animal Behaviour Science* 32, 139–155.

Partridge, G.C., Fisher, J., Gregory, H. and Prior, S.G. (1992) Automated wet feeding of weaner pigs *vs* conventional dry diet feeding: effect on growth rate and food consumption. *Animal Production* 54, 484.

Phillips, P.A. and Fraser, D. (1991) Discovery of selected water dispensers by newborn pigs. *Canadian Journal of Animal Science* 71(1), 233–236.

Pluske, J.R., Williams, I.H. and Aherne, F.X. (1996) Maintenance of villous height and crypt depth in piglets by providing continuous nutrition after weaning. *Animal Science* 62(1), 131–144.

Pluske, J.R., Hampson, D.J. and Williams, I.H. (1997) Factors influencing the structure and function of the small intestine in the weaned pig: a review. *Livestock Production Science* 51(1–3), 215–236.

Ratcliffe, B., Cole, C.B., Fuller, R. and Newport, M.J. (1986) The effect of yoghurt and milk fermented with a porcine strain of *Lactobacillus reuteri* on the performance and gastrointestinal flora of pigs weaned at two days of age. *Food Microbiology* 3, 203–211.

Rudo, N.D., Rosenberg, I.H. and Wissler, R.W. (1976) The effect of partial starvation and glucagon treatment on intestinal villus morphology and cell migration. *Proceedings of the Society for Experimental Biology and Medicine* 152, 277–280.

Russell, P.J., Geary, T.M., Brooks, P.H. and Campbell, A. (1996) Performance, water use and effluent output of weaner pigs fed *ad libitum* with either dry pellets or liquid feed and the role of microbial activity in the liquid feed. *Journal of Science of Food and Agriculture* 72, 8–16.

Smith, H.W. and Jones, J.E.T. (1963) Observations on the alimentary tract and its bacterial flora in healthy and diseased pigs. *Journal of Pathological Bacteriology* 86, 387–412.

Smith, P. (1976) A comparison of dry, wet and soaked meal for fattening bacon pigs. *Experimental Husbandry* 30, 87–94.

Spitz, F. (1986) Current state of knowledge of wild boar biology. *Pig News and Information* 7(2), 171–175.

Steiner, M., Bourges, H.R., Freedman, L.S. and Gray, S.J. (1968) Effect of starvation on the tissue composition of the small intestine in the rat. *American Journal of Physiology* 215(1), 75–77.

Taverner, M.R., Reale, T.A. and Campbell, R.G. (1987) Nutrition of the young pig. In: Farrell. D.J. (ed.) *Recent Advances in Animal Nutrition in Australia.* University of New England, Armidale, NSW, Australia, pp. 338–346.

Thomlinson, J.R. and Lawrence, T.L.J. (1981) Dietary manipulation of gastric pH in the prophylaxis of enteric disease in weaned pigs: some field observations. *The Veterinary Record* 109, 120–122.

van Winsen, R.L., Urlings, H.A.P. and Snijders, J.M.A. (1997) Feed as a vehiculum of salmonella in pigs. In: Bech-Nielsen, S. and Nielsen, J.P. (eds) *Proceedings of the Second International Symposium on Epidemiology and Control of Salmonella in Pork. Copenhagen, Denmark, August 20–22.* Federation of Danish Pig Producers and Slaughterhouses, Copenhagen, Denmark, pp. 157–159.

Yang, T.S., Howard, B. and Macfarlane, W.V. (1981) Effects of food and drinking behaviour of growing pigs. *Applied Animal Ethology* 7, 259–270.

Yang, T.S., Price, M.A. and Aherne, F.X. (1984) The effects of level of feeding on water turnover in growing pigs. *Applied Animal Behaviour Science* 12, 103–109.

Yen, J.T., Neinaber, J.A., Hill, D.A. and Pond, W.G. (1991) Potential contribution of absorbed volatile fatty acids to whole animal requirement in conscious swine. *Journal of Animal Science* 69, 2001–2012.

9

Digestive Physiology and Development in Pigs

D. Kelly and T.P. King

Rowett Research Institute, Bucksburn, Aberdeen AB21 9SB, UK

Introduction

The intestinal tract provides a complex interface between the animal and its environment. It has to cope with abrupt dietary changes at birth and weaning. Investigations of the interactions between pre- and post-weaning nutrition, gut physiology and immunology and their relevance to bodily functions and health are fundamental to solving the problems of dietary change and post-weaning performance.

The influences of dietary change on intestinal epithelial differentiation and growth are especially marked at birth and weaning. The early weeks of neonatal life see extensive changes in gut morphology, transiently elevated protein transcytosis, sustained increases in *de novo* protein synthesis and both age-related and diet-induced changes in brush-border membrane digestive and transporter functions. During the same period, the rapidly changing mucosal surface becomes colonized by successions of gut bacterial groups. In the majority of animals, the dynamic balance between host physiology, diet and the gastrointestinal microbiota leads to the establishment of a stable microbial ecology characterized by the presence of commensal organisms that exert a positive influence in maintaining and establishing a healthy gut immune system. However, perturbation of the gut ecosystem can often occur in neonates and the pre-weaning period still represents the time of greatest pig morbidity and mortality.

The growth, development and intrinsic differentiation of the digestive tract in neonatal pigs are profoundly influenced through interaction with dietary constituents and the flora. Colostrum and milk contain high levels of growth factors that accelerate proliferation and maturation of the gut in neonatal animals. Variations in milk composition during lactation are associated with alterations in the synthesis and intracellular processing of brush-border hydrolases and increased transport of nutrients and electrolytes in the suckled animal. Extensive nutrient–gene interactions underpin the successful adaptation of neonates to non-liquid, lower fat, higher carbohydrate weaner diets. These changes in gene expression and the processing of gene products are entrained through diet-induced events that begin very early in postnatal life.

Intestinal Development in the Fetal Pig

Functional background

The mammalian small intestine has a remarkable capacity to respond and rapidly adapt to a diverse array of endogenous and exogenous stimuli. Vital to this adaptive capacity is a complex epithelial surface, which is continually undergoing regeneration and differentiation. Early in fetal development, cellular proliferation occurs on the surfaces of the nascent villi as well as in the intervillus regions of the epithelium. As villus formation ensues, cell divisions become increasingly confined to early intervillus crypt zones. Crypt formation is first observed in the fetal pig intestine at around day 60 of gestation (Dekaney *et al.*, 1997). The partitioning of crypt regions in the developing fetal intestinal epithelium is an important landmark that reflects the creation of cellular machinery that will regulate proliferation, differentiation and renewal of the intestinal epithelium throughout life.

Morphological development of the fetal intestine

The gestational period for the pig is 115 ± 2 days. During fetal development the small intestine develops from a simple tube of pseudostratified epithelium to a monolayer of simple columnar epithelium comprising villus and intervillus regions. Early morphogenesis of the developing intestine proceeds as a proximal to distal wave of cytodifferentiation. Several of the characteristic features of advancing epithelial morphogenesis are first observed in the duodenum and upper jejunal regions of the developing fetal intestine. The sequence of structural and ultrastructural changes that take place in the developing fetal pig intestine was analysed by Dekaney *et al.* (1997). At 30 days gestation, the immature porcine intestine consists of a simple epithelial tube, surrounded by a layer of mesenchymal cells. Primitive villi are present at day 35 of development; by day 40 of gestation, further cytodifferentiation is evident with the presence of goblet cells and enteroendocrine cells in the proximal intestine. The conversion of the pseudostratified epithelium to simple columnar type during fetal development involves the establishment of distinct apical and basolateral membrane domains, separated from one another by tight junctions. Differentiation and migration of epithelial cells are influenced by ontogenic changes in the composition of extracellular matrix and the nature of mesenchymal–epithelial crosstalk (Pacha, 2000).

By day 45 of gestation, the proximal to distal gradient in cytodifferentiation is less marked. As development progresses, microvilli lengthen and, by day 60, components of an apical endocytic complex are observed. This organelle system is characterized by invaginations of the apical membrane and a system of apical tubules and vesicles. In other species, the complex has been shown to have a role in the uptake, sorting and lysosomal degradation of hormones, growth factors and other macromolecules from fluid swallowed during gestation (Wilson *et al.*, 1991; Trahair, 1993; Trahair *et al.*, 1995). By day 110 of gestation the villi are well developed, absorptive cells of the duodenum and proximal jejunum are not vacuolated and large lysosomal vacuoles are observed in the distal jejunum.

Functional development of the fetal intestine

Functional differentiation of the intestine occurs concomitantly with the structural development. The intestinal brush-border membrane consists of the apical microvillous border and associated glycocalyx. Functional differentiation of this membrane, particularly the prenatal development of brush-border hydrolases, is an important component in the preparation for life and nutrition *ex utero* (Trahair and Sangild, 1997). In comparison with pre- and post-weaned animals, however, the levels of brush-border enzymes in the fetal intestine are relatively low (Danielsen *et al.*, 1995). Among the major enzymes, only lactase phlorizin hydrolase and aminopeptidase-N activities reach substantial levels before birth (Buddington and Malo, 1996; Trahair and Sangild, 1997). Aminopeptidase-N gene expression is detected in the early fetal pig intestine at a time coincident with the formation of nascent intestinal villi (Perozzi *et al.*, 1993). Cytochemical analysis of the enzyme activity in mid-gestational pigs showed a substantial proportion of aminopeptidase-N located intracellularly, within the tubules and vacuoles of the apical endocytic complex, where it is thought to function in the intracellular degradation of meconium (Danielsen *et al.*, 1995).

Intestinal ion and nutrient transport develops during early fetal life and the absorptive capacity for specific components is maximal at birth and decreases postnatally (Buddington and Diamond, 1989; Buddington *et al.*, 2001). Fetal swallowing of amniotic fluid starts early in gestation, thus nutrient transporters may enable the absorption of amniotic fluid components (Trahair and Sangild, 1997). The fetal pig intestine has an ability to absorb protein macromolecules. Intestinal macromolecule absorption was recently investigated in fetal pigs after infusion of colostrum *in utero* (Sangild *et al.*, 1999). The results showed that the prenatal pig intestine is similar to the neonatal pig intestine in that colostrum stimulates both macromolecular absorption and the cessation of macromolecular uptake (intestinal closure). At 90% of term (102 days gestation) the intestine of pigs is largely but not completely developed (Buddington *et al.*, 2001).

Intestinal Development in the Suckling Pig

Functional background

The perinatal period is attended by important modifications in energy metabolism (Girard *et al.*, 1992). *In utero*, the fetus receives a continuous intravenous supply of substrates for growth and oxidative metabolism. At birth, the maternal supply of substrates ceases abruptly and the newborn has to withstand a brief period of starvation before being fed at intervals on relatively low-carbohydrate and high-fat colostrum and milk (Koletzko *et al.*, 1998). Thus the newborn gastrointestinal tract has to support a shift from mainly parenteral nutrition to enteral nutrition and intestinal absorptive function becomes a critical factor in the maintenance of physiological well-being (Stoll *et al.*, 2000). In the perinatal period the maturational programme of the intestinal epithelium is influenced by a complex interplay of local, systemic and luminal factors (Trahair and Sangild, 1997).

Digestive physiology and gut development are profoundly influenced by the many and varied host responses to colonizing microorganisms. Pathogens such as enterotoxigenic *Escherichia coli* and rotavirus are major causes of mucosal damage and contribute to scours or diarrhoea in suckling pigs. The colonization of the gut by non-pathogenic microorganisms also profoundly modifies gut structure and function. Their stimulation of the intestinal immune system results in a constitutive, low-level inflammation and epithelial changes that can have both negative and positive effects on nutrient and energy absorption in the young (Gaskins, 1997; Anderson *et al.*, 2000; Kelly and King, 2001).

Effect of colostral growth factors on intestinal structure and function

The onset of suckling stimulates rapid growth of the neonatal intestine and this is supported by a high rate of protein synthesis (Burrin *et al.*, 1992). The ingestion of colostrum stimulates crypt cell proliferation (Zhang *et al.*, 1997). The non-nutritive colostral factors that elicit faster intestinal growth include both immunoglobulins and biologically active substances, including insulin-like growth factor-I (IGF-I) (Xu and Wang, 1996; Zhang *et al.*, 1998). Colostrum and milk feeding have been shown to promote the maturation of the developing intestinal epithelium (Kelly and King, 1991; Burrin *et al.*, 1992; Kelly *et al.*, 1993; Wang and Xu, 1996).

The insulin-like growth factors (IGF-I and IGF-II), and their soluble membrane-associated and extracellular matrix-associated binding proteins and receptors, constitute an endocrine/autocrine/paracrine-acting system that mediates growth, differentiation and apoptosis of vertebrate cells (reviewed by Simmen *et al.*, 1998). High levels of IGFs and their binding proteins are present in pig mammary secretions (Donovan and Odle, 1994; Donovan *et al.*, 1994; Morgan *et al.*, 1996). Since both mammary tissues and intestinal tissues express IGF binding proteins and IGF receptors, it is possible that the IGFs contribute to both mammogenesis and lactogenesis in the sow and also play a role in regulating postnatal development and gastrointestinal maturation in the suckling young (Morgan *et al.*, 1996; Shen and Xu, 2000). Experiments have shown that feeding colostrum along with IGF-I increases growth factor stability in the intestinal lumen (Shen and Xu, 2000). Colostrum feeding also enhances macromolecular absorption and increases circulating IGF-I levels in newborn pigs (Wester *et al.*, 1998; Sangild *et al.*, 1999). Evidence from several nutritional investigations suggests that oral administration of physiological concentrations of IGF-I in formula has little effect, if any, on intestinal mucosal growth (Burrin, 1997; Houle *et al.*, 1997; Alexander and Carey, 1999). The small intestine of healthy neonates with adequate nutrition may already be achieving maximal growth rates and this may be why IGF-I supplementation is unable to stimulate additional growth (Alexander and Carey, 1999). Although oral IGF-I has little effect on intestinal growth or structure, the peptide significantly enhances intestinal epithelial Na^+ and Na^+-coupled nutrient absorption (Alexander and Carey, 1999). Houle *et al.* (2000) demonstrated that administration of 33 and 65 nmol l^{-1} and higher levels of IGF-I to neonatal piglets significantly upregulated intestinal lactase phlorizin hydrolase mRNA abundance and enzyme activity. Although the administered concentrations of IGF-I were approximately equivalent to those found in sow colostrum (Donovan *et al.*, 1994) it is possible that the induction of lactase activity was pharmacological rather than physiological, because the dosage was administered for 14 days. The administration of lower milk levels of IGF-I (1–3 nmol l^{-1}; Donovan *et al.*, 1994) may not have a similar effect on epithelial differentiation.

Transforming growth factor-alpha (TGFα) may be an exogenous ligand for the epidermal growth factor (EGF) receptor in the suckling porcine intestine (Kelly *et al.*, 1992a,b). The expression of TGFα within the intestinal crypts of suckled pigs during the first 3 weeks of life suggests that this growth factor is involved in remodelling and maturation of the neonatal mucosa (Jaeger, 1996).

There is an increasing awareness that the biological efficacy and potency of maternal colostrum is related to previously unidentified proteins defined as non-nutrient components (Burrin *et al.*, 1995; Fiorotto *et al.*, 2000). One such protein has been successfully isolated and sequenced from porcine colostrum and appears to be a potent stimulus for porcine, human and rat intestinal cells (Kelly and Coutts, 1997; Kelly, 1998).

Macromolecular uptake and 'closure' in the intestine of the suckled pig

Not all of the increased protein content in the intestine of the suckling pig is newly synthesized. A significant proportion of mucosal protein is derived from endocytosed colostral immunoglobulins (Kiriyama, 1992) and non-selective endocytosis of luminal solutes (Ekstrom and Westrom, 1991). Intact protein absorption is further enhanced by the presence of protease inhibitors in colostrum (Lindberg, 1982).

Large proportions of the total protein in colostrum comprise immunoglobulins but milk-specific proteins (casein, β-lactoglobulin and α-lactalbumin) are also present. In newborn pigs high levels of these proteins are transferred into the circulation within the first few hours of colostrum ingestion (Westrom *et al.*, 1985; Sangild *et al.*, 1997). In pigs, there is no transplacental transfer of immunoglobulins and the newborn animals acquire passive immunity by absorbing intact immunoglobulins from colostrum. The levels of these maternally derived antibodies are highest at day 1 post-farrowing and then decline to very low levels by the time the pig reaches 3 weeks of age. In newborn piglets, the intestinal epithelium has the ability to take up large quantities of intact protein from colostrum by the process of endocytosis. This macromolecular uptake is non-selective and several colostral components, including immunoglobulins and trophic factors may be rapidly absorbed through the epithelium and transported into the circulation.

Cessation of macromolecular transfer across the intestine in piglets begins approximately 6–12 h after colostrum feeding and progresses rapidly thereafter to completion at 24–36 h (Westrom *et al.*, 1984; Sangild *et al.*, 1999). The cellular basis of this so-called 'closure' is not fully understood but there is evidence that factors responsible for mediating closure are present in the whey fraction of colostrum (Sangild *et al.*, 1999). Absorption of macromolecules into the circulation depends on three processes: the endocytosis of macromolecules at the apical surface of the absorptive enterocyte; the intracellular transport and processing of the macromolecules within an endosomal/lysosomal system; and the exocytosis and release of macromolecules across the basolateral surfaces of the absorptive cells. These processes, which together constitute transcytosis, are variably expressed in the fetal-like enterocytes that persist on the villus surfaces for the first 1–2 weeks of life. Intestinal closure is not due to decreased endocytic capacity of the enterocytes or to a higher lysosomal degradation rate within these cells, but is thought to result from a decreased transfer of the intestinal macromolecules into the blood (Ekstrom and Westrom, 1991). Diminished exocytosis may be influenced by structural or functional modulation of basolateral membranes signalled by elevated levels of colostrum-derived bioactive molecules in the interstitial spaces between enterocytes. This hypothesis has not been tested but investigations have shown that circulating glucocorticoids may regulate the uptake of macromolecules in the newborn pig. Sangild *et al.* (1997) showed that vaginally delivered piglets have a higher absorption of intact proteins (IgG, albumin) at birth than corresponding littermates delivered by Caesarean section (high vs. low plasma cortisol at birth), suggesting that cortisol stimulates macromolecular uptake in the perinatal period. The process leading to termination of absorptive capacity in the newborn pig is thought, to a marked degree, to be controlled by a decrease in glucocorticoids (Bate *et al.*, 1991).

Some macromolecular uptake into the circulatory system occurs in suckling pigs after the process of intestinal closure. For example, Xu and Wang (1996) reported positive absorption of IGF-I into the blood of 3-day-old piglets and suggested that this post-closure absorption occurs via a receptor-mediated pathway. Specific IGF-I receptors have been reported in the small intestine of neonatal pigs (Schober *et al.*, 1990; Morgan *et al.*, 1996).

Replacement of fetal-like enterocytes with more adult-type cells occurs over the first 2 weeks of life. The replacement cells exhibit much lower endocytic activity (Smith and Jarvis, 1977; Smith and Peacock, 1980). The cell replacement proceeds in a proximal–distal direction along the intestine, being complete in the proximal section by the time the pig is 6 days of age (Ekstrom and Westrom, 1991). Distally there is still uptake of macromolecules, which terminates by the time the pigs are 4–8 weeks of age and is concomitant with the complete replacement of the fetal-type cells.

Structure and function of the intestine in the suckling pig

Rapid increases in intestinal dimensions occur during the early postnatal period, particularly the first 6 h of suckling (Zhang *et al.*, 1997). Compared with unsuckled newborns, the intestines of pigs 24 h old were 29% longer and 86% heavier, with 130% more mucosa (Buddington *et al.*, 2001).

The enterocytes of the small intestine of newborn suckling piglets possess complex apical endosomal systems comprising tubular endosomes and associated vesicles that are involved in the extensive uptake of maternal immunoglobulins and other colostral constituents (Murata and Namioka, 1977; Komuves and Heath, 1992). Research on other species has characterized a glycoprotein called endotubin, which is located in the apical endosomal tubules in developing intestinal cells (Trahair *et al.*, 1995; Allen *et al.*, 1998). The function of endotubin remains unknown but its expression in the early postnatal intestines of several species suggests that it is critical for development of the intestinal epithelium (Trahair *et al.*, 1995).

The enterocytes in the small intestine of suckled newborn piglets possess a complex apical endosomal system of tubules and vesicles and also large subnuclear granules that contain colostral constituents, including IgG (Murata and Namioka, 1977; Komuves and Heath, 1992). At the time of intestinal closure (1–2 days *post partum*), large crystalloid inclusions appear in the subnuclear granules (Komuves *et al.*, 1993; Ma *et al.*, 1997). Immunogold cytochemistry has shown that the crystalloid inclusions are highly enriched in maternal immunoglobulins (Komuves *et al.*, 1993).

The activity of lactase-phlorizin hydrolase, the brush-border α-glycosidase that hydrolyses milk lactose, was found to decline significantly in colostrum-fed pigs when compared with colostrum-deprived animals (Kelly *et al.*, 1991). The detection of intestinal α-glucosidase (sucrase, maltase) in the suckling pig, albeit at low activities, suggests that pigs are provided with a limited capacity to process carbohydrates other than lactose (Zhang *et al.*, 1997).

In the pig, an age-related decrease in ion and nutrient fluxes takes place in the immediate postnatal period (Sangild *et al.*, 1993). However, the total transport capacity increases with age, due to a large increase in intestinal mass (Pacha, 2000). Thus, most of the declines in carrier-mediated amino acid absorption during the first 24 h of suckling are caused not by loss of transporters but instead by the rapid increase in tissue mass that effectively 'dilutes' the transporters (Buddington *et al.*, 2001). Significantly, the declines in carrier-mediated amino acid and glucose transport also coincide with the postnatal replacement of the fetal enterocytes, leading to a redistribution of transport functions along the crypt villus axis (Buddington *et al.*, 2001).

Microbial colonization of the small intestine of the suckling pig

Several hundred microbial species have been documented as components of the suckling pig's indigenous intestinal microflora and their origin appears to be maternal and environmental (Finegold *et al.*, 1983; Conway, 1997). The pattern of colonization is similar for most animals, with lactic acid bacteria, enterobacteria and streptococci appearing first, followed by obligate anaerobes (Conway, 1997). Microbial colonization is a complex process of natural selection and ecological succession (Rolfe, 1996) and is influenced by numerous regulatory factors of both bacterial and host origin, including bacterial antagonisms, animal genotype and physiology and, importantly, nutrition (Kelly *et al.*, 1994; Conway, 1997). Several microhabitats exist within the intestines that exert a selective influence on the local composition and metabolic activity of the microflora. These microniches are found in the proximal and distal intestine associated with the villus surface, crypts, epithelial-associated mucins and luminal mucus. Variables that contribute to the regional compositional diversity include immune reactivity, the presence of gut receptors, nutrient availability and composition, the flow of digesta, pH and Eh (oxidation/reduction potential) and available molecular oxygen (Stewart *et al.*, 1993).

For growth, bacteria require energy sources and nutrients, derived either exogenously from the host diet or endogenously from sloughed-off epithelial cells, and cell secretions from the mucous blanket that coats much of the inner surface of the gut (Stewart *et al.*, 1993). Competition for substrates is a major determining factor in the composition of the intestinal microbial population. Dietary residues influence the composition and metabolic activities of gut microorganisms (Gibson and McCartney, 1998). During postnatal development, alterations in diet are believed to induce a succession of related changes in the gut microbial ecosystem (Conway, 1997).

Non-dietary or endogenous nutrient sources are recognized as important regulators of gut microbial populations (Stewart *et al.*, 1993). The oligosaccharide chains (glycans) attached to intestinal cell surfaces and secreted proteins and lipids mediate in many important biological roles. A large part of the observed intra- and inter-species diversity in glycans on mucosal surfaces is driven by exogenous selection pressures mediated by enteric microorganisms (Gagneux and Varki, 1999). Recent evidence suggests that, during evolution, microbial pressure has led to the diverse expression of enteric glycans as essential nutrient sources for selected commensal bacterial populations.

Enterotoxigenic *E. coli* and rotaviral infections are common causes of scours, or diarrhoea, in suckled pigs. Possibly the most important predisposing factor in susceptibility of suckling pigs to infections is insufficient uptake of colostrum in the first hours of suckling. Weak, undersized, chilled or injured piglets often have difficulty in competing for functioning teats. Sows with mastitis or other infections and injuries may be unable to produce sufficient colostrum for transfer of passive immunity to the suckling piglets. A second important predisposing factor to enteric infections is the prevalence of binding sites for pathogens on the intestinal surfaces of the suckling pig. The chemistry and distribution of bacterial and viral binding sites on gut mucosal surfaces play important roles in determining host and tissue susceptibility and in triggering host responses. This is particularly noticeable in neonates, where both beneficial and harmful swings in microbial balance can accompany epithelial differentiation (Kelly *et al.*, 1992a; Stewart *et al.*, 1993). Enteric bacterial strains that cause diarrhoea in suckling pigs have been partially classified according to the nature of their fimbrial adhesins or lectins. These lectins are constituents of proteinaceous appendages that protrude from surfaces of bacteria and recognize sugar moieties of glycoproteins and glycolipids on intestinal surfaces. The synthesis of these appendages and the production of enterotoxins are essential virulence factors that enable pathogens to compete successfully with commensals in the intestine.

Intestinal glycobiology and microbial pathogenesis in the suckling pig

The structural diversity of oligosaccharides found on intestinal membranes and mucins is theoretically enormous. Monosaccharides can be combined with each other in a variety of ways that differ not only in sequence and chain length but also in anomery (α and β), position of links and branching points (Lis and Sharon, 1993). In reality, regions of structural variation are often restricted and the assembly of oligosaccharide chains is at least partially based upon sets of structural rules. Membrane and secretory glycoconjugates are not themselves primary gene products, but are constructed in a stepwise manner as monosaccharides are added to precursor oligosaccharides via several glycosyltransferases coded for by different genes (for review, see Roth, 1997). Most glycoproteins carry oligosaccharide side-chains either *N*-glycosidically linked to the amide nitrogen of asparagine or *O*-glycosidically linked to the hydroxyl groups of the amino acids serine/threonine. A diverse group of bioactive oligosaccharides is also linked to lipids.

Many of the intestinal glycosylation patterns associated with microbial attachment have their basis in only small changes in oligosaccharide chain termination by α-linked sialic acid, galactose, *N*-acetylgalactosamine or fucose (King, 1995). These relatively simple glycosylation changes may be sufficient to create, or mask, binding epitopes for bacterial fimbriae. Variations in glycosylation within pig populations means that individual animals may be more or less susceptible to infection by selected organisms. In the wild, there are advantages to animal herds if some individuals survive epidemics. Similar arguments may hold true in pig-rearing units, where variations in host susceptibility to pathogens, and faecal shedding of commensal and pathogen strains, strongly influence the spread of enteric infections.

Terminal galactosyl moieties are common constituents of all intestinal glycoproteins and glycosphingolipids in newborn and suckling pigs (King *et al.*, 1995; King, 1998). Several of the enterotoxigenic bacterial strains associated with outbreaks of diarrhoea in pigs, including *E. coli* K88 and *E. coli* 987P, express fimbriae and/or heat labile toxins that specifically interact with galactosyl structures on intestinal mucosal surfaces (reviewed by Kelly *et al.*, 1994; King, 1998; Jeyasingham *et al.*, 1999).

Sialic acids located on the terminal position of glycoconjugates are attracting increasing interest because of their involvement in various aspects of normal and pathological cellular growth and development. These sugars play a key regulatory role in cellular and molecular recognition. In some instances they act as signals for recognition and in other situations they mask recognition sites on molecules and cell membranes (Schauer, 1991). Many microbial pathogens, including viruses, mycoplasma, bacteria and protozoa, take advantage of cell surface sialic acids to adhere to their respective host cells (Schauer *et al.*, 1995). Cytochemical and biochemical analyses have shown high membrane and mucin sialylation on pig villus surfaces during the first 2 weeks post partum (King *et al.*, 1995).

The presence of a highly sialylated intestinal epithelial surface during the first days of life may influence the establishment of the enteric flora. Attachment sites for several organisms can be masked by sialic acids, whereas other pathogenic and non-pathogenic bacterial strains secrete sialidase enzymes that enable them to overcome host defensive mechanisms and create novel binding sites for colonization. Other enteric bacteria opportunistically bind to sialylated receptors in the porcine intestine. K99 enterotoxigenic *E. coli,* which causes diarrhoea in neonatal but not adult pigs, expresses fimbrial adhesins that bind to sialylated glycoproteins and glycolipids (Mouricout and Julien, 1987; Lindahl and Carlstedt, 1990). *N*-glycolyneuraminyl-lactosyl-ceramide has been identified as a major receptor for K99. The membrane content of this ganglioside is maximal in newborn pigs and gradually decreases during development (Yuyama *et al.*, 1993).

Rotaviruses are a common cause of severe gastroenteritis in suckled pigs. Early events of virus binding and entry in epithelial cells are the critical determinants of cellular permissiveness to rotavirus replication (Willoughby, 1993). The infectivity of most animal rotaviruses is dependent on the interaction of the virus spike proteins (haemagglutinin VP4 dimers) with sialic acid-containing receptors. Kuhlenschmidt *et al.* (1997) chemically characterized a biologically relevant porcine enterocyte receptor for group A porcine rotavirus. The receptor is a family of two GM3 gangliosides, one containing Neu5Gc (Neu5GcGM3) and the other Neu5Ac (Neu5AcGM3). Significantly, the authors found that the amounts of the sialylated receptors in intestinal membranes decreased rapidly during the first month of life. This factor, in addition to the absence of an effective immune response, may help to explain age-sensitivity of piglets to severe rotavirus diarrhoea. Similarly, the expression of galactosylated receptors in the intestine of weaned pigs is linked to the susceptibility of individual animals to K88 *E. coli* infections (Jeyasingham *et al.*, 1999).

The development of mucosal immunity in the suckling pig

It is evident that the neonatal pig is highly susceptible to infectious diseases. There is a widely held view that the cellular immune system in the young animal is underdeveloped compared with that in the adult. In the first weeks of life, as the cellular machinery required to mount active immune responses is expanded, the suckling young receive protective immunoglobulins in maternal colostrum and milk.

The initial phase of passive immune protection involves the uptake of high levels of colostral immunoglobulins in the early hours of life. The level of protection afforded to the suckling animals is limited by the quantity and quality of antibodies in colostrum and by the amount that the neonate is able to consume and absorb. At optimal uptake of colostral immunoglobulin, serum antibody titres of the piglet are similar to those of the sow, within 24 h of birth (Holland, 1990). However, the predominant immunoglobulin isotype in colostrum is IgG. Although maternal IgG is protective against many systemic pathogens, most pathogens encountered by the piglet are found at the mucosal surfaces, where IgG antibodies are rare and largely ineffective (Gaskins, 1998). A second, longer phase of passive protection, occurring as colostrum formation ends and lactation proceeds, sees IgG concentrations decrease quickly as IgA becomes the major immunoglobulin in sow milk (Gaskins, 1998). This maternal IgA provides short-term intestinal protection by neutralizing viruses, inhibiting bacterial attachment and opsonizing or lysing bacteria (Porter, 1986; Gaskins, 1998). Although suckling piglets receive partial protection against those antigens to which the sow has previously developed immunity, they have little or no protection against new infectious agents that may be introduced to rearing units.

The most abundantly produced immunoglobulin in mammals is IgA, which is secreted mainly across mucous membranes. Conventional immune responses leading to production of IgA involve two principal players, the T and B lymphocytes. IgA is synthesized by B (B2 type) lymphocytes that are first exposed to antigens in Peyer's patches. Luminal antigens are transported through specialized epithelial cells (membranous or M-cells) overlying Peyer's patches, into an interfollicular area where they are presented by resident antigen-presenting cells (macrophages and dendritic cells) to helper T (TH) lymphocytes. The TH cells, in turn, secrete cytokines that stimulate B lymphocytes that produce IgA. After leaving Peyer's patches and passing through the systemic circulation, IgA^+ lymphocytes migrate back to the lamina propria, where they differentiate into plasma cells capable of secreting large amounts of antibody. Upon reintroduction of the antigen, plasma cells secrete antigen-specific IgA, which is then transported back toward the intestinal lumen (see reviews by Gaskins, 1997, 1998).

There is general consensus, albeit based on little quantitative data, that although newborn and suckling pigs possess some of the effector B cell machinery to initiate immune responses, they do not develop a fully functioning T cell repertoire until the late suckling or early weaning periods. Before birth, the spleen, lymph nodes, Peyer's patches and thymus contain detectable levels of immunoglobulin-containing cells. Just after birth, the incidence of IgM^+ B cells is increased and this is followed by an increase in either IgG^+ or IgA^+ B lymphocytes, depending on the tissue evaluated (Bianchi *et al.*, 1992). In at least the thymus, this B cell isotype switching to IgG- and IgA-secreting cells is not influenced by external antigenic stimuli of conventional microflora (Cukrowska *et al.*, 1996). The numbers of B and T lymphocytes present in the small intestine lamina propria doubles during the first 4 weeks after birth (Bianchi *et al.*, 1992). Over the same period there occurs a marked change in the differentiation of the T lymphocyte population; $CD4^+$ T cells increase dramatically in number during the first postnatal week, while the number of $CD8^+$ cells is low at birth and increases only moderately by 5–7 weeks (Stokes *et al.*, 1992; Gaskins, 1998).

Although the T cell machinery in neonatal pigs may be functionally underdeveloped, it is possible that the young animal can produce some form of defensive intestinal IgA barrier. Recent investigations in mice have shown that a large proportion of the intestinal IgA against cell wall antigens and proteins of commensal bacteria is specifically induced in response to their presence in the microflora, but independently of T cells or germinal centre formation (Macpherson *et al.*, 2000). The cells responsible for producing this IgA originate from so-called B1 lymphocytes found in the peritoneal and pleural cavities. These B1 derived cells recognize ubiquitous bacterial antigens such as phosphoryl choline as well as self-antigens such as Ig, DNA and membrane proteins on erythrocytes and thymocytes (Bao *et al.*, 1998; Fagarasan and Honjo, 2000). IgA antibodies produced by B1 cells prevent systemic penetration of commensal bacteria (Fagarasan and Honjo, 2000). Porcine B1 cells have many characteristics in common with those of other mammalian species, including the expression of the transmembrane glycoprotein CD5, which mediates in intracellular signalling events (Appleyard and Wilkie, 1998). The B1 system is believed to be a primitive form of specific immune defence that evolved before the T cell-dependent B cell (B2) systems. The B1 subset of lymphocytes arises early in ontogeny and is a major component of the neonatal B cell system (Wuttke *et al.*, 1997). Further work is required on B1 cell function and specific intestinal mucosal IgA induction against antigens of the changing commensal intestinal microflora in the suckling pig.

Influence of the microbiota on intestinal immune development

Very little antigen exposure occurs *in utero*. Hence, at birth the immune system of a healthy neonate, from an immunological standpoint, is naïve. As described earlier, during the birth process and early postnatal life, microbes from the mother and surrounding environment colonize the gastrointestinal tract of the infant. Exposure to this microbiota is a major predisposing factor in the anatomical and functional expansion of the intestinal immune system. Bacterial antigens play a very significant role in the proliferation and development of the gut-associated lymphoid tissue (Brandtzaeg, 1996; Helgeland *et al.*, 1996). This feature has been highlighted by comparative investigations on gnotobiotics and animals harbouring a conventional flora. It is noteworthy that germ-free animals, exposed only to dietary antigens but not bacterial antigens, possess only a rudimentary immune system.

In the neonate, it is generally accepted that the generation of appropriate immune responses and the development of immune regulatory networks are dependent upon the development of a normal/optimum intestinal flora and the exposure to dietary antigen (Brandtzaeg, 1996). Furthermore, the immunological outcome following exposure to antigen is determined by a number of variables, including genetic background and the nature, timing and dose of administered antigen (Strobel and Mowat, 1998).

In the pig, there are approximately 30 discrete Peyer's patches in the jejunum and upper ileum and one long continuous patch in the terminal ileum (Binns,

1982). The long ileal Peyer's patch is a major antigen-independent site for the generation of the repertoire of primary immunoglobulins and consequent production of the systemic B cell pool (Andersen *et al.*, 1999). The organized germinal centres of jejunal and upper ileal Peyer's patches are important sites of antigen-specific B cell production where the collaboration of epithelial cells with antigen-presenting and lymphoid cells is highly developed. The postnatal development of these two components of Peyer's patch system is variably influenced by the presence of live microbial antigens. The jejunal patches were found to be significantly larger in specified pathogen-free (SPF) and conventional pigs than in germ-free animals (Barman *et al.*, 1997). The same study showed that ileal Peyer's patch follicles in germ-free pigs increased in size between the first and second month, whereas the equivalent follicles in SPF and conventional pigs remained the same size.

The mechanisms by which microbes influence the phenotype and function of lymphoid cells are largely unknown but are likely to involve complex events that are probably triggered following the 'normal' route of antigen uptake and processing. Commensal bacteria and pathogens can also directly influence intestinal cytokine profiles (D. Kelly *et al.*, unpublished observations, Wilson *et al.*, 1998) and these signalling molecules can have very dramatic effects on immune parameters such as the polarization of the immune response and TH cell subset development (Delespesse *et al.*, 1998).

There is a tendency to distinguish between different types of IgA produced on the intestinal mucosal surfaces of neonates. This somewhat arbitrary categorization is based on experimental detection or non-detection of IgA/antigen interactions. Antigen-specific IgA is easily defined but use of the term 'natural IgA' is more confusing. In essence, natural IgA designates all IgA produced for which specificity cannot be determined. The production of most natural IgA is thought to be stimulated, not necessarily specifically, by the presence of members of the normal gut flora (Cebra, 1999). The fact that major components of the gut flora consist of obligate anaerobes that are difficult (or currently impossible) to culture *in vitro* may explain the difficulties in fully defining natural IgA responses.

Commensal enteric bacteria engender a self-limiting humoral mucosal immune response while permanently colonizing the intestine. It has been proposed that successful secretory IgA responses to commensal bacteria can attenuate chronic stimulation of Peyer's patch germinal centre reactions even though the bacteria persist in the gut (Shroff *et al.*, 1995; Talham *et al.*, 1999). The remaining bacteria in the gut are continuously coated with IgA (Schroff *et al.*, 1995). In the suckling pig it is likely that IgA coating of gut commensal bacteria is supplemented with maternal antibodies. Indeed, maternally derived IgA coating may shield the neonatal immune system from microbial antigen to the point where it delays active development of natural IgA responses (Cebra, 1999). As inferred earlier, sustained low-level mucosal IgA responses to commensal bacteria may be driven by the more primitive T cell-independent pathways that (unlike T cell-dependent pathways) continually resupply the lamina propria with specific anti-commensal IgA plasma cells in the face of self-limiting germinal centre reactions (MacPherson *et al.*, 2000).

Intestinal Development in the Weaned Pig

Functional background

Over the past 50 years, the weaning age of piglets has been decreased from 10–12 weeks to current ages of 3–5 weeks (Nabuurs, 1998). Weaning is a stressful time in a pig's life where it has to adapt rapidly to major changes in environment and nutrition. The weaning transition is commonly accompanied by adverse changes in intestinal morphology, including reduced villus height, increased villus width, increased crypt depth and reduced absorptive capacity and brush-border enzyme activity (McCracken *et al.*, 1999). A major challenge for the pig industry is to formulate economically viable growth-promoting diets to ease the transition from sow's milk to nursery diets (Thacker, 1999). In addition to satisfying the nutritional requirements of weaned pigs, such diets are increasingly assessed for their ability to modulate microbial succession, stabilize the commensal microbiota, improve immune function and enhance disease resistance in the young animal.

Intestinal structure and function in the weaner pig

Several factors may be implicated in weaning-associated morphological changes in the pig intestine. Although post-weaning villus atrophy is partly caused by the stress of separating the pigs from the sow and moving them to other pens, there is increasing evidence that the degree of villus atrophy is more closely associated with diminished level of feed intake over the weaning transition (Beers-Schreurs *et al.*, 1998). These changes may predispose the weaner pig to malabsorption, possible dehydration, diarrhoea and enteric infections (Cera *et al.*, 1988; McCracken *et al.*, 1999; Zijlstra *et al.*, 1999). The introduction of creep feeding before weaning appears to have some benefit in limiting the detrimental changes in gut morphology and function over the weaning transition (Makinde *et al.*, 1997). Giving pigs supplementary feed during the suckling period has been shown to reduce the degree of villus shortening after weaning (Nabuurs *et al.*, 1993).

The pre- and post-weaning periods are characterized by major developmental and diet-induced changes in the expression of intestinal brush-border enzymes. Thus the specific activity (activity per unit mass of tissue) of lactase-phlorizin hydrolase, the brush-border α-glycosidase, which hydrolyses milk lactose, reaches a peak in 3-week-old pigs but declines rapidly over the weaning period. Intestinal growth in the weaner pig compensates for the loss of lactase specific activity and the total digestive capacity of the pig intestine exceeds the normal dietary load (Kelly, 1998). In contrast to the decline in lactase activity, enzymes such sucrase and maltase, needed to break down the carbohydrates found in cereal grains, are expressed at low levels at birth and steadily increase with age. Levels of such enzymes are often less than 50% of the value at digestive maturity (Jensen *et al.*, 1997). The precise mechanisms underlying transcriptional regulation of lactase and other microvillar enzymes in the pig intestine have still not been defined. There is, as yet, no satisfactory explanation of the factors regulating the reciprocal expression of lactase and sucrase isomaltase observed in the weaning period.

The composition of the starter diet distributed after weaning has at least some capacity to develop carbohydrase and protease activity in the pancreas and small intestine (Kelly, 1998; Aumaître, 2000). In protein-malnourished pigs fed high-carbohydrate diets, abundance of lactase-phlorizin hydrolase mRNA is significantly lower than in well-nourished pigs (Dudley *et al.*, 1997).

A wide range of feeding strategies has been introduced to maintain food intake and preserve optimal structure and function of the small intestine of the piglet during weaning. Phased-feeding strategies have been proposed to convert the young pig gradually from a high-fat, high-lactose liquid milk diet prior to weaning to a low-fat, low-lactose, high-carbohydrate diet comprising cereal grains and soybean meal. The addition of milk protein to starter diets provides a continuing supply of highly digestible and balanced nutrients and helps to diminish the impact of weaning anorexia. Piglets fed cow milk after weaning were found to maintain villus height and crypt depth and were capable of enhanced disaccharide digestion and monosaccharide absorption (Pluske *et al.*, 1996). In the USA and some other regions, spray-dried animal plasma has been introduced as a protein source for weaner pigs. It is claimed that pig plasma significantly increases food intake, growth rate and, potentially, food conversion efficiency. Supplementing spray-dried plasma in weaner diets is also believed to suppress the local proinflammatory response associated with weaning and thereby to reduce leukocytic infiltration into the mucosal lamina propria (Jiang *et al.*, 2000). The adverse health risks associated with widespread application of serum products have not been comprehensively investigated.

After weaning, lactose is usually replaced by complex carbohydrates, requiring partial hydrolysis by luminal enzymes before the brush-border hydrolases can release the constituent monosaccharides (Puchal and Buddington, 1992). The intestines of weaned pigs have high sucrase and maltase activities and are characterized by correspondingly high rates of glucose and fructose transport across the apical membranes of enterocytes (Puchal and Buddington, 1992). Carrier-mediated amino acid absorption declines significantly in the post-weaned pig intestine. This decline is consistent with a shift to adult diet that is lower in protein (Buddington *et al.*, 2001). The exception is the post-weaning increase in rates of carrier-mediated absorption of lysine, the principal amino acid limiting the performance of early-weaned pigs (Thacker, 1999). In the weaner pig, specific transport systems may be regulated to match dietary inputs and requirements for amino acids (Buddington *et al.*, 2001). After weaning, the net absorption of fluid and electrolytes in the small intestine of pigs is temporarily decreased (Nabuurs *et al.*, 1994).

Microbial/mucosal interactions in the post-weaning intestine

Postnatal intestinal development involves extensive epithelial cell proliferation and cytodifferentiation, including changes in the expression of enzymes, receptors and transport systems. Age-related intestinal glycosylation changes play an important role in modifying the properties of intestinal receptors for dietary constituents as well as commensal and pathogenic bacteria (Kelly *et al.*, 1992a; Stewart *et al.*, 1993).

Enterotoxigenic *E. coli* infections are common contributing factors to post-weaning scours in pigs. Predisposing factors in such infections include: the removal of protective levels of IgA and other beneficial factors present in sow milk; inadequate feed and water intake; inadequate gastric acid secretion; unstable microbiota; and expression of membrane and mucin glycoconjugates that serve as binding sites for enteropathogens (Kelly *et al.*, 1994; Thacker, 1999; Kelly and Coutts, 2000; Kelly and King, 2001). Post-weaning malnutrition predisposes to infection by compromising the barrier and immune functions of the gut. At the same time, the infections adversely influence dietary intake and absorption and cause loss of endogenous nutrients (Calder and Jackson, 2000). Niewold *et al.* (2000) proposed that intestinal ischaemia is a key predisposing factor to post-weaning diarrhoea in weaned pigs. Intestinal blood supply may be unable to meet the metabolic demands of the fast-growing intestine undergoing hyper-regenerative villus repair. It is suggested that the resultant ischaemia leads to intestinal acidosis and increased permeability for enterotoxigenic *E. coli* toxins (Niewold *et al.*, 2000).

Microbial succession in the intestine of the weaned pig is influenced by the interplay of environmental factors, dietary change, intrinsic variations in host physiology, endogenous nutrients and the composition of the microbiota (Stewart *et al.*, 1993; Mackie *et al.*, 1999). As the young animal is weaned, the obligate anaerobic bacteria become numerically dominant and *E. coli* and enterococci decrease in numbers (Conway, 1997). The increase in anaerobes is particularly marked in the hindgut of the pig but high levels of such bacteria are also found in the ileum. The microbiota in the small intestine compete with the host animal for easily digestible nutrients. As much as 6% of the net energy in the pig diet can be lost due to microbial fermentation in the stomach and small intestine (Jensen, 1998). However, from 5 to 20% of the total energy supply in the adult pig is achieved from microbial fermentation in the large intestine (Jensen, 1998; Anderson *et al.*, 2000).

For several decades, orally administered antibiotics have been used to enhance growth of livestock. Astonishingly, the precise mechanisms underlying the beneficial effects of antibiotics remain unclear. Anderson *et al.* (2000) recently proposed that the benefits of growth-promoting antibiotics result from substantial decreases in bacterial populations and consequent alterations in epithelial functions in the pig small intestine, whereas changes in large intestinal microbial populations exert less impact on whole-animal growth. The use of avoparcin and virginiamycin as growth promoters in animal feed has been associated with an increase in resistance of bacteria to therapeutic agents and a fear that this could reduce the ability to treat diseased humans (Jensen, 1998). The addition of growth-promoting antibiotics to the feed of growing pigs is now banned in many European countries. The removal of such additives will profoundly influence intestinal microbiology and physiology of weaned pigs. Research in several European centres is directed towards finding alternative methods to sustain the growth and health of the animals in pig production.

Mikkelsen and Jensen (2000) investigated the impact of fermented liquid feed on the activity and composition of the microbiota in the post-weaning pig intestine.

The feed (containing high levels of lactic acid and lactic acid bacteria) reduced gastric pH, lowered microbial activity in the small intestine and reduced the number of enterobacteria (including coliform bacteria) throughout the gastrointestinal tract. Further research is required to define the mechanisms whereby fermented liquid feed reduces microbial fermentation and helps to prevent coliform scours in the small intestine of the weaned pig.

Intestinal glycobiology and microbial pathogenesis in the weaned pig

Increased production of intestinal mucins occurs in weaned pigs compared with the levels detected in pre-weaned animals (Pestova *et al.*, 2000). Epithelial mucins are major glycoprotein components of the mucus that coats the mucosal surfaces of the gastrointestinal tract. They are believed to protect epithelial cells from infection, dehydration and physical or chemical injury, as well as to aid the passage of materials through the tract (Perez-Villar and Hill, 1999). Mucin production may therefore be considered a key innate defence mechanism of intestinal epithelial cells. A negative feature of enhanced mucin production is that it may slightly reduce nutrient absorption (Satchithanandam *et al.*, 1990).

The dietary composition and microbial flora, as well as interactions between the dietary constituents and the flora, influence the composition and functional characteristics of intestinal mucins (Sharma *et al.*, 1995). Degradation of the carbohydrate chains of mucin glycoproteins involves glycosidases and glycosulphatases produced by specialized strains of normal enteric bacteria, resulting in the release of component monosaccharides that can be used as a source of nutrition by other, larger populations. Such functional specialization provides an ecological niche for enzyme-producing specialists and is likely to be a contributing factor to microbial diversity in enteric bacterial ecosystems. Bacterial/mucosal cross-talk may lead to changes in gene expression for mucin peptides. For example, the ability of selected probiotic strains of *Lactobacillus* to inhibit the adherence of attaching and effacing bacteria is mediated through their ability to increase expression of MUC2 and MUC3 intestinal mucins (Mack *et al.*, 1999).

Glycosylation is a major factor governing the adherence of bacteria to intestinal mucins. In weaned pigs, the major glycosylation patterns of secreted mucins vary according to the AO histo-blood group secretor status of the individual animals. Thus, high levels of terminal α-linked fucose characterize mucin oligosaccharides in O-secretor pigs. A-secretor pigs have the same levels of fucose but this sugar is extensively masked by terminal α-linked *N*-acetylgalactosamine (King, 1995; King *et al.*, 1995). Within individuals there are often conspicuous differences in mucin glycosylation in different regions of the small intestine and indeed on the same intestinal villi. Immature goblet cells deep within the crypts produce neutral mucins containing little sialic acid. As they mature and migrate to the villus tip, the mucins become sialylated (King, 1995). It is highly likely that animal-to-animal and site-to-site variation in mucin subtypes is reflected in the composition of the mucin-associated microbiota. In addition to the commensal flora, several enterotoxigenic *E. coli* species are known to adhere to mucin glycoproteins or glycolipids in weaner pigs (Blomberg *et al.*, 1995; Dean-Nystrom and Samuel, 1994). The precise significance of these associations is uncertain but the emerging view is that the mucus barrier reduces the pathogen colonization of villus membranes (Pestova *et al.*, 2000).

Increased levels of membrane and mucin glycoprotein fucosylation occur in the intestines of pigs and other mammalian species at the time of weaning (King, 1995). Available evidence suggests that such glycosylation events are partly pre-programmed but are also sensitive to changes in dietary regime and to weaning (Kelly and King, 1991; Kelly *et al.*, 1993). Supplementing the diet with galactose results in modification of mucin glycosylation in weaned pigs. This change may limit microbial degradation of the mucin (Pestova *et al.*, 2000). Experimental evidence from other species suggests that intestinal fucosylation may also be enhanced by the presence of selected commensal bacteria. Bry *et al.* (1996) compared genetically identical germ-free mice with mice raised with a functional microbiota and determined that the production of fucosylated glycoconjugates, appearing in the intestine and colon after the age of weaning, requires components of the microbiota. Fucoconjugates were largely absent from weaned germ-free mice; inoculation with the commensal bacterium *Bacteroides thetaiotaomicron* restored the same fucosylation pattern as in conventional mice and induced the accumulation of α1,2-fucosyltransferase RNA (Bry *et al.*, 1996). Molecular studies on *B. thetaiotaomicron* have identified a transcriptional repressor that serves as a molecular sensor of fucose availability that coordinates bacterial fucose metabolism and host fucosylated glycan production (Hooper *et al.*, 2000). The identity and mode of action of the fucosylation-inducing signal produced by the commensal has not been determined (Hooper *et al.*, 2000). The signals may be polyamines. In weaned rats, *B. thetaiotaomicron* contributes high amounts of putrescine and spermidine in the caecum and ileum of pectin-fed gnotobiotic rats (Noack *et al.*, 2000). Polyamines have recently been shown to be potent maturation factors implicated in the expression of increased α1,2-fucosylation in the rat gut at the time of weaning (Greco *et al.*, 1999, 2000). Endogenous sources of polyamines may also be of importance for maturation of the post-weaning intestine. Plasma glucocorticoids are markedly increased in pigs during weaning. This cortisol surge has an essential role in enhancing polyamine synthesis, which may be of physiological importance for intestinal adaptation and remodelling (Wu *et al.*, 2000).

The mucosal immune system in the weaned pig

Some investigators have suggested that immune responses to dietary antigen, especially those derived from soybean protein, are an important cause of local inflammation and result in villus atrophy (Li *et al.*, 1990, 1991; Bailey *et al.*, 1993; Miller *et al.*, 1994). McCracken *et al.* (1999) concluded that soybean-induced inflammation, if present, is likely to compromise intestinal morphology, due to local inflammation caused by anorexia in the immediate post-weaning period. Much more research is required into the possible cellular mechanisms linking hypersensivity and the shaping of the villus epithelium at weaning. Although hypersensitivity is a contentious issue, there are good nutritional reasons why soybean protein should not be used at too high a concentration in weaner diets. Pigs 2–3-weeks old cannot effectively utilize bean meal, because they lack adequate levels of the digestive enzyme systems needed to break down complex proteins and carbohydrates.

The banning of growth-promoting antibiotics in European pig production means that new disease management practices are required, to minimize the impact of infectious organisms in commercial piggeries. There will be little point in withdrawing antibiotic growth promoters only to see an escalation in the use of such products for clinical purposes. The perceived advantage of growth-promoting antibiotics is that they decrease the energetic costs associated with constitutive low-level inflammation caused by bacteria in the gut (Anderson *et al.*, 2000). However, although bacteria play an important role in the development of the intestinal immune system (Gaskins, 1997, 1998; MacDonald and Pettersson, 2000), some may actually also help to reduce maintenance costs of the gastrointestinal system (Kelly and King, 2001). Commensal and pathogenic bacteria have evolved a diverse range of mechanisms that promote their survival within the gut ecosystem. Bacteria can produce a vast array of cytokine-inducing or cytokine-modulating molecules that will regulate or direct the host response. Certain of these factors may promote the virulence and pathogenic potential of bacteria but others, paradoxically, may facilitate maintenance of the indigenous microflora by beneficially regulating the immuno-inflammatory status of the gut (Kelly and King, 2001).

Exposure to bacterial antigen is now recognized to be of immense importance, both in early life, in order to prime the immune system in the correct way, and throughout life, to maintain a functional immune system (Kelly and Coutts, 2000; Kelly and King, 2001). The idea of an 'optimum' functional immune system has now crept into the equation, along with the challenge to identify bacteria and bacterial antigens that can be employed to provide appropriate and optimal stimulation of the immune system. In addition to stimulation of protective immunity, the immunomodulatory potential of bacteria or bacterial antigens may be used to adjust or correct immune dysfunction or hyperfunction.

Conclusions

The development of the gut and its associated immune system, under the protection afforded by maternal passive immunity, occurs in a precise and highly regulated manner and results in an optimally primed system capable of providing nutrients for normal bodily functions and growth and also for immune protection, all essential for the survival of the young animal. However, current high-health intensive farming systems are designed to provide a high throughput of animals and rely strongly on the use of both oral and injectable antibiotics to promote the growth and disease resilience of animals. This practice is now threatened by legislation that prohibits the use of those antimicrobials associated with health risks and the emergence of antibiotic resistance.

The conflict between animal production systems and the health and welfare of animals and humans provides a real dilemma, particularly for those in the farming community and those involved in setting national food policy. The alternative, such as production of food based on the organic tradition, is unlikely to be universally acceptable or to meet market requirements. Hence, the development of new strategies for food production that provide protection to animals and humans is now essential. Information on the 'signals and inputs' required in early life to promote optimum immune function and disease resistance, and the energetics of these processes, may enable the manipulation of the nutrition and the environment (microbial exposure) of the young animal to satisfy both welfare and production criteria.

Acknowledgement

The authors' research is funded by the Scottish Executive Environment Rural Affairs Department.

References

Alexander, A.N. and Carey, H.V. (1999) Oral IGF-1 enhances nutrient and electrolyte absorption in neonatal piglet intestine. *American Journal of Physiology* 277, G619–G625.

Allen, K., Gokay, E., Thomas, M.A., Speelman, B.A. and Wilson, J.M. (1998) Biosynthesis of endotubin: an apical early endosomal glycoprotein from developing intestinal epithelial cells. *Biochemical Journal* 330, 367–373.

Andersen, J.K., Takamatsu, H., Oura, C.A.L., Brookes, S.M., Pullen, L. and Parkhouse, R.E.M. (1999) Systematic characterization of porcine ileal Peyer's patch. I. Apoptosis-sensitive immature B cells are the predominant cell type. *Immunology* 98, 612–621.

Anderson, D.B., McCracken, V.J., Aminov, R.I., Simpson, J.M., Mackie, R.I., Verstegen, W.A. and Gaskins, H.R. (2000) Gut microbiology and growth-promoting antibiotics in swine. *Nutrition Abstracts and Reviews. Series B : Livestock Feeds and Feeding* 70, 101–108.

Appleyard, G.D. and Wilkie, B.N. (1998) Characterization of porcine CD5 and CD5+ B cells. *Clinical Experimental Immunology* 111, 225–230.

Aumaître, L.A. (2000) Adaptation and efficiency of the digestive process in the gut of the young piglet: consequences for the formulation of a weaning diet. *Asian-Australasian Journal of Animal Sciences* 13, 227–242.

Bailey, M., Miller, B.G., Telemo, E., Stokes, C.R. and Bourne, F.J. (1993) Specific immunological unresponsiveness following active primary responses to proteins in the weaning diet of piglets. *International Archives Allergy Immunology* 101, 266–271.

Bao, S., Beagley, K.W., Murray, A.M., Caristo, V., Matthaei, K.I., Young, I.G. and Husband, A.J. (1998) Intestinal IgA plasma cells of the B1 lineage are IL-5 dependent. *Immunology* 94, 181–188.

Barman, N.N., Bianchi, A.T.J., Zwart, R.J., Pabst, R. and Rothkotter, H.J. (1997) Jejunal and ileal Peyer's patches in pigs differ in their postnatal development. *Anatomy and Embryology* 195, 41–50.

Bate, L.A., Ireland, W., Connell, B.J. and Grimmelt, B. (1991) Development of the small intestine in piglets in response to prenatal elevation of glucocorticoids. *Histology and Histopathology* 6, 207–216.

Beers-Schreurs, H.M., Nabuurs, M.J., Vellenga, L., Kalsbeek-van der Valk, H.J., Wensing, T. and Breukink, H.J. (1998) Weaning and the weanling diet influence the villous height and crypt depth in the small intestine of pigs and alter the concentrations of short-chain fatty acids in the large intestine and blood. *Journal of Nutrition* 128, 947–953.

Bianchi, A.T.J., Zwart, R.J., Jeurissen, S.H.M. and Moonenleusen, H.W.M. (1992) Development of the B-cell and T-cell compartments in porcine lymphoid organs from birth to adult life – an immunohistological approach. *Veterinary Immunology and Immunopathology* 33, 201–221.

Binns, R.M. (1982) Organization of the lymphoreticular system and lymphocyte markers in the pig. *Veterinary Immunology and Immunopathology* 3, 95–146.

Blomberg, L., Gustafsson, L., Cohen, P.S., Conway, P.L. and Blomberg, A. (1995) Growth of *Escherichia coli* K88 in piglet ileal mucus: protein expression as an indicator of type of metabolism. *Journal of Bacteriology* 177, 6695–6703.

Brandtzaeg, P. (1996) Development of the mucosal immune system in humans. In: Bindels, J.G., Goedhart, A.C. and Visser, H.K.A. (eds) *Recent Developments in Infant Nutrition.* Kluwer Academic Publishers, London, pp. 349–376.

Bry, L., Falk, P.G., Midtvedt, T. and Gordon, J.I. (1996) A model of host–microbial interactions in an open mammalian ecosystem. *Science* 273, 1380–1383.

Buddington, R.K. and Diamond, J.M. (1989) Ontogenetic development of intestinal nutrient transporters. *Annual Review in Physiology* 51, 601–619.

Buddington, R.K. and Malo, C. (1996) Intestinal brush-border membrane enzyme activities and transport functions during prenatal development of pigs. *Journal of Pediatric Gastroenterology and Nutrition* 23, 51–64.

Buddington, R.K., Elnif, J., Puchal-Gardiner, A.A. and Sangild, P.T. (2001) Intestinal apical amino acid absorption during development of the pig. *American Journal of Physiology* 280, R241–R247.

Burrin, D.G. (1997) Is milk-borne insulin-like growth factor-I essential for neonatal development? *Journal of Nutrition* 127, 975S–979S.

Burrin, D.G., Shulman, R.J., Reeds, P.J., Davis, T.A. and Gravitt, K.R. (1992) Porcine colostrum and milk stimulate visceral organ and skeletal muscle protein synthesis in neonatal piglets. *Journal of Nutrition* 122, 1205–1213.

Burrin, D.G., Davis, T.A., Ebner, S., Schoknecht, P.A., Fiorotto, M.L., Reeds, P.J. and McAvoy, S. (1995) Nutrient-independent and nutrient-dependent factors stimulate protein synthesis in colostrum-fed newborn pigs. *Pediatric Research* 37, 593–599.

Calder, P.C. and Jackson, A.A. (2000) Undernutrition, infection and immune function. *Nutrition Research Reviews* 13, 3–29.

Cebra, J. (1999) Influences of microbiota on intestinal immune system development. *American Journal of Clinical Nutrition* 69 (Supplement), 1046S–1051S.

Cera, K.R., Mahan, D.C., Cross, R.F., Reinhart, G.A. and Whitmoyer, R.E. (1988) Effect of age, weaning and postweaning diet on small intestinal growth and jejunal morphology in young swine. *Journal of Animal Science* 66, 574–584.

Conway, P.L. (1997) Development of intestinal microbiota. Gastrointestinal microbes and host interactions. In: Mackie, R.I., Whyte, B.A. and Isaacson, R.E. (eds) *Gastrointestinal Microbiology*, Vol. 2. Chapman & Hall, London, pp. 3–39.

Cukrowska, B., Sinkora, J., Mandel, L., Splichal, I., Bianchi, A.T., Kovaru, F. and Tlaskalova-Hogenova, H. (1996) Thymic B cells of pig fetuses and germ-free pigs spontaneously produce IgM, IgG and IgA: detection by ELISPOT method. *Immunology* 87, 487–492.

Danielsen, E.M., Hansen, G.H. and Nielschristiansen, L.L. (1995) Localisation and biosynthesis of aminopeptidase-N in pig fetal small intestine. *Gastroenterology* 109, 1039–1050.

Dean-Nystrom, E.A. and Samuel, J.E. (1994) Age-related resistance to 987P fimbria-mediated colonization correlates with specific glycolipid receptors in intestinal mucus in swine. *Infection and Immunity* 62, 4789–4794.

Dekaney, C.M., Bazer, F.W. and Jaeger, L.A. (1997) Mucosal morphogenesis and cytodifferentiation in fetal porcine small intestine. *Anatomical Record* 249, 517–523.

Delespesse, G., Yang, L.P., Ohshima, Y., Demeure, C., Shu, U., Byun, D.G. and Sarfati, M. (1998) Maturation of human neonatal CD4+ T lymphocytes into TH1/TH2 effectors. *Vaccine* 16, 1415–1419.

Donovan, S.M. and Odle, J. (1994) Growth factors in milk as mediators of infant development. *Annual Reviews in Nutrition* 14, 147–167.

Donovan, S.M., McNeil, L.K., Jimenez-Flores, R. and Odle, J. (1994) Insulin-like growth factors and insulin-like growth factor binding proteins in porcine serum and milk throughout lactation. *Pediatric Research* 36, 159–168.

Dudley, M.A., Wykes, L., Dudley, A.W.J., Fiorotto, M., Burrin, D.G., Rosenberger, J., Jahoor, F. and Reeds, P.J. (1997) Lactase phlorizin hydrolase synthesis is decreased in protein-malnourished pigs. *Journal of Nutrition* 127, 687–693.

Ekstrom, G.M. and Westrom, B.R. (1991) Cathepsin B and D activities in intestinal mucosa during postnatal development in pigs. Relation to intestinal uptake and transmission of macromolecules. *Biology of the Neonate* 59, 314–321.

Fagarasan, S. and Honjo, T. (2000) T-independent immune response: new aspects of B cell biology. *Science* 290, 89–92.

Finegold, S.M., Sutter, V.L. and Mathisen, G.E. (1983) Normal indigenous intestinal flora. In: Hentges D.J. (ed.) *Human Intestinal Microflora in Health and Disease.* Academic Press, New York, pp. 3–31.

Fiorotto, M.L., Davis, T.A., Reeds, P.J. and Burrin, D.G. (2000) Nonnutritive factors in colostrums enhance myofibrillar protein synthesis in the newborn pig. *Pediatric Research* 48, 2000.

Gagneux, P. and Varki, A. (1999) Evolutionary considerations in relating oligosaccharide diversity to biological function. *Glycobiology* 9, 747–755.

Gaskins, H.R. (1997) Immunological aspects of host/microbiota interactions at the intestinal epithelium In: Mackie, R.I., Whyte, B.A. and Isaacson, R.E. (eds) *Gastrointestinal Microbiology*, Vol. 2. Chapman and Hall, London, pp. 537–587.

Gaskins, H.R. (1998) Immunological development and mucosal defence in the pig intestine. In: Wiseman, J., Varley, M.A. & Chadwick, J.P. (eds) *Progress in Pig Science.* Nottingham University Press, Nottingham, UK, pp. 81–101.

Gibson, G.R. and McCartney, A.L. (1998) Modification of the gut flora by dietary means. *Biochemical Society Transactions* 26, 222–228.

Girard, J., Ferre, P., Pegorier, J.P. and Duee, P.H. (1992) Adaptations of glucose and fatty acid metabolism during perinatal period and suckling–weaning transition. *Physiological Reviews* 72, 507–562.

Greco, S., George, P., Hugueny, I., Louisot, P. and Biol, M.C. (1999) Spermidine-induced glycoprotein fucosylation in immature rat intestine. *Comptes Rendus de l'academia des Sciences, Serie III*, 322, 543–549.

Greco, S., Hugueny, I., George, P., Perrin, P., Louisot, P. and Biol, M.C. (2000) Influence of spermine on intestinal maturation of the glycoproteins glycosylation process in neonatal rats. *Biochemical Journal* 345, 69–75.

Helgeland, L., Vaage, J.T. and Rolstad, B. (1996) Microbial colonization influences composition and T-cell receptor V beta repertoire of intraepithelial lymphocytes in rat intestine. *Immunology* 89, 494–501.

Holland, R.E. (1990) Some infectious causes of diarrhea in young farm animals. *Clinical Microbiology Reviews* 3, 345–375.

Hooper, L.V., Falk, P.G. and Gordon, J.I. (2000) Analyzing the molecular foundations of commensalism in the mouse intestine. *Current Opinion in Microbiology* 3, 79–85.

Houle, V.M., Schroeder, E.A., Odle, J. and Donovan, S.M. (1997) Small intestinal disaccharidase activity and ileal villus height are increased in piglets consuming formula containing recombinant human insulin-like growth factor-I. *Pediatric Research* 42, 78–86.

Houle, V.M., Park, Y.K., Laswell, S.C., Freund, G.G., Dudley, M.A. and Donovan, S.M. (200) Investigation of three doses of oral insulin-like growth factor-I on jejunal lactase phlorizin hydrolase activity and gene expression and enterocyte proliferation and migration in piglets. *Pediatric Research* 48, 497–503.

Jaeger, L.A. (1996) Immunohistochemical localization of transforming growth factor-alpha in suckling porcine intestine. *Acta Anatomica (Basel)* 155, 14–21.

Jensen, B.B. (1998) The impact of feed additives on the microbial ecology of the gut in young pigs. *Journal of Animal and Feed Sciences* 7, 45–64.

Jensen, M.S, Jensen, S.K. and Jakobsen, K. (1997) Development of digestive enzymes in pigs with emphasis on lipolytic activity in the stomach and pancreas. *Journal of Animal Science* 75, 437–445.

Jeyasingham, M.D., Butty, P., King, T.P., Begbie, R. and Kelly, D. (1999) *Escherichia coli* K88 receptor expression in intestine of disease-susceptible weaned pigs. *Veterinary Microbiology* 68, 219–234.

Jiang, R., Chang, X., Stoll, B., Fan, M.Z., Arthington, J., Weaver, E., Campbell, J. and Burrin, D.G. (2000) Dietary plasma protein reduces small intestinal growth and lamina propria cell density in early weaned pigs. *Journal of Nutrition* 130, 21–26.

Kelly, D. (1998) Gut development and regulation. In: Wiseman, J., Varley, M.A. and Chadwick, J.P. (eds) *Progress in Pig Science.* Nottingham University Press, Nottingham, UK, pp. 103–120.

Kelly, D. and Coutts, A.G. (1997) Biologically active peptides in colostrum. In: Laplace, J.P., Fevrier, C. and Barbeau, A. (eds) *Digestive Physiology in Pigs. Proceedings of the VII International Symposium.* EAAP Publication no. 88, Saint Malo, France, pp. 163–170.

Kelly, D. and Coutts, A.G. (2000) Early nutrition and the development of immune function in the neonate . *Proceedings of the Nutrition Society* 59, 177–185.

Kelly, D. and King, T.P. (1991) The influence of lactation products on the temporal expression of histo-blood group antigens in the intestines of suckling pigs: lectin histochemistry and immunohistochemical analysis. *Histochemical Journal* 23, 55–60.

Kelly, D. and King, T.P. (2001) Luminal bacteria: regulation of gut function and immunity. In: Piva, A., Bach Kudsen, K.E and Lindberg, J.E. (eds) *Manipulation of the Gut Environment in Pigs.* Nottingham University Press, Nottingham, UK, pp. 113–131.

Kelly, D., King, T.P., McFadyen, M. and Travis, A.J. (1991) Effect of lactation on the decline of brush-border lactase activity in neonatal pigs. *Gut* 32, 386–392.

Kelly, D., Begbie, R. and King, T.P. (1992a) Postnatal intestinal development. In: Varley, M.A, Williams, P.E.V. and Lawrence, T.L.J. (eds) *Neonatal Survival and Growth.* BSAP Occasional Publication Number 15, British Society of Animal Production, Edinburgh, pp. 63–79.

Kelly, D., McFadyen, M., King, T.P. and Morgan, T.P. (1992b) Characterisation and autoradiographical localisation of the epidermal growth factor receptor in the jejunum of neonatal pigs. *Reproduction Fertility and Development* 4, 183–191.

Kelly, D., King, T.P., McFadyen, M. and Coutts, A.G.P. (1993) Effect of pre-closure colostrum intake on the development of the intestinal epithelium of artificially-reared piglets. *Biology of the Neonate* 64, 235–244.

Kelly, D., Begbie, R. and King, T.P. (1994) Nutritional influences on interactions between bacteria and the small intestinal mucosa. *Nutrition Research Reviews* 7, 233–257.

King, T.P. (1995) Lectin cytochemistry and intestinal epithelial cell biology. In: Pusztai, A. and Bardocz, S. (eds) *Lectins: Biomedical Perspective.* Taylor and Francis, London and Bristol, pp. 183–210.

King, T.P. (1998) The carbohydrate biology of intestinal surfaces: interactions with dietary and microbial constituents. In: *Proceedings of the 19th Western Nutrition Conference*, Saskatoon, Saskatchewan, pp. 69–88.

King, T.P. and Kelly, D. (1991) Ontogenic expression of histo-blood group antigens in the intestines of suckling pigs: lectin histochemistry and immunohistochemical analysis. *Histochemical Journal* 23, 43–54.

King, T.P., Begbie, R., Slater, D., McFadyen, M., Thom, A. and Kelly, D. (1995) Sialylation of intestinal microvillar membranes newborn, suckling and weaned pigs. *Glycobiology* 5, 525–534.

Kiriyama, H. (1992) Enzyme-linked immunoabsorbent assay of colotral IgG transported into lymph and plasma in neonatal pigs. *American Journal of Physiology* 263, R976–R980.

Koletzko, B., Aggett, P.J., Bindels, J.G., Bung, P., Ferre, P., Gil, A., Lentze, M.J., Roberfroid, M. and Strobel, S. (1998) Growth, development and differentiation: a functional food science approach. *British Journal of Nutrition* 80 (Supplement 1), S5–S45.

Komuves, L.G. and Heath, J.P. (1992) Uptake of maternal immunoglobulins in the enterocytes of suckling piglets: improved detection with a streptavidin-biotin bridge gold technique. *Journal of Histochemistry and Cytochemistry* 40, 1637–1646.

Komuves, L.G., Nicols, B.L., Hutchens, T.W. and Heath, J.P. (1993) Formation of crystalloid inclusions in the small intestine of neonatal pigs: an immunocytochemical study using colloidal gold. *Histochemical Journal* 25, 19–29.

Kuhlenschmidt, M.S., Rolsma, M.D., Kuhlenschmidt, T.B. and Gelberg, H.B. (1997) Characterization of a porcine enterocyte receptor for group A rotavirus. *Advances in Experimental Medical Biology* 412, 135–143.

Li, D.F., Nelssen, J.L., Reddy, P.G., Blecha, F., Hancock, J.D., Allee, G.L., Goodband, R.D. and Klemm, R.D. (1990) Transient hypersensitivity to soybean meal in the early-weaned pig. *Journal of Animal Science* 68, 1790–1799.

Li, D.F., Nelssen, J.L., Reddy, P.G., Blecha, F., Klemm, R. and Goodband, R.D. (1991) Interrelationship between hypersensitivity to soybean proteins and growth performance in early-weaned pigs. *Journal of Animal Science* 69, 4062–4069.

Lindahl, M. and Carlstedt, I. (1990) Binding of K99 fimbriae of enterotoxigenic *Escherichia coli* to pig small intestinal mucin glycopeptides. *Journal of General Microbiology* 136, 1609–1614.

Lindberg, T. (1982) Protease inhibitors in human milk. *Pediatric Research* 16, 479–483.

Lis, H. and Sharon, N. (1993) Protein glycosylation. Structural and functional aspects. *European Journal of Biochemistry* 218, 1–27.

Ma, L., Crissinger, K.D. and Specian, R.D. (1997) Unique crystalline inclusions in neonatal pig small intestine epithelial cells. *Journal of Submicroscopical Cytology and Pathology* 29, 401–404.

MacDonald, T.T. and Pettersson, S. (2000) Bacterial regulation of intestinal immune responses. *Inflammatory Bowel Diseases* 6, 116–122.

Mack, D.R., Michail, S., Wei, S., McDougall, L. and Hollingsworth, M.A. (1999) Probiotics inhibit enteropathogenic *E. coli* adherence in vitro by inducing intestinal mucin gene expression. *American Journal of Physiology* 276, G941–G950.

Mackie, R., Sghir, A. and Gaskins, H.R. (1999) Developmental microbial ecology of the neonatal gastrointestinal tract. *American Journal of Clinical Nutrition* 69(5),1035S–1045S.

Macpherson, A.J., Gatto, D., Sainsbury, E., Harriman, G.R., Hengartner, H. and Zinkernagel, R.M. (2000) A primitive T cell-independent mechanism of intestinal mucosal IgA responses to commensal bacteria. *Science* 288, 2222–2226.

Makinde, M.O., Umapathy, E., Akingbemi, B.T., Mandisodza, K.T. and Skadhauge, E. (1997) Differential response of legumes and creep feeding on gut morphology and faecal composition in weanling pigs. *Comparative Biochemistry and Physiology A – Physiology* 118, 349–354.

McCracken, B.A., Spurlock, M.E., Roos, M.A., Zuckermann, F.A. and Gaskins, H.R. (1999) Weaning anorexia may contribute to local inflammation in the piglet small intestine. *Journal of Nutrition* 129, 613–619.

Mikkelsen, L.L. and Jensen, B.B. (2000) Effect of fermented liquid feed on the activity and composition of the microbiota in the gut of pigs. *Nutrition Abstracts and Reviews. Series B: Livestock Feeds and Feeding* 70, 919–924.

Miller, B.G., Whittemore, C.T., Stokes, C.R. and Telemo, E. (1994) The effect of delayed weaning on the development of oral tolerance to soyabean protein in pigs. *British Journal of Nutrition* 71, 615–625.

Morgan, C.J., Coutts, A.G.P., McFadyen, M., King, T.P. and Kelly, D. (1996) IGF-1 receptor expression in the small intestine of piglets during postnatal development; localisation and characterisation. *Journal of Nutritional Biochemistry* 7, 339–349.

Mouricout, M.A. and Julien, R.A. (1987) Pilus-mediated binding of bovine enterotoxigenic *Escherichia coli* to calf intestinal mucins. *Infection and Immunity* 55, 1216–1223.

Murata, H. and Namioka, S. (1977) The duration of colostral immunoglobulin uptake by the epithelium of the small intestine of neonatal piglets. *Journal of Comparative Pathology* 87, 431–439.

Nabuurs, M.J. (1998) Weaning piglets as a model for studying pathophysiology of diarrhea. *Veterinary Q* 20 (Supplement 3), S42–S45.

Nabuurs, M.J., Hoogendoorn, A., van der Molen, E.J. and van Osta, A.L. (1993) Villus height and crypt depth in weaned and unweaned pigs, reared under various circumstances in The Netherlands. *Research in Veterinary Science* 55, 78–84.

Nabuurs, M.J., Hoogendoorn, A. and van Zijderveld, F.G. (1994) Effects of weaning and enterotoxigenic *Escherichia coli* on net absorption in the small intestine of pigs. *Research in Veterinary Science* 56, 379–385.

Niewold, T.A., van Essen, G.J., Nabuurs, M.J., Stockhofe-Zurwieden, N. and van der Meulen, J. (2000) A review of porcine pathophysiology: a different approach to disease. *Veterinary Q* 22, 209–212.

Noack, J., Dongowski, G., Hartmann, L. and Blaut, M. (2000) The human gut bacteria *Bacteroides thetaiotaomicron* and *Fusobacterium varium* produce putrescine and spermidine in cecum of pectin-fed gnotobiotic rats. *Journal of Nutrition* 130, 1225–1231.

Pacha, J. (2000) development of intestinal transport function in mammals. *Physiological Reviews* 80, 1633–1667.

Perez-Vilar, J. and Hill, R. (1999) The structure and assembly of secreted mucins. *Journal of Biological Chemistry* 274, 31751–31754.

Perozzi, G., Barila, D., Murgia, C., Kelly, D., Begbie, R. and King, T.P. (1993) Expression of differentiated functions in the developing porcine small intestine. *Journal of Nutritional Biochemistry* 4, 699–705.

Pestova, M.I., Clift, R.E., Vickers, R.J., Franklin, M.A. and Mathew, A.G. (2000) Effect of weaning and dietary galactose supplementation on digesta glycoproteins in pigs. *Journal of the Science of Food and Agriculture* 80, 1918–1924.

Pluske, J.R., Thompson, M.J., Atwood, C.S., Bird, P.H., Williams, I.H. and Hartmann, P.E. (1996) Maintenance of villus height and crypt depth, and enhancement of disaccharide digestion and monosaccharide absorption, in piglets fed on cows' whole milk after weaning. *British Journal of Nutrition* 76, 409–422.

Porter, P. (1986) Immune system. In: Leman, A.D. (ed.) *Diseases of Swine*, 6th edn. Iowa State University Press, Ames, Iowa, pp. 44–57.

Puchal, A.A. and Buddington, R.K. (1992) Postnatal development of monosaccharide transport in pig intestine. *American Journal of Physiology* 262, G895–G902.

Rolfe, R.D. (1996) Colonisation resistance. In: Mackie, R.I. (ed.) *Gastrointestinal Microbiology*, Vol. 2, *Gastrointestinal Microbes and Host Interactions*. Chapman & Hall, New York, pp. 501–536.

Roth, J. (1997) Topology of glycosylation in the Golgi apparatus. In: Berger, E.G. and Roth, J. (eds) *The Golgi Apparatus*. Birkhauser Verlag, Basel, Switzerland, pp. 131–161.

Sangild, P.T., Diernaes, L., Christiansen, I.J. and Skadhauge, E. (1993) Intestinal transport of sodium, glucose and immunoglobulins in neonatal pigs. Effect of glucocorticoids. *Experimental Physiology* 78, 485–497.

Sangild, P.T., Holtug, K., Diernaes, L., Schmidt, M. and Skadhauge, E. (1997) Birth and prematurity influence intestinal function in the newborn pig. *Comparative Biochemistry and Physiology A – Physiology* 118, 359–361.

Sangild, P.T., Trahair, J.F., Loftager, M.K. and Fowden, A.L. (1999) Intestinal macromolecule absorption in the fetal pig after infusion of colostrum *in utero*. *Pediatric Research* 45, 595–602.

Satchithanandam, S., Vargofcak-Apker, M., Calvert, R.J., Leeds, A.R. and Cassidy, M.M. (1990) Alteration in gastrointestinal mucin after fiber feeding in rats. *Journal of Nutrition* 120, 1179–1184.

Schauer, R. (1991) Biosynthesis and function of N- and O-substituted sialic acids. *Glycobiology* 1, 449–452.

Schauer, R., Kelm, S., Reuter, G., Roggentin, P. and Shaw, L. (1995) Biochemistry and role of sialic acids. In: Rosenberg, A. (ed.) *Biology of the Sialic Acids*. Plenum Press, New York, pp. 7–67.

Schober, D.A., Simmen, F.A., Hadsell, D.L. and Baumrucker, C.R. (1990) Perinatal expression of type I IGF receptors in porcine small intestine. *Endocrinology* 126, 1125–1132.

Schroff, K.E., Meslin, K. and Cebra, J.J. (1995) Commensal enteric bacteria engender self-limiting humoral immune response while permanently colonizing the gut. *Infection and Immunity* 63, 3904–3913.

Sharma, R., Schumacher, U., Ronaasen, V. and Coates, M. (1995) Rat intestinal mucosal responses to a microbial flora and different diets. *Gut* 36, 209–214.

Shen, W.H. and Xu, R.J. (2000) Stability of insulin-like growth factor I in the gastrointestinal lumen in neonatal pigs. *Journal of Pediatric Gastroenterology and Nutrition* 30, 299–304.

Simmen, F.A., Badinga, L., Green, M.L., Kwak, I., Song, S. and Simmen, R.C. (1998) The porcine insulin-like growth factor system: at the interface of nutrition, growth and reproduction. *Journal of Nutrition* 128, 315S–320S.

Smith, M.W. and Jarvis, L.G. (1977) Villus growth and cell replacement in the small intestine of the neonatal pig. *Experientia* 33, 1587–1588.

Smith, M.W. and Peacock, M.A. (1980) Anomalous replacement of foetal enterocytes in the neonatal pig. *Proceedings of the Royal Society of London, B, Biological Sciences* 206, 411–420.

Stewart, C., Hillman, K., Maxwell, F., Kelly, D. and King, T.P. (1993) Recent advances in probiosis in pigs: observations on the microbiology of the pig gut. In: Garnsworthy, P.C. and Cole, D.J.A. (eds) *Recent Advances in Animal Nutrition.* Nottingham University Press, Nottingham, pp. 197–219.

Stokes, C.R., Vega-Lopez, M.A., Bailey, M., Telemo, E. and Miller, B.G. (1992) Immune development in the gastrointestinal tract of the pig. In: Varley, M.A., Williams, P.E.V. and Lawrence, T.L.J. (eds) *Neonatal Survival and Growth.* BSAP Occasional Publication No. 15, British Society of Animal Production, Edinburgh, pp. 9–12.

Stoll, B., Chang, X., Fan, M.Z., Reeds, P.J. and Burrin, D.G. (2000) Enteral nutrient intake level determines intestinal protein synthesis and accretion rates in neonatal pigs. *Amercian Journal of Physiology* 279, G288–G294.

Strobel, S. and Mowat, A. (1998) Immune responses to dietary antigens: oral tolerance. *Immunology Today* 19, 173–181.

Talham, G.L., Jiang, H.Q., Bos, N.A. and Cebra, J.J. (1999) Segmented filamentous bacteria are potent stimuli of a physiologically normal state of the murine gut mucosal immune system. *Infection and Immunity* 67, 1992–2000.

Thacker, P.A. (1999) Nutritional requirements of early weaned pigs: a review. *Pig News and Information* 20, 13N–24N.

Trahair, J.F. (1993) Is fetal enteral nutrition important for normal gastrointestinal growth? A discussion. *Journal of Parenteral and Enteral Nutrition* 17, 82–85.

Trahair, J.F. and Sangild, P.T. (1997) Systemic and luminal influences on the perinatal development of the gut. *Equine Veterinary Journal Supplement* 24, 40–50.

Trahair, J.F., Wilson, J.M. and Neutra, M.R. (1995) Identification of a marker antigen for the endocytic stage of intestinal development in rat, sheep, and human. *Journal of Pediatric Gastroenterology and Nutrition* 21, 277–287.

Wang, T. and Xu, R.J. (1996) Effects of colostrum feeding on intestinal development in newborn pigs. *Biology of the Neonate* 70, 339–348.

Wester, T.J., Fiorotto, M.L., Klindt, J. and Burrin, D.G. (1998) Feeding colostrum increases circulating insulin-like growth factor I in newborn pigs independent of endogenous growth hormone secretion. *Journal of Animal Science* 76, 3003–3009.

Westrom, B.R., Svendsen, J., Ohlsson, B.G., Tagesson, C. and Karlsson, B.W. (1984) Intestinal transmission of macromolecules (BSA and FITC-labelled dextrans) in the neonatal pig. Influence of age of piglet and molecular weight of markers. *Biology of the Neonate* 46, 20–26.

Westrom, B.R., Ohlsson, B.G., Svendsen, J., Tagesson, C. and Karlsson, B.W. (1985) Intestinal transmission of macromolecules (BSA and FITC-labelled dextrans) in the neonatal pig. Enhancing effect of colostrum, proteins and proteinase. *Biology of the Neonate* 47, 359–366.

Willoughby, R.E. (1993) Rotaviruses preferentially bind O-linked sialylglycoconjugates and sialomucins. *Glycobiology* 3, 437–445.

Wilson, J.M., Whitney, J.A. and Neutra, M.R. (1991) Biogenesis of the apical endosome–lysosome complex during differentiation of absorptive epithelial cells in rat ileum. *Journal of Cell Science* 100 (Part 1), 133–143.

Wilson, M., Seymour, R. and Henderson, B. (1998) Bacterial perturbation of cytokine networks. *Infection and Immunity* 66, 2401–2409.

Wu, G., Flynn, N.E., Knabe, D.A. and Jaeger, L.A. (2000) A cortisol surge mediates the enhanced polyamine synthesis in porcine enterocytes during weaning. *American Journal of Physiology* 279, R554–R559.

Wuttke, N.J., Macardle, P.J. and Zola, H. (1997) Blood group antibodies are made by CD5+ and by CD5– B cells. *Immunology and Cell Biology* 75, 478–483.

Xu, R.J. and Wang, T. (1996) Gastrointestinal absorption of insulinlike growth factor-I in neonatal pigs. *Journal of Pediatric Gastroenterology and Nutrition* 23, 430–437.

Yuyama, Y., Yoshimatsu, K., Ono, E., Saito, M. and Naiki, M. (1993) Postnatal change of pig intestinal ganglioside bound by *Escherichia coli* with K99 fimbriae. *Journal of Biochemistry (Tokyo)* 113, 488–492.

Zhang, H., Malo, C. and Buddington, R.K. (1997) Suckling induces rapid intestinal growth and changes in brush-border digestive functions of newborn pigs. *Journal of Nutrition* 127, 418–426.

Zhang, H.Z., Malo, C., Boyle, C.R. and Buddington, R.K. (1998) Diet influences development of the pig (*Sus scrofa*) intestine during the first 6 hours after birth. *Journal of Nutrition* 128, 1302–1310.

Zijlstra, R.T., McCracken, B.A., Odle, J., Donovan, S.M., Gelberg, H.B., Petschow, B.W., Zuckermann, F.A. and Gaskins, H.R. (1999) Malnutrition modifies pig small intestinal inflammatory responses to rotavirus. *Journal of Nutrition* 129, 838–843.

Enteric Immunity and Gut Health 10

M. Bailey[1], M.A. Vega-Lopez[2], H.-J. Rothkötter[3], K. Haverson[1], P.W. Bland[1], B.G. Miller[1] and C.R. Stokes[1]

[1]*Division of Molecular and Cellular Biology, Department of Clinical Veterinary Sciences, University of Bristol, Langford House, Langford, Bristol BS40 5DU, UK;* [2]*CINVESTAV-IPN, Experimental Pathology Department, Av. IPN 2508, Mexico 07360 DF, Mexico;* [3]*Department of Functional Anatomy, Hannover Medical School, 4120 Carl-Neuberg Strasse, 30625 Hannover, Germany*

Introduction

The concept that control of mucosal immune responses to food antigens contributes to 'enteric health' has been considered within the limited field of mucosal immunology for three decades. Although work in this field contributed to a shift away from early weaning of human infants, early weaning of domestic species is still widely practised. However, in the last 5 years or so it has begun to be appreciated that a stable mucosal immune system is necessary for controlling responses not only to these food antigens but also to commensal environmental bacteria. In addition, exposure to environmental antigens, particularly bacteria, is necessary for the development and maintenance of a stable mucosal immune system.

The interactions between the adaptive immune system, diet and microflora are extremely complex and are the product of millions of years of coevolution of the higher vertebrates and their environments. Once this is appreciated, it becomes apparent that changes in lifestyle are likely to stress the stability of the interactions.

The so-called 'hygiene hypothesis', discussed later, proposes that the increasing incidence of allergies in human infants in industrialized countries is associated with changes in the patterns of colonization by harmless and pathogenic microorganisms. In commercial pig production, exposure to food antigens and microorganisms is also very different from the conditions under which stable mucosal immune systems developed. Given the complex requirements of a mucosal immune system, it is no longer adequate to ask simply, 'Can a neonatal or weaned pig respond actively to antigen?' Instead, it is necessary to examine the functions of the mucosal immune system and the effect of food and bacterial antigens on these functions in order to predict the short- and long-term effects of weaning.

The increasing interest in this area as a determinant of human health means that, in the next decade, dramatic changes can be expected in our understanding of the role of enteric immunity in intestinal and systemic health. There is much greater control over the rate at which neonates of the domestic species become exposed to microbial and food antigens: it should be possible not only to translate results rapidly from human studies to pigs, but also to contribute to the understanding of these interactions in human infants.

Mucosal Tolerance and Active Responses

The immune system is commonly viewed as having evolved as a protective mechanism to allow specific adaptive responses to potential pathogens. While this is undoubtedly true, it has led to the concept that the immune system is normally in a static, 'antigen-free' state, capable of responding to any 'non-self' antigen. However, studies in humans, rodents and pigs have clearly demonstrated that the immune system is continuously exposed to non-self antigens derived from the environment. Small quantities of harmless but antigenic proteins are absorbed from the intestine. While nutritionally insignificant, these proteins are enough to trigger strong and damaging allergic immune responses in the intestine. In pigs, considerable attention has focused on the immune responses to the major storage proteins of the soybean. Two days after weaning 3-week-old piglets on to soya-based diets, it is possible to detect β-conglycinin in their serum (Fig. 10.1). Detection by radioimmunoassay demonstrates that this retains structural B-lymphocyte epitopes and dialysis studies suggest that it is biochemically intact (Wilson *et al.*, 1989). The level of absorbed protein in serum (around 80 ng ml^{-1}) represents a significant antigenic load. In the case of soya feeding, this is associated with rapid serum IgG antibody responses to the soya protein (Fig. 10.2) (Wilson *et al.*, 1989; Dreau *et al.*, 1994). The magnitude of the response is comparable to that produced in a piglet of the same age after parenteral injection of soya with adjuvant, but the two responses appear to be controlled differently, since neonatal exposure to antigen depresses the response when the antigen is fed, but not when it is injected (Bailey *et al.*, 1994b).

The observation that fed proteins can be absorbed from the intestine in a form capable of triggering an immune response raises questions as to the normal function of the immune system. If the immune system simply responds to 'non-self', then it should be continuously reacting to fed proteins. Such continuous responses would have two undesirable consequences. Firstly, committing biological resources to the immune system may compromise growth and development. Secondly, expression of active responses in the intestine is associated with functional impairment (reduced ability to digest and absorb nutrients). This has been formally demonstrated in experimental rodent models, and diarrhoea in 3-week-old piglets weaned on to soya has long been associated with the observed transient hypersensitivity to soya proteins. Firm evidence for the association between gut damage and soya hypersensitivity is difficult in a multifactorial system such as weaning, but the two occur concurrently and mathematical models have shown that measures of immune responses to soya proteins are strong predictors of average daily liveweight gain after weaning (Li *et al.*, 1991). Interestingly, in these mathematical models, the strength of the cellular immune response, measured by skin thickness after challenge, was a strong negative predictor of weight gain, while antibody levels were positively related. This observation is consistent with rodent and human data, where cellular (or T-helper type 1) immune responses are primarily involved in mucosal inflammation. Further circumstantial evidence comes from the observation that diets based on egg proteins do not trigger active immune responses and are less likely to cause post-weaning diarrhoea (Telemo *et al.*, 1991).

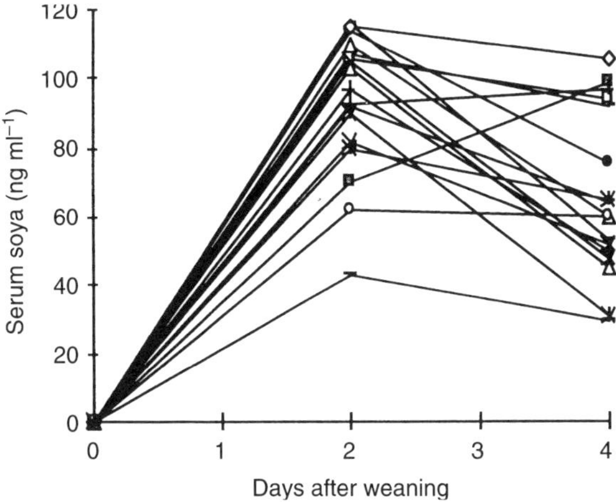

Fig. 10.1. Serum β-conglycinin detected by radioimmunoassay in piglets weaned on to soya-based diets at 3 weeks old. Each line represents results from a single piglet.

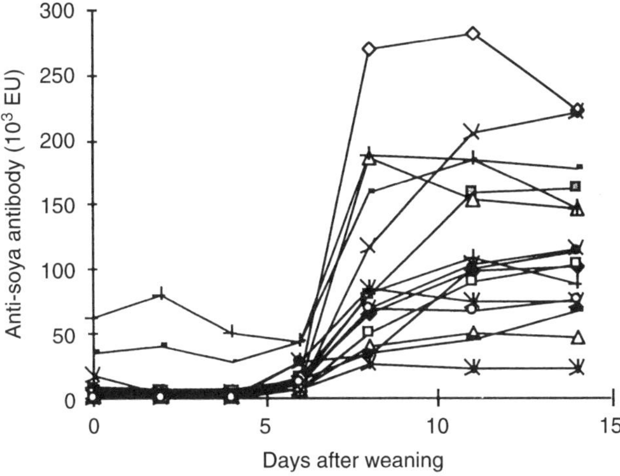

Fig. 10.2. Serum IgG antibody to crude soya antigens (mixed glycinin and β-conglycinin) measured by ELISA in piglets weaned on to soya-based diets at 3 weeks old. Each line represents results from a single piglet.

Normally, these responses are actively prevented by the phenomenon of 'oral tolerance'. In rodents, tolerance to fed proteins is easily demonstrated by the failure to mount antibody or cellular responses to injected antigen in animals previously exposed to the antigen by feeding. Tolerance to fed proteins has now been demonstrated in a wide range of animal species, including humans. Tolerance to soya proteins has been more difficult to demonstrate in piglets around the age of weaning, due to the initial primary response to the fed protein. However, comparison of secondary responses to injected soya in piglets that are either soya-naïve, or weaned on to soya (primary response to fed soya) or injected with soya (primary response to injected soya) has shown that weaning is followed by tolerance, rather than priming (Fig. 10.3) (Bailey *et al.*, 1993).

Since allergic responses in humans could be identified relatively easily, early studies on tolerance induction focused on responses to fed protein. However, it is now apparent that tolerance is also commonly generated to antigens derived from commensal luminal bacterial, and that breakdown of tolerance may be the basis of inflammatory bowel diseases in humans. T cells from normal mice failed to make responses *in vitro* to antigens from their own bacterial flora, but did make responses to antigens from flora of other mice (Duchmann *et al.*, 1996b). A similar phenomenon has been demonstrated for humans but, importantly, this tolerance is lost in individuals with inflammatory bowel disease (Duchmann *et al.*, 1995). T cells with specificity for bacterial antigens are present in increased frequency in the intestines of colitic mice, and cloned bacteria-specific T cells from colitic mice can directly transfer disease to recipient mice (Duchmann *et al.*, 1996a). Taken together, and with the results from reconstituted immunodeficient mice discussed later, it is apparent that antigens of commensal bacteria can be targets for mucosal immune responses, and that such responses can be associated with disease.

Given the continuous level of exposure to antigens associated with food and commensal bacteria, the immune system probably generates tolerance, rather than active responses, to the majority of 'non-self' antigens. That is, tolerance, rather than active immune responses, is the default response to 'non-self' antigens. However, the mucosa also retains the ability to respond actively to potential pathogens. In order to balance these effector and regulatory functions, the immune system must be continuously making value judgements about the antigens to which it is exposed and making the appropriate immune response – are they 'harmless' or 'dangerous'? The correct decisions will allow rapid control of pathogens when necessary without unduly compromising intestinal function. The wrong decisions will result in loss of intestinal function as a result of inadequate control of pathogens or allergic responses to food. Gut health can be defined in immunological terms, therefore, as the ability to discriminate between harmful and harmless antigens efficiently and accurately and to express appropriate responses to each.

Development of the Mucosal Immune System

In the mature pig, there is a high degree of immunological organization both within the Peyer's patches and outside in the villi and crypts. The architecture of the Peyer's patches is to some extent analogous to that of the lymph nodes and spleen and includes antigen-sampling dome regions, containing dendritic cells (DC), overlying follicles flanked by T cell areas. The organized tissues of the Peyer's patches and mesenteric lymph nodes are the site at which primary contact with novel antigens takes place. Primed T and B lymphocytes spread from these tissues back to the intestinal lamina propria to express function. The intestinal mucosa also has a major immunological component, which also shows a striking level of organization (Vega-Lopez *et al.*, 1993). There is clear spatial separation of T cells in the villi and plasma cells around the crypts: within the villi, $CD8^+$ (cytotoxic) T cells are in and immediately underlying the epithelium, while $CD4^+$ (helper) T cells lie deeper in the lamina propria. Expression of major histocompatibility complex (MHC) class II antigens (necessary for recognition of antigen by T cells) occurs on stromal cells (predominantly endothelial cells) and on dendritic cells in the villi (Wilson *et al.*, 1996; Haverson *et al.*, 2000). Comparable expression of MHC class II (on stromal and dendritic cells) is seen in human and rodent intestine but the degree of spatial organization of T cells in the pig intestine is unusually high.

Until recently, it had been assumed that the immunological mechanisms present in the intestinal mucosa were primarily directed at expression of active immune responses to pathogens. Studies of the phenotype of mucosal T cells initially supported this view (James and Zeitz, 1994). All leucocytes express on their surface a

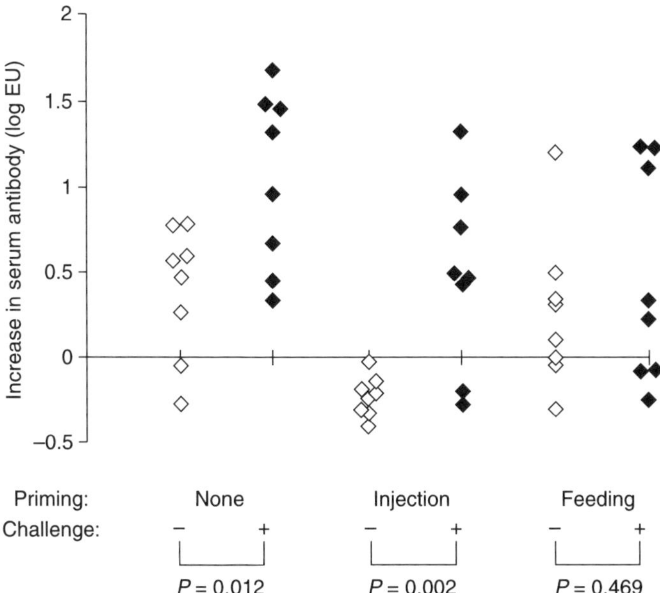

Fig. 10.3. Induction of tolerance in weaned piglets (modified from Bailey *et al.*, 1993). Two groups of piglets were weaned on to either soya-free fishmeal diets at weaning (soya-naïve), soya-free fishmeal diets but injected with soya protein in adjuvant (primed by injection), or soya-based diets (primed by feeding) at 3 weeks old. One of each pair of groups was injected with soya protein in adjuvant at 9 weeks old. The increase in antibody between 9 and 11 weeks old was determined by ELISA.

large, heavily glycosylated molecule termed CD45 (or leucocyte common antigen). In T lymphocytes, transcripts encoding the CD45 molecule are variably spliced, resulting in expression of truncated forms of the molecule. Monoclonal antibodies directed against these variably expressed regions, termed CD45RA, CD45RB and CD45RC, discriminate subsets of T cells. Antigen-inexperienced T cells express all three epitopes, while exposure to antigen results in progressive loss of these high molecular weight forms (Akbar *et al.*, 1988; Plebanski *et al.*, 1992). Reversion from CD45RABC$^-$ to CD45RABC$^+$ can also occur at low levels in the absence of antigen. Studies on human, rodent and pig lamina propria T cells have shown that they are primarily CD45RABC$^-$, consistent with their being recently activated memory or effector cells (Halstensen *et al.*, 1990; Bailey *et al.*, 1998).

Accumulating evidence suggests that the mucosa may be a heavily regulated and regulating environment. All memory/effector T lymphocytes become increasingly susceptible to programmed cell death (apoptosis) and increasingly dependent on their microenvironment for survival, making them extremely good targets for regulation (Salmon *et al.*, 1994). This is particularly true of lamina propria T cells, in which activation results in secretion of an unusual range of regulatory cytokines and is followed by apoptosis (Bailey *et al.*, 1994a, 1998; Boirivant *et al.*, 1996). The T cell types in the intestinal mucosa are, therefore, potential effector, regulatory and regulated cells, although whether these are functions of discrete cell types or of the same cell type under different microenvironmental conditions is unclear. This combined immunological architecture – the Peyer's patches and intestinal mucosa – maintains a balance between effector and regulatory functions and is well organized to make appropriate decisions regarding the potential danger associated with each environmental antigen. In this model of the function of the mucosal immune system, the large number of T cells present in the intestinal mucosa represents a heavy commitment to the maintenance of tolerance to harmless food and commensal bacterial antigens, rather than to active responses to pathogens.

An efficient, decision-making mucosal immune system is a particular requirement in neonatal piglets. After birth and after weaning, the young piglet is exposed to a huge array of novel antigens associated with food and with commensal, opportunist and pathogenic microorganisms. These pigs must make the correct decisions about all of these antigens and mount the appropriate responses. Failure to make the correct decision will compromise intestinal function, as a result either of failure to mount active responses to pathogens, or of expression of damaging immune responses to harmless food antigens.

Studies in a number of laboratories have shown that, in addition to the absence of endogenous antibody, piglets have essentially no active mucosal immune system at birth. Importantly, this system develops after birth in a series of well-defined phases (Bianchi *et al.*, 1992; Vega-lopez *et al.*, 1995; Pabst and Rothkötter, 1999). Essentially four stages can be distinguished by immunohistology.

1. The newborn piglet has very few lymphocytes in its intestinal epithelium or lamina propria. Clusters of lymphocytes are present in the mucosa, in the areas that will subsequently develop into Peyer's patches, but these clusters have no clear immunological structure comparable with the follicles and T cell areas of mature animals.

2. In the first 2 weeks of life, the intestine rapidly becomes colonized with lymphoid cells. These cells express the CD2 surface marker but do not coexpress CD4 or CD8. Although initially assumed to be an unusual T cell subset, their identity is now in question since naïve (IgM^+) pigs can express surface CD2 at low intensity (Sinkora *et al.*, 1998). The Peyer's patches begin to organize during this time, reaching a relatively normal architecture at 10–15 days.
3. In piglets 2–4 weeks old, the intestinal mucosa becomes colonized by $CD4^+$ T cells, primarily in the lamina propria. $CD8^+$ cells are still largely absent (Fig. 10.4). Small numbers of B cells appear, preferentially expressing IgM.
4. From the age of 5 weeks onwards, $CD8^+$ cells begin to appear in the intestinal epithelium and around the epithelial basement membrane. In the crypt areas many IgA^+ B cells are appearing. By 7 weeks the architecture of the intestine is comparable to that of a mature animal.

Studies using germ-free animals have clearly demonstrated the importance of an intestinal bacterial flora on this phased development of the mucosal immune system. In germ-free piglets, the early appearance of $CD2^+$ cells has been observed, but the later appearance of conventional $CD4^+$ and $CD8^+$ T cells does not occur (Rothkötter *et al.*, 1991). Similarly, the Peyer's patches are smaller in germ-free piglets than in conventional or SPF animals (Pabst *et al.*, 1988; Barman *et al.*, 1997) and contain primarily IgM^+, rather than IgA^+ B cells. The requirement for microbial stimulation for expansion of the B lymphocyte system is consistent with recent data on generation of the antibody repertoire in germ-free piglets compared with gnotobiotic animals colonized with a benign exclusion flora. The ability of immunoglobulin molecules to bind a wide range of antigens occurs as a result of genetic rearrangements in which one of a large number of V gene-segments are linked to one of a similarly large number of D and J gene-segments to produce a large number of possible V-D-J regions encoding the antigen-binding site. However, unlike rodents and humans, pigs appear to use only a single J gene-segment, providing an initial restriction on diversity. Further, immunoglobulin gene rearrangement in fetal and postnatal germ-free animals uses almost exclusively four V gene-segments and two D gene-segments and therefore produces only limited diversity of the immunoglobulin heavy-chain variable region. This means that the piglet can recognize only a limited range of antigenic epitopes (Sun *et al.*, 1998). In colonized piglets, mucosal immunoglobulin transcripts used a wider range of V and D gene-segments and contained higher levels of point mutations in regions coding for antigen-binding sites, suggesting the ability to recognize a much greater range of antigenic epitopes (Butler *et al.*, 2000). Importantly, this effect was much more apparent in mucosal than in systemic tissues and was present both in IgM and IgA transcripts. This observation demonstrates that the increasing presence of IgM^+ B cells in the intestines during the first 5 weeks represents a specific antigen- or mitogen-driven expansion before switching to IgA production.

The requirement for microbial flora for development of a mucosal immune system can be interpreted in two ways. Firstly, it may represent accumulation of effector cells in response to challenge with true or facultative pathogens, independent

(a)

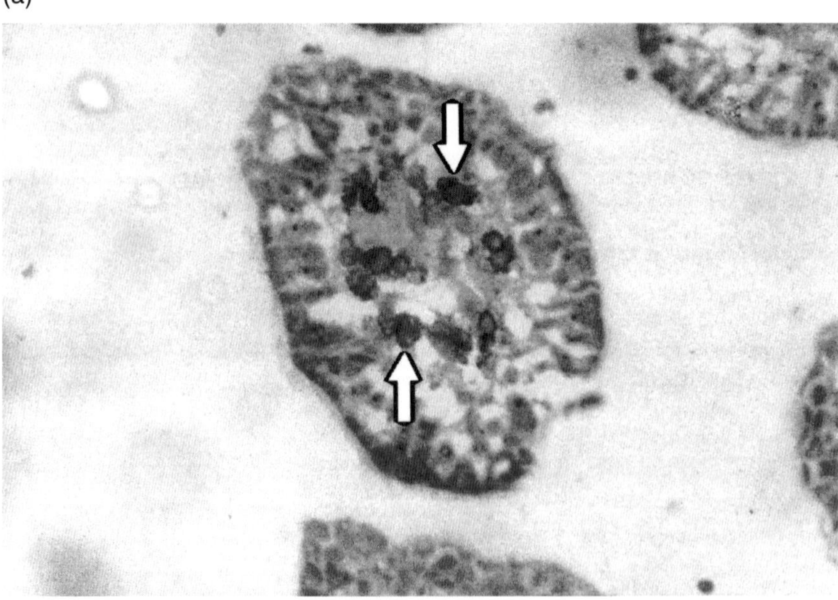

(b)

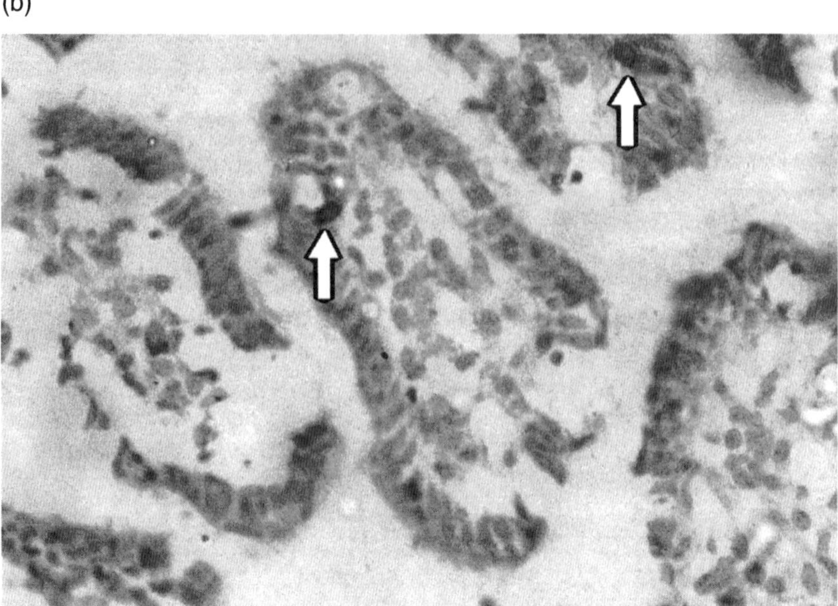

Fig. 10.4. Subsets of T cells in intestinal villi of 4-week-old unweaned piglets. Samples of intestine were snap-frozen in liquid nitrogen, sectioned and stained with (a) anti-CD4 or (b) anti-CD8. $CD4^+$ T cells are evident within the intestinal lamina propria, while $CD8^+$ T cells are still scarce in the epithelium.

of regulatory mechanisms. While possible, this ignores the clear need to minimize expression of active immune responses in the intestine and the immunoregulatory role of the mucosa. Secondly, and more likely, it may represent gradual development of a mucosal immune system capable of expression of effector and regulatory immune responses. Although the mucosal immunological architecture passes through defined developmental phases, successive waves of T cell types do appear to maintain a balance between regulatory and effector function. At 3 weeks old, the small number of T cells in the intestinal lamina propria already have the unusual cytokine profile seen in mature pigs (Bailey *et al.*, 1994a), although mucosal immunization at this age can be effective (Bertschinger *et al.*, 2000). A model for this development is proposed in Fig. 10.5a, in which regulatory and effector function develop together after birth, and intestinal integrity depends on the balance between these two functions being maintained within certain acceptable limits. Mucosal damage will result from excessive regulation or excessive effector function. This model allows predictions of the impact of immunological manipulation on neonatal piglets.

The Impact of Weaning

The period of weaning has been associated with intestinal and immunological alterations in young piglets. Some villus shortening can be apparent within 24 h of weaning but the most severe morphological changes in the intestine do not occur until 4–5 days after weaning (Hampson, 1986). At this stage the numbers of $CD2^+$ leucocytes in the intestine is significantly increased (Dreau *et al.*, 1995; Vega-Lopez *et al.*, 1995). In addition, the rate of crypt cell production falls transiently 1–2 days after weaning, before subsequently increasing (B.G. Miller, unpublished). Shortly after weaning, piglets showed reduced ability to react to the lymphocyte mitogen

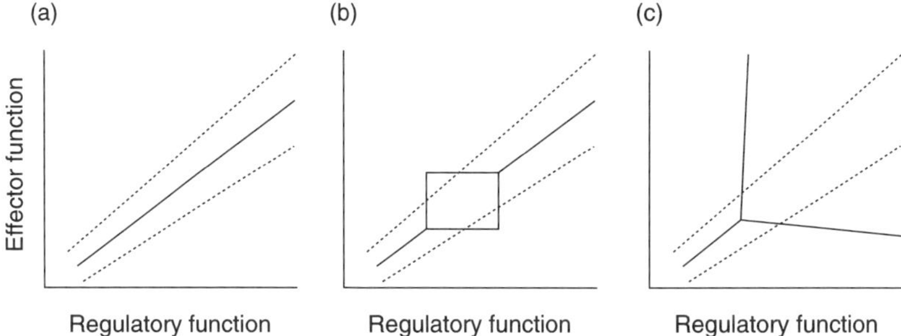

Fig. 10.5. Models for the development of the mucosal immune system in pigs. (a) Normal development (solid line) is within the 'safe' limits of a balance between effector and regulatory function indicated by the dotted line. (b) Weaning can push the balance in either direction outside the safe limits, resulting in transient loss of gastrointestinal function. (c) Under some undefined conditions (genetic, environmental) balance is not restored, resulting in chronic inability to regulate appropriate responses to pathogens or to food and commensal bacterial antigens.

phytohaemagglutinin either given intradermally or when isolated lymphocytes were cultured *in vitro* (Blecha *et al.*, 1983). Production of specific antibody was also reduced when sheep red blood cells were given to piglets 1 day after weaning compared with 2 weeks after weaning (Blecha and Kelley, 1981). This may be associated with the reduced ability of systemic T cells to secrete interleukin-2 (IL-2), a major T cell growth factor, despite expressing receptors for IL-2 (Bailey *et al.*, 1992; Wattrang *et al.*, 1998). Our recent studies (unpublished) suggest that similar disturbances occur in T cells from the intestinal lamina propria of weaned piglets. Production of IL-2 by activated spleen cells was depressed in mice undergoing graft-versus-host reactions (Levy *et al.*, 1990), and shortening of intestinal villi and increased crypt cell production rate in such mice appear to be a consequence of local immune responses (Mowat and Ferguson, 1982).

Many factors active at weaning are likely to affect the ability of piglets to mount effector immune responses or tolerance to environmental antigens. The interaction between serum antibody to soya proteins and growth has been discussed. Cold- or draught-stressors have been reported to influence T cell responses to non-specific mitogens (Blecha and Kelley, 1981; Scheepens *et al.*, 1994). Similarly, mixing and housing changes have been reported to affect parameters of immune responses (Hicks *et al.*, 1998; Kelly *et al.*, 2000). Restraining 8-week-old piglets resulted in increased serum levels of cortisol and decreased responses to intradermal phytohaemagglutinin similar to the effects of weaning (Westly and Kelley, 1984). IL-2 production, but not receptor expression, was reduced when T cells were activated in the presence of dexamethasone (Larsson, 1981; Reem and Yeh, 1984). The reduced IL-2 production in weaned animals may therefore be associated with expression of active immune responses in the intestine or consequent upon stress-induced release of endogenous steroids or, similarly, of histamine (Rezai *et al.*, 1990).

Caution must be used in attributing too many effects of weaning to cortisol, since some studies have not shown weaning-associated changes in cortisol levels (Carroll *et al.*, 1998). Other factors, such as transient anorexia, may also contribute to mucosal inflammation (McCracken *et al.*, 1999). Regardless of the cause, it is apparent that there are disturbances in the development of the mucosal immune system associated with weaning. This will affect the crucial balance between regulatory and effector function and the ability of the mucosal immune system to discriminate between harmless and harmful antigens.

The effect of weaning on the young piglet is clearly multifactorial, involving withdrawal of maternal antibodies, mixing stress, reduction in gut contents, change in intestinal microflora and inappropriate responses to environmental antigens. However, weaning on to soya protein in the early neonatal period, in the absence of a mucosal immune system, does not have a severe impact on gut morphology (McCracken *et al.*, 1998). This and studies in rodents suggest that the final morphological and functional alterations are a result of expression of immune mechanisms. The model that we propose predicts that this occurs because of disruption of the balance between effector and regulatory function, resulting either in failure to control pathogens (excess regulatory function) or in allergic reactions to diet (excess effector function) (Fig. 10.5b). Under normal circumstances, the balance between regulatory and effector function could be expected to re-establish, with only point-of-weaning growth disturbance. Studies in pigs (discussed) and humans (see later) now suggest that the developing gut bacterial flora in the neonate is critical in establishing normal immune function in later life. It is likely, therefore, that disruption of the flora and the immune system at the point of weaning may have long-lasting effects on mucosal function. That is, the development of excessive regulatory or effector function as a result of weaning insults leaves a long-term inability to mount appropriate responses to mucosal antigens (Fig. 10.5c). The emerging phenomenon of 'non-specific colitis' in growing pigs may reflect a chronic inability to make appropriate responses to mucosal antigen, and may well be a consequence of weaning practice.

Models of Enteric Health in the Pig

The problem for the neonatal piglet, as described here, is that it must populate its intestine with 'memory' cells while maintaining a balance between regulatory and effector cells. At birth essentially all its recirculating cells are antigen-naïve and it must therefore control the transition from naïve to memory cell status to maintain this effector/regulatory balance. This requirement for the control of T cell differentiation during population of the intestinal mucosa is also seen in a well-defined model of colitis in immunodeficient mice that may be relevant to the developing piglet (Groux and Powrie, 1999). Colitis occurs in a number of strains of mice in which specific immunologically relevant genes have been deleted. However, these models are difficult to interpret, since this may reflect failure to control pathogens effectively or inappropriate responses to commensals. More relevant are the so-called *scid* (severe combined immunodeficient) mice. Defects in these mice result in an absence of both T and B lymphocytes. Despite the complete lack of any specific adaptive component of the immune system, these mice cope perfectly well in clean SPF environments. Under these conditions, they have relatively normal colonic microflora and the implication is that a specific adaptive immune system is not necessary for control of true commensal microorganisms. In this respect, they are comparable to the neonatal pig, which also lacks a mucosal immune system at birth. The difference is that the neonatal piglet must rapidly develop a system capable of dealing with true or facultative pathogens in its less controlled environment. Of importance, therefore, is the effect of re-constituting the *scid* mice with low levels of a normal, functional immune system to allow slow mucosal colonization in a manner comparable to that in the neonatal piglet. Transfer of all the components of the immune system results in the development of a healthy, functional immune system and the maintenance of 'enteric health'. However, reconstitution with antigen-naïve T cells, in the absence of antigen-experienced regulatory T cells, results in the appearance of colitis after 6–12 weeks, characterized by infiltration of the colon with effector T lymphocytes and macrophages. This colitis does not occur when the mice are concurrently repopulated with regulatory cells (Powrie *et al.*, 1993), or kept germ-free or treated with gut-acting antibiotics, demonstrating a dependence on the intestinal microflora. Importantly, this is the same microflora with which unreconstituted mice cope perfectly well in the absence of an immune system.

The implications of this model are that specific adaptive immune responses to commensal bacteria are not necessary and, when unregulated, can be extremely damaging. Elegant studies (Groux *et al.*, 1997) have shown that regulatory T cells recognizing food antigens can prevent the development of colitis, providing support for the idea of development requiring a balance between effector and regulatory function. These studies also demonstrated that regulation could be essentially antigen-independent, in that regulatory T cells with specificity for ovalbumin could prevent colitis caused by immune responses to bacterial antigens.

Although used primarily as a model for inflammatory bowel disease in humans, this system may also be a valuable model for the gradual development of mucosal immune function in the neonatal pig and the effect of failure to regulate immune responses. In particular, it may provide insights into the aetiology and pathogenesis of non-specific colitis in growing pigs.

Further stimulus to research interactions between the immune system, intestinal microflora and food antigens can be expected from human medicine, as a result of the concern over the rapidly increasing incidence of childhood allergies in Western populations. While initial epidemiological studies implicated antigenic loads and environmental pollutants, many of these (with the exception of passive smoking) have been discounted (von Mutius, 2000; Pearce *et al.*, 2000). The establishment and later removal of the 'iron curtain' in Germany meant that genetically comparable populations lived under different lifestyles for one generation and were then reunited. These large natural 'experiments' in humans have been analysed in the last 10 years and clearly demonstrate a strong effect of lifestyle on the frequency of respiratory allergy in children (von Mutius *et al.*, 1994). Frequency of allergy is not higher in areas of high traffic or industrial pollution: lower frequencies in rural areas appear to be associated with higher levels of association with animals (Braun-Fahrlander *et al.*, 1999). More surprisingly, the detection of antibody to two or more enteric pathogens (*Toxoplasma*, *Helicobacter*, hepatitis A virus) was associated with lower risk of allergy (Matricardi *et al.*, 2000). Studies of this nature have led to the 'hygiene hypothesis', in which controlled exposure to microbial antigens is proposed to be necessary for development of immune system capable of discriminating between 'harmless' and 'dangerous' antigens (Wold, 1998). This proposal has clear parallels and implications for the developing piglet. The mucosal immune system of the newborn piglet is much less developed than the newborn human. In addition, it must develop to an adult capability in a much shorter time. The requirement for microbial exposure for development of the piglet mucosal immune system has been discussed. The conditions under which human infants and piglets are kept in the developing world differs markedly from the conditions under which interactions between the immune system, microbial flora and food evolved. It would not be surprising, therefore, if the allergic diseases seen in humans were to occur in pigs. In humans, such allergies develop over a number of years, starting from birth, and early life events in piglets may also be expected to influence subsequent immune responses to food, commensal microorganisms and pathogens.

Conclusions

The contribution of the immune system to gut health is the ability to mount appropriate immune responses to harmful and harmless antigens, rather than the simple ability to mount active responses to non-self antigens. The mechanisms involved in making the appropriate decisions are not entirely clear, but lie in the organized mucosal lymphoid tissue (Peyer's patches and mesenteric lymph nodes) and the diffuse immunological architecture of the intestinal villi. These structures are poorly developed in the neonatal piglet and do not fully develop until 5–7 weeks old, implying that the ability of the neonatal and weaned piglet to make these immunological decisions may be poor and requires exposure to microbial flora for its development. At weaning, a range of factors can affect immune responses, further compromising the ability to make appropriate responses to antigens associated with food, commensal and pathogenic microorganisms. Inappropriate responses will contribute to impairment of gut digestive and absorptive function at the time of weaning, and such environmentally triggered errors in immune development continue to have long-term effects. Given the increasing evidence for immunological immaturity at 3 weeks old, it seems wise to delay weaning of piglets until 5 weeks old.

References

Akbar, A.N., Terry, L., Timms, A., Beverley, P.C.L. and Janossy, G. (1988) Loss of CD45R and gain of UCHL1 reactivity is a feature of primed T cells. *Journal of Immunology* 140, 2171–2178.

Bailey, M., Clarke, C.J., Wilson, A.D. and Stokes, C.R. (1992) Depressed potential for interleukin-2 production following early weaning of piglets. *Veterinary Immunology and Immunopathology* 34, 197–207.

Bailey, M., Miller, B.G., Telemo, E., Stokes, C.R. and Bourne, F.J. (1993) Specific immunological unresponsiveness following active primary responses to proteins in the weaning diet of pigs. *International Archives of Allergy and Immunology* 101, 266–271.

Bailey, M., Hall, L., Bland, P.W. and Stokes, C.R. (1994a) Production of cytokines by lymphocytes from spleen, mesenteric lymph node and intestinal lamina propria of pigs. *Immunology* 82, 577–583.

Bailey, M., Miller, B.G., Telemo, E., Stokes, C.R. and Bourne, F.J. (1994b) Altered immune response to proteins fed after neonatal exposure of piglets to the antigen. *International Archives of Allergy and Applied Immunology* 103, 183–187.

Bailey, M., Plunkett, F., Clarke, A., Sturgess, D., Haverson, K. and Stokes, C. (1998) Activation of T cells from the intestinal lamina propria of the pig. *Scandinavian Journal of Immunology* 48, 177–182.

Barman, N.N., Bianchi, A.T.J., Zwart, R.J., Pabst, R. and Rothkötter, H.J. (1997) Jejunal and ileal Peyers patches in pigs differ in their postnatal development. *Anatomy and Embryology* 195, 41–50.

Bertschinger, H.U., Nief, V. and Tschape, H. (2000) Active oral immunisation of suckling piglets to prevent colonisation after weaning by enterotoxigenic *Escherichia coli* with fimbriae F18. *Veterinary Microbiology* 71, 255–267.

Bianchi, A.T., Zwart, R.J., Jeurissen, S.H. and Moonen-Leusen, H.W. (1992) Development of the B- and T-cell compartments in porcine lymphoid organs from birth to adult life: an immunohistological approach. *Veterinary Immunology and Immunopathology* 33, 201–221.

Blecha, F. and Kelley, K.W. (1981) Effects of cold and weaning stressors on the antibody-mediated immune response of pigs. *Journal of Animal Science* 53, 439–447.

Blecha, F., Pollman, D.S. and Nicholls, D.A. (1983) Weaning pigs at an early age decreases cellular immunity. *Journal of Animal Science* 56, 396–400.

Boirivant, M., Pica, R., Demaria, R., Testi, R., Pallone, F. and Strober, W. (1996) Stimulated human lamina propria T-cells manifest enhanced FAS-mediated apoptosis. *Journal of Clinical Investigation* 98, 2616–2622.

Braun-Fahrlander, C., Gassner, M., Grize, L., Neu, U., Sennhauser, F.H., Varonier, H.S., Vuille, J.C. and Wuthrich, B. (1999) Prevalence of hay fever in allergic sensitization in farmers children and their peers living in the same rural community. *Clinical and Experimental Allergy* 29, 28–34.

Butler, J.E., Sun, J., Weber, P., Navarro, P. and Francis, D. (2000) Antibody repertoire development in fetal and neonatal piglets. III. Colonisation of the gastrointestinal tract selectively diversifies the preimmune repertoire in mucosal lymphoid tissues. *Immunology* 100, 119–130.

Carroll, J.A., Veum, T.L. and Matteri, R.L. (1998) Endocrine responses to weaning and changes in post-weaning diet in the young pig. *Domestic Animal Endocrinology* 15, 183–194.

Dreau, D., Lalles, J.P., Philouzerome, V., Toullec, R. and Salmon, H. (1994) Local and systemic immune responses to soybean protein ingestion in early-weaned pigs. *Journal of Animal Science* 72, 2090–2098.

Dreau, D., Lalles, J.P., Toullec, R. and Salmon, H. (1995) B-lymphocyte and T-lymphocyte are enhanced in the gut of piglets fed heat-treated soybean proteins. *Veterinary Immunology and Immunopathology* 47, 69–79.

Duchmann, R., Kaiser, I., Hermann, E., Mayet, W., Ewe, K. and zum Buschenfelde, K.-H.M. (1995) Tolerance exists towards resident intestinal flora but is broken in active inflammatory bowel disease (IBD). *Clinical and Experimental Immunology* 102, 448–455.

Duchmann, R., Hermann, R.M. and zum Buschenfelde, K.H.M. (1996a) Bacteria-specific T cell clones are selective in their reactivity towards different enterobacteria or *H. pylori* and increased in inflammatory disease. *Scandinavian Journal of Immunology* 44, 71–79.

Duchmann, R., Schmitt, E., Knolle, P., zum Buschenfelde, K.H.M. and Neurath, M. (1996b) Tolerance towards resident intestinal flora in mice is abrogated in experimental colitis and restored by treatment with interleukin-10 or antibodies to interleukin-12. *European Journal of Immunology* 26, 934–938.

Groux, H. and Powrie, F. (1999) Regulatory T cells and inflammatory bowel disease. *Immunology Today* 20, 442–446.

Groux, H., O'Garra, A., Bigler, M., Rouleau, M., Antonenko, S., deVries, J.E. and Roncarolo, M.G. (1997) A $CD4^+$ T-cell subset inhibits antigen-specific T cell responses and prevents colitis. *Nature* 389, 737–742.

Halstensen, T.S., Scott, H. and Brandtzaeg, P. (1990) Human $CD8^+$ intraepithelial T lymphocytes are mainly $CD45RA^-RB^+$ and show increased co-expression of CD45R0 in celiac disease. *European Journal of Immunology* 20, 1825–1830.

Hampson, D.J. (1986) Alterations in piglet small intestinal structure at weaning. *Research in Veterinary Science* 40, 32–40.

Haverson, K., Singha, S., Stokes, C.R. and Bailey, M. (2000) Professional and non-professional antigen-presenting cells in the porcine small intestine. *Immunology* 101, 1–13.

Hicks, T.A., McGlone, J.J., Whisnant, C.S., Kattesh, H.G. and Norman, R.L. (1998) Behavioural, endocrine, immune and performance measures for pigs exposed to acute stress. *Journal of Animal Science* 76, 474–483.

James, S.P. and Zeitz, M. (1994) Human gastrointestinal mucosal T cells. In: Ogra, L., Mestecky, J., Lamm, M.E., Strober, W., McGhee, J.R. and Bienenstock, J. (eds) *Handbook of Mucosal Immunology*. Academic Press, San Diego, California, pp. 275–285.

Kelly, H.R.C., Bruce, J.M., Edwards, S.A., English, P.R. and Fowler, V.R. (2000) Limb injuries, immune response and growth performance of early-weaned pigs in different housing systems. *Animal Science* 70, 73–83.

Larsson, E.L. (1981) Mechanisms of T cell activation II. Antigen- and lectin-dependent acquisition of responsiveness to TCGF is a non-mitogenic active response of resting T cells. *Journal of Immunology* 126, 1323.

Levy, R.B., Jones, M. and Cray, C. (1990) Isolated peripheral T cells from GvHR recipients exhibit defective IL-2R expression, IL-2 production and proliferation in response to activation stimuli. *Journal of Immunology* 145, 3998–4005.

Li, D.F., Nelssen, J.L., Reddy, P.G., Blecha, F., Klemm, R. and Goodband, R.D. (1991) Interrelationship between hypersensitivity to soybean proteins and growth performance in early-weaned pigs. *Journal of Animal Science* 69, 4062–4069.

Matricardi, P.M., Rosmini, F., Riondino, S., Fortini, M., Ferrigno, L., Rapicetta, M. and Bonini, S. (2000) Exposure to foodborne and orofecal microbes versus airborne viruses in relation to atopy and allergic asthma: an epidemiological study. *British Medical Journal* 320, 412–417.

McCracken, B.A., Zilstra, R.T., Donovan, S.M., Odle, J., Lien, E.L. and Gaskins, H.R. (1998) Neither intact nor hydrolysed soy proteins elicit intestinal inflammation in neonatal pigs. *Journal of Parenteral and Enteral Nutrition* 22, 91–97.

McCracken, B.A., Spurlock, M.E., Roos, M.A., Zuckermann, F.A. and Gaskins, H.R. (1999) Weaning anorexia may contribute to local inflammation in the piglet small intestine. *Journal of Nutrition* 129, 613–619.

Mowat, A.M. and Ferguson, A. (1982) Intraepithelial count and crypt hyperplasia measure the mucosal component of the graft-versus-host reaction in mouse small intestine. *Gastroenterology* 83, 417–423.

von Mutius, E. (2000) The environmental predictors of allergic disease. *Journal of Allergy and Clinical Immunology* 105, 9–19.

von Mutius, E., Martinez, F.D., Fritzsch, C., Nicolai, T., Roell, G. and Thiemann, H.H. (1994) Prevalence of asthma and atopy in two areas of West and East Germany. *American Journal of Respiratory and Critical Care Medicine* 149, 358–364.

Pabst, R. and Rothkötter, H.-J. (1999) Postnatal development of lymphocyte subsets in different compartments of the small intestine of pigs. *Veterinary Immunology and Immunopathology* 72, 167–173.

Pabst, R., Geist, M., Rothkötter, H.J. and Fritz, F.J. (1988) Postnatal development and lymphocyte production of jejunal and ileal Peyers patches in normal and gnotobiotic pigs. *Immunopathology* 64, 539–544.

Pearce, N., Douwes, J. and Beasley, R. (2000) Is allergen exposure the major primary cause of asthma? *Thorax* 55, 424–431.

Plebanski, M., Saunders, M., Burtles, S.S., Crowe, S. and Hooper, D.C. (1992) Primary and secondary human *in vitro* T-cell responses to soluble antigens are mediated by subsets bearing different CD45 isoforms. *Immunology* 75, 86–91.

Powrie, F., Leach, M.W., Mauze, S., Caddle, L.B. and Coffman, R.L. (1993) Phenotypically distinct subsets of $CD4^+$ T-cells induce or protect from chronic intestinal inflammation in C.B-17 scid mice. *International Immunology* 5, 1461–1471.

Reem, G.H. and Yeh, N.-H. (1984) Interleukin 2 regulates expression of its receptor and synthesis of gamma interferon by human T lymphocytes. *Science* 225, 429–430.

Rezai, A.R., Salazar-Gonzalez, J.F., Martinez-Maza, O., Afrasiabi, R. and Kermani-Arab, V. (1990) Histamine blocks interleukin 2 (IL-2) gene expression and regulates IL-2 receptor expression. *Immunopharmacology and Immunotoxicology* 12, 345–362.

Rothkötter, H.J., Ulbrich, H. and Pabst, R. (1991) The postnatal development of gut lamina propria lymphocytes: number, proliferation and T and B cell subsets in conventional and germ-free pigs. *Pediatric Research* 29, 237–242.

Salmon, M., Pilling, D., Borthwick, N.J., Viner, N., Janossy, G., Bacon, P.A. and Akbar, A.N. (1994) The progressive differentiation of primed T cells is associated with an increasing susceptibility to apoptosis. *European Journal of Immunology* 24, 892–899.

Scheepens, C.J.M., Hessing, M.J.C., Hensen, E.J. and Henricks, P.A.J. (1994) Effect of climatic stress on the immunological reactivity of weaned pigs. *Veterinary Quarterly* 16, 137–143.

Sinkora, J., Rehakova, Z. and Sinkora, M. (1998) Expression of CD2 on porcine B lymphocytes. *Immunology* 95, 442–449.

Sun, J., Hayward, C., Shinde, R., Christensen, R., Ford, S.P. and Butler, J.E. (1998) Antibody repertoire development in fetal and neonatal piglets. I. Four V_H genes account for 80% of V_H usage during 84 days of fetal life. *Journal of Immunology* 161, 5070.

Telemo, E., Bailey, M., Miller, B.G., Stokes, C.R. and Bourne, F.J. (1991) Dietary antigen handling by mother and offspring. *Scandinavian Journal of Immunology* 34, 689–696.

Vega-Lopez, M.A., Telemo, E., Bailey, M., Stevens, K. and Stokes, C.R. (1993) Immune cell distribution in the small intestine of the pig: immunohistological evidence for an organized compartmentalization in the lamina propria. *Veterinary Immunology and Immunopathology* 37, 49–60.

Vega-Lopez, M.A., Bailey, M., Telemo, E. and Stokes, C.R. (1995) Effect of early weaning on the development of immune cells in the pig intestine. *Veterinary Immunology and Immunopathology* 44, 319–327.

Wattrang, E., Wallgren, P., Lindberg, A. and Fossum, C. (1998) Signs of infections and reduced immune functions at weaning of conventionally reared and specific pathogen free pigs. *Journal of Veterinary Medicine B* 45, 7–17.

Westly, H.J. and Kelley, K.W. (1984) Physiologic concentrations of cortisol suppress cell-mediated immune events in the domestic pig. *Proceedings of the Society for Experimental Biology and Medicine* 177, 156–164.

Wilson, A.D., Stokes, C.R. and Bourne, F.J. (1989) Effect of age on absorbtion and immune responses to weaning or introduction of novel dietary antigens in pigs. *Research in Veterinary Science* 46, 180–186.

Wilson, A.D., Haverson, K., Bland, P.W., Stokes, C.R. and Bailey, M. (1996) Expression of class II MHC antigens on normal pig intestinal endothelium. *Immunology* 88, 98–103.

Wold, A.E. (1998) The hygiene hypothesis revised: is the rising frequency of allergy due to changes in intestinal flora? *Allergy* 53 (Supplement 46), 20–25.

11 Enteric and Respiratory Diseases in the Young Weaned Piglet

Stan Done

Veterinary Laboratories Agency, Woodham Lane, Addlestone, Surrey KT15 3BP, UK

Introduction

The major enteric and respiratory diseases are listed in Tables 11.1 and 11.2. This chapter will not mention obscure or uncommon conditions but proposes only to describe changes that have occurred in the last decade and to refer readers to the standard texts. Taylor (1999) is sufficient but for a reference text see Straw *et al.* (1999).

Approximately 70% of all losses (mortality, morbidity, culling, loss of production, increased hygiene and veterinary costs and abattoir losses) are incurred as a result of enteric and respiratory diseases. The neonatal pig is particularly prone to viral/parasitic alimentary disease (Table 11.3). At weaning, the young pig is confronted for the first time by artificial food and feeding in an artificial environment, with the result that alimentary and respiratory disease is significant.

There have been more changes in the area of weaner pig production than in any other area of pig production and some of these factors are listed in Table 11.4. Most are detrimental and many serious, with only new methods of rearing and vaccination as really positive items. One of the most serious changes is that, with the shrinking industry, who is going to want (let alone pay for) diagnosis of disease? A dead pig, until fairly recently, has in fact saved money by reducing feeding and labour costs – a new concept of economic death. The effects of this situation are shown in Table 11.5, which lists the pig submissions to the Veterinary Laboratories Agency in the first quarter of each year from 1995 to 2000. The submission level in 1999 was half the level in 1995 and is probably now so low (and perhaps biased) that what is actually going on in the pig world may be unknown.

The farrowing house must be a relatively safe place compared with the nursery house. The most physiologically threatening moment in the life of the piglet is weaning. At this point, maternal antibodies are at the lowest levels; the piglet is not

Table 11.1. Enteric disorders of the weaned piglet.

Class	Disorder
Notifiable exotic	Classical swine fever African swine fever
Important	Transmissible gastroenteritis Epidemic diarrhoea I, II Salmonellosis
Common	Rotaviruses Swine dysentery Intestinal spirochaetosis *Escherichia coli* Spirochaetal diarrhoea Porcine proliferative enteropathy
Worrying	Cryptosporidia Gastric ulceration *Yersinia enterocolitica* *Clostridium perfringens* (A or B?)

Table 11.2. Respiratory disorders of the weaned piglet.

Class	Disorder	
Notifiable exotic	Aujeszky's disease Blue-eye	
Potentially zoonotic but rare	Chlamydia Salmonellosis	
Common	Pleuropneumonia	– *Actinobacillus pleuropneumoniae*
	Atrophic rhinitis	– *Bordetella bronchiseptica* – *Pasteurella multocida*
	Enzootic pneumonia	– *Pasteurella multocida* – *Mycoplasma hyopneumoniae*
	Glasser's disease (and pneumonia) *Haemophilus parasuis*	
	Porcine reproductive and respiratory syndrome	
	Swine influenza	
	Porcine respiratory coronavirus	
Unknown status	CAR-bacillus	– *Pneumocystis carinii*

Table 11.3. Neonatal disorders/diseases that occur before weaning and subsequently affect the weaner.

Disorders	Disease	
Starvation	Rotavirus	Villus atrophy
Poor bodyweight	Transmissible gastroenteritis	Villus atrophy
Low daily gain	Coccidia	
Poor maternal antibody provision and absorption	*Clostridium perfringens* type A	
	Escherichia coli	

Table 11.4. Other major factors contributing to alimentary and respiratory disease losses.

Production

1. There are more changes occurring in nursery pig production and management than anywhere else in the production cycle.
2. Outdoor production now probably 25–30% of sows and produces weaners.
3. Welfare constraints and changes have removed stalls and may well in the future eventually affect crates, floors, slats or no-slats, straw or not and also tail docking and age of weaning.
4. A tendency to lower weaning age.
5. More techniques that involve mixing/moving/batching according to weight, even on day 1.
6. New systems, e.g. SEW, 2–3-site production, isowean, etc.
7. Continual expansion of unit size – more hazardous, more severe when an exotic disease hits, and sub-populations may be formed; i.e. pathogenic and non-pathogenic load is increasing within a unit.
8. Economic situation → no profit → no investment
   (a) No repair of existing facilities/waiting to close/poor welfare
   (b) No surplus labour
      – reduced cleaning/disinfection
      – reduced welfare attention
      – reduced care of large numbers of piglets

Disease

1. Effects of removal of growth promoters and future potential for pathogenic intestinal agents to gain the upper hand – process may continue.
2. Threats to prescribing may not help future control of disease
3. Vaccination, particularly for respiratory disease, has proved extremely beneficial but alimentary disease vaccines have not advanced far.
4. Other diseases, particularly of septicaemic origin, may affect the respiratory and alimentary tracts, e.g. Erysipelas, salmonellosis, *Streptococcus suis* of many types and *Actinobacillus suis.*
5. Identification of 'immunosuppressive' agents – starting with PCV-2 (1980s), PRRS (1992), swine influenza (1986, 1987, 1991 and 1996).
6. The structure of veterinary farm animal practice is changing, even disappearing.
7. The corporate farmer will employ the corporate vet.
8. Information will become company property.
9. The Investigative Service will not be perceived to be necessary for a shrinking industry and will decline further.
10. The research basis, albeit already very poor in animal health, will reduce even further.
11. Veterinary schools will eventually not teach pig medicine and there will be a need for a centre of excellence in one place, with a full cadre of people (preferably whilst they still exist).
12. An expanding EU will not only bring problems of cheap pork but possibly also foreign disease, particularly Aujeszky's disease, CSF, ASF and foot-and-mouth disease.
13. New diseases will always appear, e.g. Nipah Virus, PRRS virus, new strains of swine influenza virus (Done, 2000).

Table 11.5. Porcine submissions to the VLA 1995–2000 (first quarter each year).

Year	No. submissions
1995	1097
1996	1105
1997	943
1998	907
1999	505
2000	637

properly adjusted to dry food; pathogens picked up from the sow are beginning to establish, colonize and cause disease; and the active processes of cell-mediated immunity and active humoral immunity are still impaired. Added to these already detrimental factors are the further challenges of moving, mixing, regrouping and transport to a new environment with new floors, heating, ventilation and many more animals in one air space, and the fact that piglets themselves are reservoirs of viral and bacterial pathogens. Weaning is, therefore, likely to be followed by a wealth of intestinal disorders associated with these instantaneous changes and by a multitude of respiratory disorders that are associated with a high stocking density, adverse aerial environment and transmission of respiratory pathogens from contemporaries or, more seriously, from older pigs in the same building.

It cannot be guaranteed that, during life in the farrowing house, all pigs will receive maternal antibodies to all diseases equally and for the same period. Nor does the presence of maternal antibodies always mean protection and there is often a long period, before titres ultimately disappear, in which protection is not provided by the falling levels. This has far-reaching consequences in that there may be infected and non-infected litters or individuals within the overall group after weaning. Thus there are increased possibilities for the lateral spread of pathogens between sub-populations and the development of acute disease outbreaks in immature animals.

The diseases listed in Tables 11.1 and 11.2 will be considered briefly, with comments on any significant changes and developments. Initially it is important to consider some general changes that have occurred in the last 10 years, including: (i) growth promoters and their potential loss; (ii) modern methods of early weaning and separate site production; (iii) the use of vaccines; (iv) feedback techniques; and (v) post-weaning respiratory disease and the appearance of viral diseases.

Growth Promoters

The use of growth promoters began in about 1950 (Jukes *et al.*, 1950). They fall into one of four groups (Table 11.6), all affecting Gram-positive organisms except Olaquindox, which is broad spectrum. Currently only Avilamycin and Salinomycin are allowed.

The use of growth promoters has generally increased feed intake, and has improved average daily liveweight gain and food conversion efficiency in each species. The effects of the removal of growth promoters may include: (i) fall in daily gain; (ii) deterioration in food conversion efficiency; (iii) increased diarrhoea; (iv) possibly more specific disease (*Lawsoniana intracellularis*, *Serpulina pilosicoli* and *S. hyodysenteriae*); (v) reduced food intake; (vi) deteriorating hygiene; (vii) increased output of excreta; (viii) declining air quality; and (ix) poorer carcass quality.

Early Weaning and Sites of Production

If conditions are good in the first place, there is often no benefit in separate site production (i.e. an off-site nursery), though there may be more benefit if conditions are very bad. The system separates pathogens and pigs and avoids exposure of recently weaned pigs to the endemic non-indigenous pathogens of older pigs (Drum *et al.*, 1998). Maternal antibody does not necessarily prevent infection but it will inhibit colonization and thereby reduce clinical signs.

Essentially, pigs are removed when their maternal antibody levels are high (i.e. when the piglets are aged less than 21 days) and when pigs are free from infection. Most pathogens (e.g. common bacterial diseases) cannot cross the placenta though some viruses can, including classical swine fever (CSF), porcine cytomegalic virus and particularly porcine reproductive and respiratory syndrome (PRRS) and porcine circovirus type 2 (PCV-2) (congenital tremor aII virus?). Thus a piglet under 21 days of age is potentially free from most pathogens, except those excreted by the sow around farrowing (probably *Streptococcus suis*, *Actinobacillus suis*, *Haemophilus parasuis*, *Pasteurella multocida*, *Bordetella bronchiseptica* and *Actinobacillus pleuropneumoniae*). Even these pathogens are probably picked up in only small numbers whilst maternal antibody provides cover; major acquisition of infection occurs in the nursery.

Medicated early weaning (MEW) was first described by Alexander *et al.* (1980) and further described by Connor (1992). Segregated early weaning (SEW) encourages low pathogen load without the use of the medication and sanitation used in MEW (Clark *et al.*, 1994; Dritz, 1994; Drum *et al.*, 1998). The clean new nursery site should provide no challenge. In the UK a new nursery room with separate ventilation and effluent control operated on an 'all-in/all-out' basis should achieve the

Table 11.6. Classification of antimicrobial growth factors.

Class	Growth factors
A. Inhibitors of pertidoglycan synthesis	Bacitracin
	Avoparcin
B. Inhibitors of protein synthesis	Avilamycin
	Tylosin
	Spiromycin
	Virginiamycin
C. Metabolites inhibit DNA synthesis	Olaquidox
D. Inhibition of ion transport	Salinomycin

same control as SEW and MEW. *S. suis*, PRRS and *H. parasuis* passed on from the sow, however, may never be controlled using any of these methods.

Harris (1990; 1993) argued that earlier weaning ages were necessary to prevent spread of organisms (Table 11.7) and suggested that MEW is much more successful (fewer organisms) the younger the pig. On the other hand, Pijoan (1995) suggested that this philosophy may be flawed in that a small proportion of pigs is always infected before weaning and these pigs have clinical disease when in a nursery. Consequently huge numbers of bacteria are released into the environment. Infection spreads slowly in the nursery and subsequently a larger proportion of pigs express disease until there is enough active immunity to prevent the clinical disease (Moore, 1995; Clark, 1998). *S. suis* is often recovered in early-weaned pigs and it is possible that these organisms are accordingly able to colonize the animal.

In pathological terms the results of cytokine-mediated physiological responses are visible in the comparison of specific pathogen free (SPF) (low antigenic load) or MEW/SEW pigs with conventional pigs, when the thymus and lymph nodes are examined. Enlarged thymuses and increased thymic/bodyweight ratios are found in SPF animals (there is no need to populate peripheral lymph nodes with T lymphocytes, as these pigs are not on the immunological threshold: Waxler and Drees, 1972; Tsay *et al.*, 1989; Wiseman *et al.*, 1992). Animals that are infected early in life soon show a reduction in thymus size and an increase in lymph node size. The role of cytokines in pig growth, production and disease is being studied (Fossum, 1998), as is the role of acute inflammatory reactions (Heegard *et al.*, 1998).

The Use of Vaccines

The 1990s saw the development of several respiratory disease vaccines that have revolutionized the control of respiratory diseases. Two of these vaccines are for *Mycoplasma hyopneumoniae*, one for *A. pleuropneumoniae*, one for atrophic rhinitis (*P. multocida* and *B. bronchiseptica*) and one for the control of Glasser's disease (*H. parasuis* serotypes 4 and 5).

Progressive atrophic rhinitis

All pigs probably carry *B. bronchiseptica* in the nasal cavity and, in uncomplicated infections, produce a catarrhal rhinitis and inflammation of the epithelial surfaces. The bordetella component of these vaccines is always killed. A number of formalin-killed alum adjuvanted vaccines have been produced. Most have been used in the control of progressive atrophic rhinitis and are accompanied by *P. multocida* toxoid. They are used to immunize sows 6 weeks and 2 weeks prior to farrowing and protect piglets from bronchopneumonia and rhinitis by means of maternal antibody (passive). The piglets can then be vaccinated at 7 and 8 days to stimulate active immunity. Clinical signs are reduced and the carriage of the organism is reduced.

Haemophilus parasuis

Recently the increase in *Haemophilus* infections has, to some extent, been halted by the appearance of *H. parasuis* vaccine that contains serotypes 4 and 5 and is apparently very effective in controlling these serotypes, which are principally involved in Glasser's disease. Only time will tell if they are also helpful in controlling the other 13 serovars of *H. parasuis* that have been discovered. If not, the time-honoured method of control by autogenous bacterins will be continued.

Mycoplasma hyopneumoniae

The *M. hyopneumoniae* vaccines apparently work in the face of maternal antibodies. In studies of vaccinated vs. non-vaccinated pigs, daily gain was improved by 20–60 g day^{-1} and feed conversion ratio (FCR) by 0.1–0.7. Carcass quality may also be better in vaccinated animals. Most herds also show a large reduction in average lung scores (% pneumonia) following vaccination. Field observations have shown benefits in reduced coughing, less reliance on in-feed antibiotics, reduction in the use of injectable antibiotics, improved batch uniformity, improved daily liveweight gain and FCR, and possibly reductions in vice due to a general improvement in health status (Thacker *et al.*, 1998a; Fernald and Clyde, 1970).

Actinobacillus pleuropneumoniae

Current vaccines are not perfect but they usually prevent death and severely curtail the lung lesions. Eventually sub-unit vaccines, which work well experimentally, will become available commercially.

Table 11.7. Weaning ages for MEW freedom (Harris 1993).

Disease	Weaning age (days)
Aujeszky's disease	< 21
APP	< 21
TGE	< 21
HPS	< 14
Salmonella cholerae-suis	< 12
Mycoplasma hyopneumoniae	< 10
Pasteurella multocida	< 10
PRRS	< 10

Feedback Techniques

Several authorities have suggested that feedback is useful in preventing disease, but it is not. Madsen (1995) described a case of *Clostridum perfringens* type A infection in a large group of suckling pigs following the use of feedback to control an outbreak of transmissible gastroenteritis (TGE) in the USA. Feedback is commonly used for parvovirus, other enteroviruses, *Escherichia coli* and TGE. It may be useful to prevent one disease but spreading another disease whilst protecting against the first disease is not recommended. In this instance, there would have been considerable mortality if it had been *Cl. perfringens* type C. Other diseases known to be spread by feedback include PRRS (Christianson *et al.*, 1992) and also *Erysipelas insidiosa, Brachyspira hyodysenteriae* and salmonellosis.

Post-weaning Respiratory Diseases

In 1990 the UK was effectively free of viral respiratory disease (no Aujeszky's, no African swine fever (ASF), no CSF, little porcine cytomegalic virus (PCMV)) but this is no longer the case (see Table 11.8). The nature of porcine respiratory disease has changed since 1991 and the term enzootic pneumonia has been replaced by terms such as porcine respiratory disease complex (PRDC), the '18-week wall' or the post-weaning respiratory syndrome in the pig.

Post-weaning respiratory syndrome

Usually the clinical signs occur 8–10 weeks after weaning and possibly 70% of lung may be consolidated in some cases. Typically, viral agents such as PRRS, swine influenza and porcine respiratory coronavirus are involved. Also present are bacterial pathogens, including *M. hyopneumoniae, P. multocida, S. suis* and *A. pleuropneumoniae* (Done and White, 1995).

Quite often piglets sneeze, cough and have eye discharge, either before weaning or up to 7–10 days after weaning. Frequently, affected piglets are also pyrexic and hyperpnoeic. Following weaning, anorexia leads to rapid weight loss. Morbidity follows quickly, with up to 25% also developing a hairy coat, anaemia, occasionally jaundice and a pot-bellied appearance. The post-mortem appearance is usually one of a pneumonia, often with pleurisy/pericarditis, and the aetiological components include PRRS, swine influenza and a range of secondary bacterial infections (e.g. *H. parasuis, A. pleuropneumoniae, P. multocida, B. bronchiseptica, Eperythrozoon suis, E. coli, M. hyopneumoniae*). This syndrome appeared with the onset of viral respiratory pathogens in 1991, the first of which was PRRS.

Porcine reproductive and respiratory syndrome (PRRS)

The European strains of PRRS virus (PRRSV) are much milder than the strains that are found in the USA, and thankfully have proved to be much more stable (Christianson *et al.*, 1992; Collins *et al.*, 1992). PRRSV is now recorded as existing in two types: Type 1 are the European strains and Type 2 the North American strains. This suggests a different evolutionary path as the Type 1 and Type 2 strains had a common origin, but now Type 1 and 2 are very different. For example, for many years the UK strain would grow only on alveolar macrophages and not in tissue culture. Type 1 strains have shown little evolution over the period since occurring in 1991, whereas Type 2 strains continue to evolve. Many strains of varying pathogenicity exist in the USA; some will kill a susceptible pig within 12 h but other strains do not produce clinical signs. The exact virulence factors are still not clear. The sow abortion and mortality syndrome (SAMS) is an example of acute PRRS, as it is now called, in which up to 40% of sows may die.

The respiratory disease in the UK in its simple form is an uncomplicated interstitial pneumonia with non-collapsible rubbery lungs. It is, however, often accompanied by secondary bacterial infection due to the adverse effects of the virus on alveolar macrophages and ciliated epithelium producing a failure of mucociliary clearance, particularly at the alveolar level, until macrophage recovery takes place at about day 28 after infection. There is a simultaneous leukopenia and lymphocyte depletion in lymph nodes. This reaction is probably not restricted to alveolar macrophages as even conditions as unusual as cryptosporidiosis, *Pneumocystis carinii* (most commonly found in AIDS patients), navel bleeding, *E. suis* (a blood parasite leading to anaemia), mange and coccidiosis have been seen more commonly in the initial PRRS diseases. The major respiratory disease result of PRRS is the occurrence of large numbers of *H. parasuis* and *S. suis* infections, together with salmonellosis. Circulating PRRS virus is probably still a significant contributor to porcine respiratory disease complex (PRDC) in the post-weaning/early finisher phase.

Table 11.8. Current swine influenza viruses in the UK.

Importance	Virus
Still present but reduced importance	H1N1 classic H3N2 human-like swine viruses H1N1 avian-like swine viruses
Increasing importance	H1N2 now replacing all other viruses in UK herd
Unknown importance (only isolated once)	H3N1 (N1 of swine and H3 from H3N2 (human-like) circulating in pigs H1N7 (N7 from equine virus, largely disappeared in horse; and H1 from human virus, largely disappeared in the 1970s)
Not seen in UK	H3N2 avian-like viruses as in Asia H9N2 (Hong Kong type of situation, immediate penetration of avian viruses into humans)

Swine influenza

This condition first reached the UK in 1941, when it caused mild clinical disease, but it subsequently disappeared. It was seen again in Europe in 1976, probably arising from US imports to Italy, and reached the UK in 1986, causing mild disease and then becoming common in the UK picture, with 25–30% of pigs being seropositive. Current swine influenza viruses are shown in Table 11.8.

Bacterial Respiratory Diseases

Mycoplasma hyopneumoniae

Two features of enzootic pneumonia have become important. The first is vaccination (discussed above), which has revolutionized control of respiratory disease and has left a niche in the respiratory tract that has been taken up by cilia-associated respiratory (CAR-bacillus) (Nietfield *et al.*, 1995; Done and Higgins, 1998) but this has not caused a huge increase in respiratory disease. Secondly, and more importantly, the work of Eileen Thacker (Thacker *et al.*, 1998a) and colleagues has shown that it is not PRRS that predisposes to *M. hyopneumoniae* but it is more likely to be *M. hyopneumoniae* that predisposes to PRRS. Vaccination for *M. hyopneumoniae* may also significantly help in the control of other diseases. Otherwise little has changed in recent years in our understanding of enzootic pneumonia.

Pasteurella multocida

This is probably now at a low level of infection in the UK, as part of the atrophic rhinitis complex (toxigenic type D), but it is still a frequent cause of pneumonic pasteurellosis (non-toxigenic type A).

Actinobacillus pleuropneumoniae (APP)

This is the most common primary bacterial pneumonic infection (20% of cases) and approximately 30% of APP cases are uncomplicated (Hoefling, 1998). The usual complication is *P. multocida* (Loeffen *et al.*, 1999). The condition may be confused with *A. suis* septicaemia or *H. parasuis* acute pulmonary infections.

The condition has not changed much since 1990 but the knowledge base on virulence factors has changed considerably. There are the 12 serotypes that have been known for some time but their virulence differences may now possibly be explained by reference to their cytolysin production. Cytolysins Apx I, II, III, IVA have now been described. Serotype 1 is probably the most virulent; 4 and 8 require only 10^2 organisms experimentally to reproduce disease; 2, 5, 9, 10 and 11 may be moderate and 3, 6, 7 and 12 the least virulent. Serotypes 1, 2 and 9 are most common in Europe and still increasing. In the USA 1, 5 and 7 are most common. The suspected virulence factors and cytotoxin distribution are shown in Table 11.9.

It has always been assumed that APP lives only in the pig, in the nasopharynx and tonsil, and is not transmitted except by pig-to-pig contact (nostril), but recently experimental aerosol spread of at least 1 m has been reported (Torremorell *et al.*, 1997; Jobert *et al.*, 2000). Larsen (1998) suggested that airborne APP may cause infection at a distance of at least 500 m and fomites (trucks, clothing) have been suggested as other possible methods of transmission (Loeffen *et al.*, 1999; Jobert *et al.*, 2000). APP can also be found in cattle, deer and lambs. Maternal antibodies are often quite high but do not persist for long (limit of protection 4 weeks) and totally disappear by 9 weeks. It may also be the last colonizer of the pig respiratory tract, as the maternal antibody stops major colonization early on, though infection is probably picked up from the sow.

The condition is very dose dependent, with lower exposure resulting in seroconversion without clinical disease, whilst a slightly higher level of exposure results in fatal infections. Growing animals are exposed to increased doses of APP, which may help to explain why outbreaks tend to occur in pigs 12–16 weeks old. APP was described adhering to tonsils (Chiers *et al.*, 1999). The significant improvement in diagnosis has been the use of tonsil screening followed by culture and/or immunofluorescence and/or PCR (Gram *et al.*, 1996; Gagne *et al.*, 1997). The combination of tonsil screening, PCR, monoclonal antibody and immunomagnetic separation has led to a 1000-fold increase in sensitivity compared with culture.

Infection with APP has recently been shown to be responsible for cytokine release, including IL-1, TNFα, IL-8 (Huang *et al.*, 1998a, b; Paradis *et al.*, 1999) and IL-6 (Fossum *et al.*, 1998) and a prolonged surge in cortisone production (Heegard *et al.*, 1998).

Table 11.9. Virulence factors.

				Cytotoxins	
Group	Function	Serotypes	Apx I	Apx II	Apx III
1. IgA protease	Hydrolyses porcine IgA and porcine IgG	• 1	√	√	
	(Negrete-Abascal *et al.*, 1998)	2		√	√
2. Capsule	Resists phagocytosis	3		√	√
3. Endotoxin	Cause of death?	4		√	√
	Haemolysin	• 5	√	√	
	Cytotoxins	6		√	√
4. Exotoxins	4 (part of the RTX group)	7		√	
	Apx I Cytotoxic	8		√	√
	Apx II Haemolytic	• 9	√	√	
	Apx III Cytotoxic	• 10	√		
		• 11	√	√	
		12		√	

• Most virulent groups (1,5,9,11)
∴ Apx I and II found in most virulent serotypes.
Many serotypes cross-react, possibly because of these toxins (e.g. 8 with 3 and 6; 9 with 1; 5 with 7; and 7 with 4).

Haemophilus parasuis

Polyserositis is also associated with *E. coli* and *S. suis*. There has been a huge increase in the number of *H. parasuis* cases over the last few years. This has been due in part to the immunosuppressive effects of PRRS and swine influenza predisposing to *H. parasuis* secondary infections, and to the reduction of *M. hyopneumoniae* after vaccination creating a vacuum in the respiratory tract, but in many ways is also a reflection of improved culture methods and the advent of serotyping. It is a great opportunist and can involve any mesothelial surface, such as pleurae, pericardial sac, peritoneum, meninges and synovial membrane (Glasser's disease), as well as being a significant pulmonary pathogen. The virulence factors for *H. parasuis* are unknown and the reason why Glasser's disease is associated with serotypes 4 and 5 and why serotypes 1–3 and 6–14 are linked to pneumonia and pleurisy is also unknown.

The agent is particularly problematic in repopulated herds, in minimal-disease/high-herd-health herds and multiple-site production systems, where it is also often associated with sudden death.

Streptococcus suis

Since 1991 *S. suis* infections have been at the forefront of pig diseases (septicaemia, Type 2 meningitis, polyserositis, endocarditis and pneumonia) because they are caught from the sow's nose (tonsil) during birth, usually before 7 days (Clark *et al.*, 1994), or *in utero*. They are not easily controlled by early-weaning systems (Wiseman *et al.*, 1992; Clark *et al.*, 1994) because they are ubiquitous and stress related, and because they are predisposed to by swine influenza and PRRS (Amass *et al.*, 1998). Experimentally, *S. suis* meningitis is still the only disease that has been shown to be more easily produced following PRRS. *S. suis* typing has increased interest in porcine streptococci and so in the UK it is known that, as well as types 2, 1 and 1/2, there are also significant problems with types 9 and 14.

Actinobacillus suis

Like all other bacterial pathogens, *A. suis* has increased in incidence following influenza and PRRS. Additionally, typing has become possible for types 1 and 2 and this always stimulates an upsurge of interest in a pathogen. Recently weaned suckling pigs are the major group affected (Sanford *et al.*, 1990; Sandford, 1995) and several pigs in a litter or litters may be affected. High-herd-health pigs are particularly badly affected.

Enteric Diseases

Progress in enteric disorders of pigs has not been as significant as that in respiratory disease. The disease agents as yet do not seem to be subject to mutational changes.

Diets probably have not changed significantly; weaning was always the biggest stress to the pig; there are no significant vaccines developed; and perhaps the only major factor is the withdrawal of growth promoters. At the same time there are the preoccupation with antibiotics, the search for satisfactory replacements (e.g. organic acids, probiotics, enzymes) and the possibility of food scares, particularly salmonellosis. Those alimentary disorders that may originate in the weaner pig, continue with the grower and extend into the finishing period are shown in Table 11.10.

Transmissible gastroenteritis

This is the major viral disease affecting the weaner pig, because it is the most catastrophic and causes the most severe form of villus atrophy, where villus height is reduced and therefore the absorptive surface is greatly diminished. Fortunately the UK pig herd, in keeping with most others in Europe, has been vaccinated by the occurrence of porcine respiratory coronavirus, the respiratory variant of porcine coronavirus transmissible gastroenteritis (TGE). The occasional new strain has appeared but the old strains are still present on the farms where these diseases occurred in the 1960s and 1970s.

Porcine epidemic diarrhoea

This condition is usually seen after sale or purchase of animals and may be encountered only in finishing pigs. Villus atrophy is severe and occurs quickly.

Rotavirus

Under normal circumstances this is the most important cause of villus atrophy and only since the mid-1980s has knowledge been extended to demonstrate that there are seven distinct serogroups of rotaviruses (A–G). Rotaviruses A, B, C and E are found in pigs. It is ubiquitous and it is impossible to keep units clean of rotavirus, which is extremely stable and resistant. Rotavirus A (40%) predominates in nursery pigs but, in post-weaning diarrhoea, 44% were A, 18% were B and 22% were C (Janke, 1990). It is often combined with *Isospora suis*. It has been shown that there is a cell receptor for a monosialoganglioside or a family of these with sialic acid as an important epitope for virus binding (Rolsma *et al.*, 1994), with a sugar-binding specificity (Mendez *et al.*, 1999), and that this may occur more frequently in the distal one-half to two-thirds of the small intestine (Collins *et al.*, 1989; Stevenson *et al.*, 1990).

Escherichia coli

Since 1990 our understanding of *E. coli* has developed considerably. Acute post-weaning mortality and morbidity is emerging as a more common disease (*E. coli*

Table 11.10. Differential diagnosis of the common intestinal disorders of finishing pigs.

Disease	Ileum	Colon	Ileocaecocolic nodes	Extra-intestinal signs/lesions
Salmonellosis	Mild inflammation	Focal to severe deep necrosis	Always enlarged (2–5 times normal)	Often generalized infarction Interstitial pneumonitis
Swine dysentery	–	Superficial diffuse/moderate blood leakage mucus	Normal, possibly slightly enlarged	None
Porcine proliferative enteropathy	Haemorrhagic necrosis or proliferation	Milder than in the ileum? Usually the proximal spiral colon	Variable with stage of disease	None
Whipworms (*Trichuris*)		May be mucohaemorrhagic colitis		
Non-specific colitis	–	Congested slight haemorrhagic mucosa. Watery mucoid faeces. Mild, not like SD		
Yersiniosis	Catarrhal enteritis	With mucus/blood button-ulcers?	Microabscesses in nodes?	

conditions are subject to significant change). It is becoming an increasing problem in high-herd-health units in the USA, particularly those using SEW and MEW techniques that do not eliminate *E. coli* (Winkelman, 1995).

E. coli disrupts the intestinal mucosal barrier, allowing bacteria to migrate from mucosae to the blood and internal organs, causing septicaemia. Diets that burden the absorptive capacity of the gut may make the small intestine more prone to colonization and subsequent diarrhoea. Highly digestible *ad libitum* diets are less likely to cause trouble, and creep feeding before weaning also helps. Lowering the pH of the pig gastrointestinal tract has been the basis of some control strategies. Similarly, if significant amounts of protein enter the large intestine, there is a change in the balance of the hindgut leading to looseness or diarrhoea. This change may explain the large numbers of *Campylobacter* spp. and *Balantidium coli* found in *E. coli*-affected pigs.

Colibacillosis is a major cause of illness and death in recently weaned pigs. Usually it is a consequence of enterotoxigenic strains of *E. coli* (ETEC) producing enterotoxins (heat-labile toxin LT) or VTEC types producing heat-stable toxins (STa/Stb) which act totally in the gut and stimulate hypersecretions of mucus and electrolytes.

The organism also produces fimbrial adhesions, which mediate the adherence of bacteria to the mucosal surface (Table 11.11). Fimbriae produced include K88 (F4), K99 (F5) and 987P (F6); F41 and F18 (F107 and 2134P) are less common but some may produce a shiga toxin (stx 2E).

Age and genetic background seem to determine the inherent susceptibility of piglets to *E. coli*. Pigs are resistant to F18^{+} *E. coli* at birth but become susceptible after several weeks (Imberechts *et al.*, 1997), whereas K99 or 987P resistance is substantially complete by 2 weeks (Runnels *et al.*, 1986; Dean *et al.*, 1989). Resistance is achieved by failure to produce the receptor on epithelial brush-border membranes to which the fimbriae adhere (Francis *et al.*, 1998).

Heritable resistance to colibacillosis caused by K88 and F18 ETEC is well documented but has not been reported with regard to *E. coli*, which produces other adherence fimbriae (Bertschinger *et al.*, 1993; Francis *et al.*, 1998).

Enterotoxigenic *E. coli* expressing K88 and F18 fimbrial adhesions account for nearly all cases of post-weaning colibacillosis in pigs (Wilson and Francis, 1986; Hide *et al.*, 1995; Wittig *et al.*, 1995). About 50% of pigs in common breeds inherit resistance to K88 organisms (Baker *et al.*, 1997). Pigs genetically susceptible to *E. coli* K88^{+} account for a high proportion of all colibacillosis and so selective breeding for resistance to K88^{+} and perhaps F18 could have a significant effect on the economic losses due to *E. coli* infections.

Susceptibility to *E. coli* infection is dominant over resistance (Francis *et al.*, 1998). Sows with the virulent phenotype do not produce K88 antibody subsequent to oral exposure with K88^{+} ETEC or K88 antigens (van der Broueck *et al.*, 1999) but they probably produce circulating IgG anti-K88 antibody following parenteral vaccination.

A high concentration of dietary zinc for 2 weeks after weaning (Holm and Poulsen, 1996) controls *E. coli* by preventing proliferation of pathogenic strains in the lumen of the gut, thereby reducing spread of *E. coli* from the small intestine to the ileal lymph nodes. (Fewer organisms are cultured from the ileal lymph nodes of pigs receiving zinc oxide.)

Table 11.11. Serogroups, fimbriae, enterotoxins and haemolysins of nursery and weaned pigs.

Sero-group	Fimbriae	Enterotoxin	Haemoly-sins	Group of pigs
0149	K88	LT and/or STa or STb	+	Nursery and weaned
0157	K88 or F18	LT and/or STa or STb or Stx2E	+/–	Nursery and weaned
08	K88 or K99	LT and/or STa or STb	+	Nursery and weaned
0138	F18	LT and/or STa or STb or Stx2E	+	Weaned
0138	F18	LT and/or STa or STb or Stx2E	–	Weaned
0141	987P	STa	–	Nursery
020	987P	STa	–	Nursery
09	987P or K99, F41	STa	–	Nursery
0101	K99, F41	STa	–	Nursery
045	–	–	–	Weaned

Salmonellosis

In the USA *Salmonella cholerae-suis* (host-adapted species) persists as the major disease-producing agent, whereas in the UK it has virtually disappeared (one in 1997) from its major position in the 1960s. In 1997, 62% were *Salmonella typhimurium* and 12% were *Salmonella derby* (both non-host adapted species). The concern is that the proliferation of salmonella may be perceived to be a farm problem. On the farm, numbers are probably only small (in Iowa, a survey found 11% in 9000 carcasses), but in abattoirs up to 48% of carcasses are infected and up to 30% of retail pork might be contaminated, i.e. it is a problem of contamination, whereas salmonellosis as an on-farm disease is much less common (Davies *et al.*, 1999). Phage types DT104, DT193 and DT208 are the most common types of *S. typhimurium* and many show antibiotic resistance, particularly to tetracyclines; that is, on-farm salmonellosis is rare but salmonella infection in pigs, pork and pork products may be more commonly found.

Davies *et al.* (1999) isolated salmonella from 7.0% of 2219 carcass swabs and 11.0% of 2205 samples of large intestine contents, with many resistant to antibiotics. Keeping pigs in the lairage and subjecting them to stress probably increased the recovery but all is relative, as Davies *et al.* (1999) showed that overnight lairage reduced salmonella compared with slaughter within 2–3 h of arrival; they also demonstrated little further spread after evisceration.

One infected pig can transmit a considerable number of organisms: a concentration of 10^6–10^7 g^{-1} of faeces is possible and a concentration of only 10^8–10^{11} is required to promote an infection (Gray *et al.*, 1995, 1996a,b). Infection may be acquired through dust or as an aerosol and there is always the possibility that other organisms (cats, birds, rodents and flies) may carry salmonellae. Meat meals and other similar foods (e.g. fishmeal) also carry a high risk of infection. Disinfection does not always help (Gray *et al.*, 1995; Wray and Woodward, 1997).

Little is known about virulence factors, of which there may be over 200, but *S. typhimurium* may adhere to the mucosa via mannose-resistant adhesins. The pathogen is usually responsible for disease in weaned pigs from 6 to 12 weeks of age and produces yellow watery diarrhoea with a low mortality and inappetance, pyrexia and lethargy. Morbidity can be quite high with a diffuse necrotizing colitis and typhlitis. Often the luminal contents are gritty, blood-stained or bile-stained, all of which are a reflection of the histological diagnosis of coagulative necrosis. In the future, competitive exclusion techniques may be useful (Fedorka-Cray *et al.*, 1997). Major improvements in diagnosis of salmonellosis have occurred, including: (i) conventional isolation; (ii) conductance; (iii) DNA-based methods, including PCR and multiplex PCR; (iv) immunomagnetic methods; (v) ELISAs; (vi) methods of typing; and (vii) serology.

Swine dysentery

Perhaps the only major development in swine dysentery in recent years has been the change in name from *T. hyodysenteriae* to *B. hyodysenteriae*. The condition is still responsible for economic loss and suffering. It has been controlled by hygiene and preventive medication but some resistance is developing, particularly to lincocin and tiamulin, and the removal of growth promoters may encourage the re-emergence of the disease.

Porcine proliferative enteropathy

This is a name given to a group of acute and chronic conditions that are caused by the obligate intracellular anaerobic bacterium *Lawsoniana intracellularis*. The range of conditions includes necrotic enteritis, regional ileitis, proliferative enteritis, porcine intestinal adenomatosis and porcine haemorrhagic enteropathy. As a clinical condition, it is either an acute haemorrhagic syndrome in young mature or adult animals or a non-haemorrhagic syndrome resulting in chronic weight loss in growing pigs (Gebhart *et al.*, 1993; Lawson and McOrist, 1993).

It is not possible to reproduce the disease in gnotobiotics and the implication is that it must require other organisms. The organism responsible is frequently found with other pathogens. It is often precipitated by stressors since many herds are sub-clinically infected and many other species can probably harbour the organism.

Porcine colonic spirochaetosis

This is one of the conditions that has been examined in detail since 1995. It has other names, including non-specific colitis and spirochaetal diarrhoea. It is caused by *Brachyspira pilosicoli*, one of the five species of Brachyspirae. The clinical signs and lesions are seen in Table 11.12.

Thomson *et al.* (1998) examined 85 herds in Scotland and found *B. hyodysenteriae* as the sole pathogen in 33% of cases and present in 52% of cases (the individual infections are shown in Table 11.13. Shedding pigs are the worst risk and the organism may possibly survive for months under wet conditions. Not all infected pigs develop diarrhoea. In experimental infections only 30–70% develop diarrhoea, although all pigs may be colonized. These inoculated pigs may shed for 6 weeks. *B. brachyspirae* may also be cultured from humans, non-human primates, dogs, commercial chickens, waterfowl and pheasants. There is considerable risk whenever there is frequent mixing, continuous pig flow, poor ventilation/sanitation and high stocking densities.

Nutritional colitis, or non-specific colitis, is also believed to occur independent of *Brachyspira* etc. It is caused by fermentation in the hindgut and may be associated with overcrowding and inadequate sanitation, high-energy food intake, change of ingredients (particularly low-cost rations), particle size and pelleting of feed. Feed that contains fat, more than 9 g lysine kg^{-1} and more than 200 g manioc kg^{-1} appears to predispose to this problem.

Table 11.12. Clinical signs and pathological lesions of porcine colonic spirochaetosis.

Clinical signs	
Non-fatal	No mortality if uncomplicated
6–16 weeks	Diarrhoea – soft, wet, watery
7–14 days after moving/mixing	Chronic cases: thick tags of mucus; rarely, flecks of blood
Incubation period 3–16 days	Generally alert, active, but inappetant
Usually less than 30% prevalence	
Lasts 2–6 weeks	
Lesions	
Pigs gaunt, rough haircoat	Colonic lymph nodes are enlarged
Spiral colon flaccid, enlarged, watery contents	Mucosal surfaces are normal/may be erosions
Mesentery and serosa thickened by oedema in acute cases and fibrous connective tissue in chronic cases	Fibrinonecrotic debris Blood and mucus in lumen

Table 11.13. Isolation of agents from enteric disorders in Scotland (Thomson *et al.*, 1998).

Pathogen	Primary agent %	Mixed infections %
Brachyspira pilosicoli	18	24
Brachyspira hyodysenteriae	13	16
Atypical brachyspirae	8	4
Y. pseudotuberculosis	4	10
Salmonella spp.	6	12
Lawsonia intracellularis	10	15
Trichuris suis	2	–
No infectious agent	8	

Porcine Dermatitis and Nephropathy Syndrome (PDNS) and Post-weaning Multisystemic Wasting Disease (PMWS)

Porcine dermatitis and nephropathy syndrome was recorded in the UK by Smith *et al.* (1993) and White and Higgins (1993). Between 1993 and 1998 there have been on average nine recorded sporadic outbreaks per annum of this condition in the UK (Gresham *et al.*, 2000a,b), with low morbidity and mortality. Since the autumn of 1999 a much greater number of acute outbreaks of PDNS (first reported by Potter, 2000) across southern England and East Anglia have been investigated. The lesions resemble those of acute African swine fever and classical swine fever and the difficulties of differentiation are reviewed by Done *et al.* (2001). The subject of PDNS was reviewed by Duran *et al.* (1997), Drolet *et al.* (1999), Gresham (1999) and Madec *et al.* (2000). This chapter considers the recent outbreak, in which there has been positive immunohistochemistry for PCV-2 and also isolation of the virus (Gresham *et al.*, 2000a,b).

Many causes of post-weaning wasting have been documented (Table 11.14) and these should not be forgotten. It is always very tempting to assume a new fashionable diagnosis before the traditional diagnoses have been ruled out. Previous contributions on the specific subject of PMWS have been published (Gresham, 1999; Madec *et al.*, 2000), and the first clinical note drawing attention to this condition in the UK was published by Potter (2000).

PMWS has only been recognized recently and is believed to be associated with PCV-2 (Harding, 1999; Harding *et al.*, 1999). In fact, since examination of this condition is in its infancy, not much is yet known about PCV-1 or PCV-2 and it may be that other factors, such as other infectious agents or immune modulations, may be involved in the expression of the disease. This is most likely as many PCV-1, PCV-2 and PRRS positive herds have not experienced either PMWS or PDNS. The clinical signs of PDNS and PMWS are given in Tables 11.15 and 11.16.

Pathology similar to the classically recorded Canadian descriptions of PMWS has been seen in several countries (Spain, France, Ireland, UK and USA) but the initial descriptions of PMWS have been used as a reference. Acute severe PDNS appears to be a particular UK phenomenon in the present outbreaks, although it has been reported elsewhere (Duran *et al.*, 1997; Drolet *et al.*, 1999).

Since September/October 1999 there has been a sudden increase in the incidence of acute cases of PDNS which have often (but not always) occurred in association with PMWS. These cases are particularly important in that the major lesions are enlarged haemorrhagic lymph nodes, which are similar to those seen in either ASF or CSF. The lesions are described in Table 11.15.

The lesions of PMWS as originally described from Canada by Harding and Clark (1997), Harding (1996), Clark (1996, 1997) have been seen in the UK cases, for which there is probably less jaundice and liver atrophy.

PDNS and PMWS conclusion

The cause of PDNS is unknown but the nature of the lesions in the blood-vessel walls are suggestive of a type of hypersensitivity reaction that may be caused by the presence of a gram-negative bacteria such as *P. multocida*. The reason for the sudden upsurge in the number of cases of sporadic PDNS that occurred in the autumn of 1999 is not

Table 11.14. Causes of post-weaning wasting.

Non-infectious	Post-infection	Infection
Anorexia/starvation Nutritional inadequacy 'Wasting pig syndrome'	Persistent diarrhoea following neonatal scouring → villus atrophy (rotavirus, coccidiosis, etc.)	PRRS Swine influenza Enzootic pneumonia *Haemophilus parasuis* (Glasser's disease) Lawsoniana infection (PIA) Post-weaning colibacillosis Cryptosporidiosis

Table 11.15. Clinical signs in PDNS (features of new outbreaks that differ from previously reported cases are in bold).

Age:	Usually 12–14 weeks, range 5 weeks–9 months
Onset:	Acute
Signs:	Anorexia, dullness, pallor pyrexia, conjunctivitis, **dyspnoea**, **diarrhoea**, **emaciated**, multifocal well-demarcated slightly raised red/purple circular to irregular skin lesions Lesions may coalesce, may be any shape or size Lesions may be anywhere, but particularly loins, hindlegs and caudal/ventral abdomen and ears **Recovered pigs unthrifty. Many fading pigs are culled** Occasionally superficial lymph nodes may be palpable **Hot legs, pyoderma and lameness**
Morbidity:	**12–14% usually; varies 0.5–30% – much higher than before**
Mortality:	0.5–7% usually; **varies 0.5–14%** Case fatality rate may be **5–14%** 60–80% of pigs with cutaneous lesions die
Response to treatment:	Usually none

Table 11.16. Clinical signs in PMWS (percentage of affected pigs showing signs).

Onset:	Subtle may be acute
Age:	Affects 6–12-week-old pigs (range 5–14) 3–8 weeks after weaning Morbidity 5–20%, but may reach 100%, mortality *c.* 10%
Signs:	May be sequential development
Major:	Progressive Severe weight loss (50%) – spine becomes prominent (this is distinctive and often the first sign)
Also seen:	Dyspnoea (40%) – may be due to profound anaemia Pallor Anorexia – not always, some may still eat Cough (40%) Diarrhoea (20% or more) Jaundice (15–20% – UK < 2%) Rough hair coat (< 20%) May be a few skin haemorrhages Enlarged palpable inguinal lymph nodes

known but it can only be concluded that, in some way, this is associated with the cases of PMWS that occurred at approximately the same time. Recently, it has been shown that PCV-2 has been present in the pig population since the mid-1980s, and a great deal of work continues in an attempt to associate conditions seen since then with PCV-2 – particularly these cases, PRRS cases and others such as the chlamydia cases observed. It is important to realize that PDNS and PMWS do not always occur together; they may also occur separately, or in sequence.

How PDNS and PMWS are transmitted is not known but it is suspected to be by either *in utero* or early neonatal transmission. The distribution of the age range of cases suggests that these may follow PRRS infection. The latter occurs before 6 weeks, and PMWS after 6 weeks.

Treatment of either condition and the most effective means of control are unknown, because essential facts about the epidemiology of the diseases remain to be confirmed. What is known suggests that minimization of stress and bacterial challenge (particularly multi-sourcing) will help, and that weaning after 6–8 weeks may facilitate the disappearance of the condition. PCV-2 is believed to be associated with the occurrence of both diseases but the significance of circovirus is unknown (Ellis *et al.*, 1999).

One cannot better the words of Cottrell *et al.* (1999) to describe the present situation:

> ... like PRRS, the presence of PCV does not always lead to the presence of PMWS and other factors may be required to initiate disease.

This is particularly important in that PCV-2 is found widely in clinically normal pigs (Ellis *et al.*, 1999).

Conclusion

This chapter has described recent advances in the study of enteric and respiratory disorders of the pig. The disorders do not change; they become more complex and more difficult to control and new diseases occur all the time. As soon as a vacuum is created, something new occurs to fill that niche. Vaccination is now set to replace chemotherapy – certainly in the field of respiratory disease but as yet not, unfortunately, in enteric disorders.

Acknowledgements

Acknowledgements are due to all practitioners who have sent material; to companies, particularly PIC of Fyfield Wick and Fort Dodge of Southampton, who have organized seminars on the subject of PMWS; most importantly the Veterinary Laboratories Agency, Ministry of Agriculture Fisheries and Food, and the Scottish Agricultural Colleges, Veterinary Surveillance Division, who have provided surveillance facilities and finance to study PDNS and PMWS; staff of regional laboratories, particularly VLA-Bury St Edmunds and VLA-Winchester, for a great deal of help and effort and with post-mortem examinations.

References

Alexander, T.J.L., Thornton, K., Boon, G., Lysons, R.J. and Gush, A.F. (1980) Medicated early weaning to obtain pigs free from pathogens endemic in the herd of origin. *Veterinary Record* 106, 114–119.

Amass, S. (1998) The effect of weaning age on pathogen removal. *Compendium of Continuing Education for the Practising Veterinarian* 20 (Supplement 9), S196–S203.

Baker, D.R., Billey, L.O. and Francis, D.H. (1997) Distribution of K88 *E coli*-adhesive and non-adhesive phenotypes amongst pigs of four breeds. *Veterinary Microbiology* 54, 123–132.

Bertschinger, H.U., Stamm, M. and Vogeli, P. (1993) Inheritance of the oedema disease in the pig. Experiments with an *E.coli* strain expressing fimbriae F107. *Veterinary Microbiology* 35, 79–89.

van der Broueck, W., Cox, E. and Goddeers, B.M. (1999) Receptor-dependent immune response in pigs after oral immunisation with F4 fimbriae. *Infection and Immunity* 67, 520–526.

Chiers, K., Haesebrouck, F., Overbeke, I., van Charlier, G. and Ducatelle, R. (1999) Early *in vivo* interactions of APP with tonsils of pigs. *Veterinary Microbiology* 68, 301–306.

Christianson, W.T., Collins, J.E., Benfield, D.A., Harris, L., Gorcyca, D.E., Chladek, D.W., Morrison, R.B. and Joo, H.S. (1992) Experimental reproduction of swine infertility and respiratory syndrome in pregnant sows. *American Journal of Veterinary Research* 53, 485–488.

Clark, E.G. (1996) Pathology of the post-weaning multisystemic wasting syndrome in the pig. In: *Proceedings Western Canadian Association Swine Practitioners*, pp. 222–226.

Clark, E.G. (1997) Post-weaning multisystemic wasting syndrome. *Proceedings American Association Swine Practitioners Annual Meeting* 28, 499–501.

Clark, L.K. (1998) New rearing technologies: influence on health, growth and production economics of swine. *Proceedings of the 15th International Pig Veterinary Society Congress, Birmingham* 15 (Part 1), 281–288.

Clark, L.K., Hill, M.A., Kniffen, T.S., VanAlstine, W., Stevenson, G., Meyer, K.B., Wu, C.C., Scheidt, A.B., Knox, K.E. and Albregts, S. (1994) An evaluation of the components of medicated early weaning. *Swine Health and Production* 2, 5–11.

Collins, J.E., Benfield, D.A. and Duimstra, J.R. (1989) Comparative virulence of 2 porcine group-A rotavirus isolates in gnotobiotic pigs. *American Journal of Veterinary Research* 50, 827–835.

Collins, J.E., Benfield, D.A., Christianson, W.T., Harris, L., Hennings, J.C., Shaw, D.P., Goyal, S.M., McCullough, S., Morrison, R.B., Joo, H.S., Gorcyca, D. and Chladek, D. (1992) Isolation of swine infertility and respiratory syndrome virus (isolate ATTC-2332) in North America and experimental reproduction of disease in gnotobiotic pigs. *Journal Veterinary Diagnostic Investigation* 4, 117–126.

Connor, J. (1992) Elimination of specific agents from herds: elimination by medication. In: *Proceedings of the Minnesota Swine Conference*, pp. 157–159.

Cottrell, T.S., Friendship, R.M. and Dewey, G.E. (1999) Epidemiology of post-weaning multi-systemic wasting syndrome in Ontario. *Proceedings of the American Association of Swine Practitioners* 30, 389–390.

Davies, R.H., McClaren, I.M. and Bedford, S. (1999) Observations on the distribution of salmonella in the pig abattoir. *Veterinary Record* 145, 656–661.

Dean, D., Whipp, S.C. and Moon, H.W. (1989) Age-specific colonisation of porcine intestinal epithelium by 987P piliated *E. coli. Infection and Immunity* 37, 82–87.

Done, S.H. (2000) New and emerging diseases (how?, why?, when?) of farm animals with special reference to the pig. *Pig Journal* 45, 76–100.

Done, S.H. and Higgins, R.H. (1998) Unusual tracheal lesions in the pig: association with CAR-bacillus like organisms. *Pig Journal* 41, 127–133.

Done, S.H. and White, M.E.C. (1995) Post-weaning respiratory syndrome. *Proceedings of the Allen D. Leman Swine Conference, University of Minnesota, Minneapolis* 21, 133–136.

Done, S.H., Gresham, A., Harwood, D., Giles, N., Jackson, G., Potter, R., Chennells, D. and Thomson, J.R. (2001) Porcine dermatitis and nephropathy syndrome. An important differential diagnosis for African and classical swine fever. *State Veterinary Journal* (in press).

Dritz, S. (1994) Application of segregated early weaning technologies in the commercial swine industry. In: *Compendium of Continuing Education for the Practising Veterinarian, May 1994*, pp. 677–685.

Drolet, R., Thibault, S., D'Allaire, S., Thomson, J.R. and Done, S.H., (1999) Porcine dermatitis and nephropathy syndrome (PDNS): an overview of the disease. *Swine Health and Production* 7, 283–285.

Drum, S.O., Walker, R.D., Marsh, W.E., Mellencamp, M.M. and King, V.L. (1998) Growth performance of segregated early-weaned versus conventionally weaned pigs through finishing. *Swine Health and Production* 6, 203–210.

Duran, C.O., Ramos-Vara, J.A. and Render, J.A. (1997) Porcine dermatitis and nephropathy syndrome: a new condition to include in the differential diagnosis list for skin discolouration in the pig. *Swine Health and Production* 5, 241–244.

Ellis, J.A., Krakowka, S., Allan, G., Clark, E.G. and Kennedy, S. (1999) The clinical scope of PRRSv infection has expanded since 1997, an alternative explanation. *Veterinary Pathology* 36, 262–265.

Fedorka-Cray, P.J., Harris, D.L. and Whipp, S.C. (1997) Using isolated weaning to raise salmonella free swine. *Veterinary Medicine* 92, 375–382.

Fernald, G.W. and Clyde, W.A. (1970) Protective effects of vaccines in experimental *M. hyopneumoniae* disease. *Infection and Immunity* 1, 559–565.

Fossum, C. (1998) Cytokines as markers for infections and their effect on growth performance and well-being in the pig. *Domestic Animal Endocrinology* 15, 439–444.

Fossum, C., Wattrang, E., Fuxler, L. Jensen, K.T. and Wallgren, P. (1998) Evaluation of various cytokines (IL-6, IFNα, IFNγ, TNFα) as markers for acute bacterial infection in swine: a possible role for serum interleukins. *Veterinary Immunology and Immunopathology* 64, 161–172.

Francis, D.H., Grange, P.A., Zeman, D.H., Baker, D.R., Sur, R. and Erikson, A.K. (1998) Expression of mucin-type glycoprotein K88 receptors strongly correlates with piglet susceptibility to K88+ enterotoxigenic *E. coli*, but adhesion of this bacterium to brush borders does not. *Infection and Immunity* 66, 4050–4055.

Gagne, A., Laccouture, S., Broes, A., D'Allaire, S. and Gottschalk, M. (1997) Development of an immunomagnetic marker for selective isolation of *Actinobacillus pleuropneumoniae* serotype 1 from tonsils. *Journal of Clinical Microbiology* 36, 251–254.

Gebhart, C.J., Jones, G.F. and McOrist, S. (1993) Porcine proliferative enteropathy: aetiology and diagnosis. *Swine Health and Production* 1, 24–25.

Gram, T., Ahrens, P. and Neilsen, J.P. (1996) Evaluation of a PCR for detection of APP in mixed bacterial cultures from tonsils. *Veterinary Microbiology* 51, 95–104.

Gray, J.T., Stabel, T.J. and Fedorka-Cray, P.J. (1966a) Effect of the dose on the immune response and persistence of *Salmonella cholerae-suis* infection in swine. *American Journal of Veterinary Research* 57, 313–319.

Gray, J.T., Fedorka-Cray, P.J., Stabel, T.J. and Ackermann, M.R. (1966b) Natural transmission of *Salmonella cholerae-suis* in swine. *Applied and Environmental Biology* 62, 141–146.

Gray, J.T., Fedorka-Cray, P.J., Stabel, T.J. and Ackermann, M.R. (1995) Influence of inoculation route on the carrier state of *Salmonella cholerae-suis*. *Veterinary Microbiology* 47, 43–59.

Gresham, A. (1999) Post-weaning multi-systemic wasting syndrome in pigs: a review and assessment of the situation in the United Kingdom. *The Pig Journal* 43, 72–79.

Gresham, A., Jackson, G., Giles, N., Allan, G., McNeilly, F. and Kennedy, S. (2000a) PMWS and porcine dermatitis nephropathy syndrome in Great Britain. *Veterinary Record* 146, 143.

Gresham, A., Giles, N. and Weaver, J. (2000b) PMWS and porcine dermatitis nephropathy syndrome in Great Britain. *Veterinary Record* 147, 115.

Harding, J.C. (1996) PMWS. Preliminary epidemiology and clinical presentations. In: *Proceedings Western Canada Association Swine Practitioners*, p. 21.

Harding, J.C. (1999) PMWS. *International Pig Letter* 19, No. 3.

Harding, J.C.S. and Clark, E.G. (1997) Recognising and diagnosing PMWS. *Swine Health and Production* 5, 201–203.

Harding, J.C.S., Clark, E.G. and Ellis, J.A. (1999) The clinical expression of porcine circovirus. *Proceedings of the 1999 Allen D. Leman Swine Conference* 26, 252–255.

Harris, D.L. (1990) Isolated weaning. *Large Animal Veterinarian* May/June 1990, pp. 10–12.

Harris, D.L. (1993) Medicated early weaning. In: *Proceedings of the South-East Swine Practitioners Conference, Raleigh, North Carolina*, pp. 1–19.

Heegard, P.M.H., Klaussen, J., Nielsen, J.P., Gonzalez-Ramon, N., Pineiro, M., Lampreave, F. and Alava, M.A. (1998) The porcine acute phase protein response to infection with APP: haptoglobin, C-reactive protein, major acute phase protein, and serum amyloid A protein are sensitive indicators of infection. *Comparative Biochemistry and Physiology; B Biochemistry and Molecular Biology* 119, 365–373.

Hide, E.J., Connaughton, I.D., Driesen, S.J., Hasse, D., Monckton, R.P. and Simmons, H.G. (1995) The presence of F107 fimbriae and their association with Shiga-like toxin II in *E. coli* strains from weaned pigs. *Veterinary Microbiology* 47, 235–243.

Hoefling, D.C. (1998) Tracking the incidence of porcine respiratory diseases. *Food Animal Practice* April 1998, pp. 391–398.

Holm, A. and Poulsen, N.D. (1996) Swine production management update: zinc oxide in treating *E. coli* diarrhoea in pigs after weaning. *Compendium of Continuing Education Practising Veterinarians* 18, S26–S28.

Huang, H.-S., Potter, A.-A., Campos, M., Leighton, F.A., Willson, P.J. and Yates, W.D.G. (1998a) Pathogenesis of porcine APP: Part 1: Effects of surface components of APP *in vitro* and *in vivo. Canadian Journal of Veterinary Research* 62, 93–101.

Huang, H.-S., Potter, A.-A., Campos, M. Leighton, F.A., Willson, P.J. and Yates, W.D.G. (1998b) Pathogenesis of porcine APP: Part 2: Roles of pro-inflammatory cytokines. *Canadian Journal of Veterinary Research* 63, 69–78.

Imberechts, H., Deprez, P., Harris, D.L. and Whipp, S.C. (1997) Chicken egg-yolk antibodies against F18ab fimbriae *E. coli* inhibit shedding of F18 positive *E. coli* by experimentally infected pigs. *Veterinary Microbiology* 54, 329–341.

Janke, B.H. (1990) Relative prevalence of typical and atypical strains among rotaviruses from diarrhoeic pigs in conventional swine herds. *Journal of Veterinary Diagnostic Investigation* 2, 306–311.

Jobert, J.L., Savoye, C., Cariolet, R., Kobisch, M. and Madec, F. (2000) Experimental aerosol transmission of APP. *Canadian Journal of Veterinary Research* 64, 21–26.

Jukes, T.H., Stokstad, E.L.R., Taylor, R.R., Cunha, T.J. and Edwards, H.Y. (1950) Growth promotion. *Archives of Biochemistry* 26, 324–331.

Larsen, H. (1998) Danish SPF herds: a lesson in precautions. *Pig Progress June 1998*, pp. 46–47.

Lawson, G.H.K. and McOrist, S. (1993) The enigma of proliferative enteropathies: a review. *Journal of Comparative Pathology* 108, 41–46.

Loeffen, W.L.A., Kamp, E.M., Stockhofe-Zurweiden, N., Nieuwstadt, A.P.K.M.I., van Bongers, J.H., Hunnemann, W.A., Elbers, A.R.W., Baars, J., Nell, T. and van Zijderveld, F.G. (1999) Sources of infectious agents involved in acute respiratory disease in finishing pigs. *Veterinary Record* 145, 123–129.

Madec, F., Albina, E., Cariolet, R., Hamon, L., Mahe, D.H., Truong, C., Jestin, A., Amenna, N. and Morvan, H. (2000) Post-weaning multisystemic wasting syndrome in the pig: a new challenge for veterinary research and practice. *Pig Journal* 45, 69–75.

Madsen, D.P. (1995) Managing management-induced *Clostridium perfringens* type A infection in suckling pigs: a case study. *Swine Health and Production* 3, 207–208.

Mendez, E., Lopez, S., Cuadras, M.A., Romero, P. and Arias, C.F. (1999) Entry of rotaviruses is a multi-step process. *Virology* 263, 450–459.

Moore, C. (1995) Using high-health technology in a modern production system. *Proceedings of the Allen D. Leman Swine Conference, University of Minnesota* 22, 15–16.

Negrete-Abascal, E., Tenorio, V.R., Guerrero, L., Garcia, R.M., Reyes, M.E. and de la Gazza, M. (1998) Purification and characterisation of a protease from APP serotype1, an antigen common to all the serotypes. *Canadian Journal of Veterinary Research* 62, 183–190.

Nietfield, J.C., Franklin, C.L., Riley, L.K., Zeman, D.H. and Groff, B.T. (1995) Colonisation of the tracheal epithelium by filamentous bacteria resembling cilia-associated respiratory bacillus. *Journal of Veterinary Diagnostic Investigation* 7, 338–342.

Paradis, S.E., Dubreuil, J.D., Gottschalk, M., Archambault, M. and Jacques, M. (1999) Inhibition of adherence of APP to porcine respiratory tract cells by monoclonal antibodies directed against LPS and partial characterisation of the LPS receptors. *Current Microbiology* 39, 313–320.

Pijoan, C. (1995) Diseases of high-health pigs. Some ideas on pathogenesis. *Proceedings of the Allen D. Leman Swine Conference, University of Minnesota* 22, 16–17.

Potter, R.A. (2000) PMWS in pigs. *Veterinary Record* 146, 84.

Rolsma, M.D., Gelberg, H.R. and Kuhlenschmidt, M.R. (1994) Assay for estimation of rotavirus–cell interactions: identification of an enterocyte ganglioside fraction that mediates group A porcine rotavirus recognition. *Journal of Virology* 68, 258–268.

Runnels, P.L., Moon, H.W. and Schneider, R.A. (1986) Development with host age of resistance to adhesion of 99+ *E. coli* to isolated intestinal cells. *Infection and Immunity* 28, 298–300.

Sanford, S.E. (1995) *Actinobacillus suis*: an overview of an emerging disease. *Proceedings of the American Association of Swine Practitioners* 26, 425–428.

Sanford, S.E., Josephson, G.K.A., Rehmetulla, A.J. and Tilker, A.M. (1990) *Actinobacillus suis* infection in pigs in SW Ontario. *Canadian Veterinary Journal* 31, 443–447.

Smith, W., Thomson, J.R. and Done, S.H. (1993) Dermatitis/nephropathy syndrome of pigs. *Veterinary Record* 132, 47.

Stevenson, G.W., Paul, P.S. and Andrews, J.J. (1990) Pathogenesis of a new serotype of porcine group A rotavirus in the small intestine mucosa of neonatal gnotobiotic and weaned conventional pigs. In: *Proceedings of the American Association of Veterinary Laboratory Diagnosticians*, p. 12.

Straw, B.E., D'Allaire, S., Mengeling, W.L. and Taylor, D.J. (eds) (1999) *Diseases of Swine*, 8th edn. Iowa State University Press, Ames, Iowa.

Taylor, D.J. (1999) *Pig Diseases*, 6th edn. Published by the author.

Thacker, E.L., Halbur, P.G., Ross, R.F., Thanawongnuwech, R. and Thacker, B.J. (1998a) *M. hyopneumoniae* potentiation of PRRS virus induced pneumonia. *Journal of Clinical Microbiology* 37, 620–627.

Thacker, E.L., Thacker, B.J., Boettcher, T.B. and Jayappa, H. (1998b) Comparison of antibody production, lymphocyte stimulation, and protection induced by four commercial *Mycoplasma hyopneumoniae* bacterins. *Swine Health and Production* 6, 107–112.

Thomson, J.R., Smith, W.J. and Murray, B.P. (1998) Investigation into field cases of porcine colitis with particular reference to infections with *Serpulina pilosicoli*. *Veterinary Record* 142, 235–239.

Torremorell, M., Pijoan, C., Janni, K., Walker, R. and Joo, H.S. (1997) Airborne transmission of APP and PRRS virus in nursery pigs. *American Journal of Veterinary Research* 58, 828–832.

Tsay, C.E., Weng, C.N., Pang, F. and Chen, P.C. (1989) A comparison of the immune system and physiological organs between SPF and conventional pigs. *Report of the Pig Research Institute, Taiwan*, pp. 69–74.

Waxler, G.L. and Drees, D.T. (1972) Comparison of body weights, organ weights, and histological features of selected organs of gnotobiotic, conventional, and isolator reared contaminated pigs. *Canadian Journal of Comparative Medicine* 36, 265–274.

White, M.E.C. and Higgins, R.J. (1993) Dermatitis/nephropathy syndrome of pigs. *Veterinary Record* 132, 199.

Wilson, R.A. and Francis, D.H. (1986) Fimbriae and enterotoxin associated with *E. coli* serotypes isolated from clinical cases of porcine colibacillosis. *American Journal of Veterinary Research* 47, 213–217.

Winkelman, N.L. (1995) *E. coli* septicaemia. *Proceedings of the Allen D. Leman Swine Conference, University of Minnesota* 22, 59–60.

Wiseman, B.S., Morrison, R.B., Dial, G.D., Molitor, T., Pijoan, C. and Bergland, M. (1992) Influence of weaning age on pathogen elimination and growth performance of comingled pigs derived by MEW. In: *Proceedings of the 12th International Pig Veterinary Society Congress*, p. 500.

Wittig, W., Klie, H., Gallien, P., Lehmann, S., Timm, N.D. and Tschape, H. (1995) Prevalence of the fimbrial antigens F18 and K88 and enterotoxins and verotoxins among *E. coli* isolated from recovered pigs. *Zentralblatt für Bakterologie* 283, 95–104.

Wray, C. and Woodward, M.T. (1997) *E. coli* infection in animals. In: Sussmann, M.J. (ed.) *E. coli: Mechanisms of Virulence*. University of Cambridge, Cambridge, UK, pp. 49–84.

12

Gut Health: Practical Considerations

P. Baynes[1] and M. Varley[2]

[1]*NuTec Ltd, Eastern Avenue, Lichfield, Staffordshire WS13 7SE, UK;* [2]*SCA Nutrition Ltd, Dalton, Thirsk, North Yorkshire YO7 3HE, UK*

Introduction

The maintenance of gut health in the young weaned piglet is of paramount importance if the young animal is to achieve its full genetic potential for growth. When piglets are weaned from the sow, there is immediate withdrawal of the passive protection afforded by the IgA class antibodies from sow milk and the piglet's active immunity processes are still under development. Furthermore, because of the stress of the weaning event itself, the ability of piglets to mount effective immune responses to antigenic challenge is compromised further.

The weaned piglets under commercial conditions are then moved to new accommodation and mixed with new pen-mates. This whole process leads to rapid changes in the microflora on the gut wall and is especially damaging to the microvilli and to the enterocytes themselves. It is well documented, for example (Miller *et al.*, 1985, 1987), that post-weaning changes in the morphology of the intestinal microvilli are quite profound and play a major role in the post-weaning growth check and high incidence of clinical disease and diarrhoea seen at this time.

Those animals that can maintain a high gut health status will be the ones that will also incur the minimum growth check after weaning; conversely, those piglets that cannot maintain gut health will be the ones that fail to grow at all beyond weaning and will be very prone to morbidity and mortality. Gut health represents a difficult balancing act between, on the one hand, the microorganisms that act as invading opportunists on the gut wall and, on the other hand, the piglet's own immune status defending the integrity of the gut and its important structures for digestion and efficient absorption of nutrients.

Undoubtedly some piglets possess genes that enable them to mount very vigorous mucosal immunity and they will continue to grow rapidly and stay healthy. Others may have genes that promote voluntary food intake and will be able to eat their way out of trouble by maintaining energy and protein intakes despite poor rates of digestibility. The average commercial piglet, however, suffers a post-weaning period of poor gut health, poor food intake patterns and poor growth for perhaps a week or 10 days, until the microflora and the gut wall have recovered. In the course of this relatively short time, significant growth is lost and it has been shown (Varley and Cole, 2000) that this growth is never recovered, all the way through to slaughter. For example, a reduction of 100 g day^{-1} in post-weaning gain may increase the overall time to slaughter by 10 days.

In commercial practice, it has been the received wisdom for many years to use in-feed agents that will help considerably to promote good post-weaning gut health. Among these agents are antimicrobial growth promoters (AGPs), copper sulphate, zinc oxide, probiotics, pre-biotics, herbal extracts and spices, and immunostimulants. Copper and zinc are still widely used in the UK and in most countries around Europe and further afield, and the AGPs have similarly been widely used to good effect. Yet in recent times there has been a rising swell of public opinion, press reports and political activity criticising the use of in-feed antimicrobial growth promoters.

British pig and poultry farmers have used antimicrobial growth promoters for some 30 years and this has facilitated significant economic benefits to both these primary industries. The benefits accrue from their predictable and generally positive influence on gut microflora. More nutrients passing down the gastrointestinal tract are therefore available for growth and for the maintenance of health status. This practice has been monitored and controlled in Great Britain by the Royal Pharmaceutical Society of Great Britain (RPSGB) and by the pharmaceutical industry itself.

Antibiotic Digestive Enhancers

Digestive enhancers have been given many names, including growth promoters, zootechnical additives, pro-nutrients and antibiotic performance enhancers. Growth promoters within the European Community are feed additives (authorized under EU Directive 701524 EEC) that improve the growth rate and efficiency of feed used in cattle, pigs and poultry. Some of the products that have been widely used are: Carbadox, Flavophospholipol, Virginiamycin, Olaquindox, Spiramycin, Zinc Baccitracin, Avilamycin, Tylosin, Salinomycin.

Feed inclusion levels have usually been very low, varying from 50 to 100 ppm (mg kg^{-1}). Typical animal responses might be a 5% improvement in feed conversion efficiency and a 3–6% improvement in growth. Table 12.1 gives data showing typical animal responses. An enormous literature over the years has demonstrated the qualitative and quantitative efficacy characteristics of growth promoters.

Table 12.1. Typical growth responses.

Species	Improvement (%)	
	Average daily gain	FCR
Piglets	5–15	5–8
Growing pigs	3–8	2–4
Finishing pigs	0–3	0–2

Potential problems of overuse of AGPs

AGPs are mild antibiotics and one of the potential pitfalls in their prolonged use is that they apply genetic selection pressure on the population of non-pathogenic gut bacteria normally resident in the gut lumen and bound to the enterocytes. Over time, this selection pressure will favour those bacterial genotypes that are best able to survive the antibiotic environment. Over long time spans, there is the possibility that extremely virulent pathogenic strains will arise out of this forced evolution and the antibiotics will no longer work as bacteriocides. In such a scenario, presumably the antimicrobial product would also lose its ability to enhance digestion and growth.

Another possible route for problems is with the phenomenon of plasmid transfer of bacterial resistance to the antibiotic to other unrelated strains of bacteria that also exist in the general environment. This could potentially be of great significance if, for example, strains of *Salmonella typhimurium* causing typhoid in humans suddenly acquired resistance to the potent antibiotics used against the disease. It is therefore important to regulate which antibiotics are used in animal feedstuffs to ensure that this does not happen. In practice, this has been rigidly applied in the industry and has been subject to rigorous legislation. The antibiotics used as growth promoters are totally separate from those used in human healthcare.

Mode of action

AGPs work in the intestine to maintain a balance between bacteria that help and those that hinder digestion. For example, many of the bacteria that impede digestion are toxin producers that damage the architecture of the villi, leading to a reduction in the absorptive surface. These toxins also increase the rate of migration of the enterocyte cells lining the villi. Many growth promoters act to prevent bacteria adhering to the gut lining by inhibiting the production of adhesive pilae.

The net result of all of these effects is that less food is wasted in maintaining the integrity of the gut lining and there is a larger surface area for absorption through a thinner gut wall (Fiems *et al.*, 1991). There are also specific effects on glucose sparing (prevention of the production of lactic acid) and amino acid sparing (prevention of the production of toxic amines, such as putrescine and cadaverine, in the caecum).

Residues in the meat and the effect on human gut flora

With the notable exception of Carbadox and Olaquindox, growth promoters administered in animal feed do not leave residues in the meat. Indeed, they can be used from birth to slaughter (i.e. they have a zero withdrawal). As there is no detectable residue in the meat, there is unlikely to be a selection pressure on human gut flora.

The use of Carbadox and Olaquindox, however, does lead to residues in meat and they both had a 28-day withdrawal period. These products were limited to use in the first 4 months of a pig's life, which, in practice, gives a minimum withdrawal period of at least 2 months.

The public health implications of the use of antibiotics in animal feeds have been examined by the Netherthorpe Committee in 1962; the Swann Committee in 1969 and the USA FDA Task Force in 1972. All of these have discussed the hazards but none have proved any risk to human health.

Benefits from using antibiotics as digestive enhancers

There are obvious economic gains to be made from production animals growing faster to slaughter weights and also in using feedstuffs far more efficiently. In addition there are welfare benefits, in that animals are healthier throughout the production cycle and will require less veterinary attention.

The increased growth rate and improved feed utilization by cattle, pigs and poultry varies enormously from farm to farm and within farms but probably lies somewhere between 3% and 10%.

In parallel with the economic arguments for the use of growth promoters runs the environmental benefits. These have been quantified in terms of the feed savings in the UK and the tonnages involved are considerable. These savings also reduce the acreage needed under tillage. Improved growth rate and feed conversion also result in reduced slurry production in pigs, cattle and poultry and a reduction in methane output (a major greenhouse gas) in cattle. They were estimated by Verbeke and Viaene (1996) for France, Germany and the UK to represent a reduction of 78,000 t day^{-1} of nitrogen, 15,000 t day^{-1} of phosphorus and 1246×10^6 m^3 day^{-1} methane released into the environment from agricultural sources.

Likely effects of antibiotic withdrawal

The consequences of the removal of AGPs are twofold: (i) a reduction in the efficiency of growth, leading to a reduction in farmer profits; and (ii) an increased amount of food and water needed to grow an animal to slaughter weight, leading to increased demand for agricultural land and drinking water. As food and water requirements rise, the impact of the animal system on the general environment will also be greater.

If antibiotic growth promoters are banned then increased slurry production is likely to increase significantly the pollution of water sources and waterways with nitrates and phosphates (OECD, 1986). Slurry, while waiting to be spread, will also produce ammonia, which is potentially toxic to housed animals, as well as contributing to acid rain. The carry-over effect of growth promoters in the faeces can reduce the ammonia output by up to 33%.

In cattle, the use of Monensin can reduce methane output by some 20%. This is of particular importance, as methane is a potent greenhouse gas and cattle are one of its major worldwide sources. The concentration of methane in the atmosphere is currently increasing at the rate of 1% per year, of which 60% is derived from animals. The US Environmental Protection Agency has recommended the use of ionophores, such as Monensin, as a method of reducing ruminant methane production. Verbeke and Viaene (1996) estimated a saving of 1246×10^6 m^3 day^{-1} from cattle in France, Germany and the UK.

The consequences of the withdrawal of growth promoters discussed in this chapter can be summarized as follows: (i) reduced profitability of local farming enterprises; (ii) increased feed requirements to produce the same weight of animal; (iii) increased number of animals to maintain same output of meat (this could be seen as increased imports if the UK industry does not respond); (iv) increased need for drinking water, much of which would be from the mains; (v) increase in slurry output and area of land needed for spreading; (vi) increased need for fossil fuels to transport feeds, greater numbers of animals and slurry; (vii) increased area of tillage to grow extra grain; (viii) increased imports of expensive protein sources; and (ix) increased environmental pollution.

Alternative Tools

There are alternatives to the use of antibiotic growth promoters for creating conditions where animals can exploit their full genetic potential for growth. Some of these have been around for many years and have been well researched. If conventional antimicrobials were to be precluded from use in the animal industries then there may be a significant increase in demand for such products. These alternatives include: (i) improved husbandry (ventilation, stocking density, grouping, all-in/all-out production flows); (ii) improving on-farm hygiene status; (iii) use of effective vaccination schemes; (iv) in-feed enzymes to promote digestion; (v) improved nutritional specification to increase nutrient and energy retention; (vi) organic acids to optimize conditions in the gut lumen microenvironment; (vii) use of herb and spice formulations, some of which have been shown to have natural growth-enhancing properties; (viii) probiotics (lactobacillus cultures that can control the conditions and microflora on the gut wall to ensure pathogenic exclusion and improved digestion); (ix) immuno-potentiating agents (either natural nutrients such as vitamin E or other compounds to improve natural mucosal immunity); and (x) prebiotics, amongst which are non-digestible sugars such as oligosaccharide molecules, which can also enhance the processes of digestion.

Thus a wide variety of tools and methods is available at present for use in animal production systems to replace existing AGPs. Amongst these, the oligosaccharides hold promise due to the fact that they are natural constituents of a whole host of plant materials. Fructo-oligosaccharides (FOS), for example, have been widely tested in animals and are already used routinely in the human food industry in many products, including yogurts. The addition of short-chain specific FOS to the diets of young growing animals can promote lactobacilli and other beneficial bacterial serotypes on the gut wall and at the same time will impede the proliferation of potential pathogenic types such as coliforms and salmonellas. This changes the pH in the gut lumen in the same way as organic acids and at the same time the lactobacillus population reduces the pathogenic loading on the animal by competitive exclusion. All of this leads to enhanced digestion and improved daily liveweight gain.

Current State of Play

There has been a chain of events since 1998 leading to the present position where the use of in-feed digestive enhancers is under threat of a blanket ban. The major important milestones in this are as follows.

The House of Lords Select Committee report (1998) called for the prudent use of antibiotics in human and animal health to ensure that they remain useful in the future (this is fully supported by the pharmaceutical industry). The report acknowledged that growth promoters represent the difference between profit and loss for the pig and poultry industry. Their removal would result in a considerable welfare issue and an increased use of therapeutic antibiotics. The report also described this experience in Sweden.

The report documented selected opinions on the possible hazards to human health of the use of growth promoters. It recommended further research in order to establish whether there is a real risk to human health. This work is in hand and needs to be assessed before any phasing out of growth promoters is undertaken.

It is important to note that growth promoters are fully licensed and have been used safely in food production for nearly 30 years. This long history of usage, in billions of animals, indicates that the possible hazards have not been translated into real risks to human health.

Future Scenarios

The pharmaceutical industry is obviously concerned about the potential withdrawal of some of its products, though in the medium term the belief is that the slack will be taken up by the application of therapeutic quantities of the same antibiotics. These will be applied to counter the huge increase in disease incidence that might be the outcome in many herds and flocks.

Companies involved in the production and distribution of antibiotic growth promoters also have a series of arguments that they believe will maintain the status quo. These arguments can be summarized as follows.

- There is no evidence over the last 25 years that bacterial serotypes have acquired changed resistance to the antibiotics currently in widespread use.
- There is only limited evidence that plasmid transfer of resistance has occurred in the bacterial populations with which humans come into regular contact.
- The alternatives to in-feed growth promoters are not effective. These may include acidulants, probiotics, cell fragments, yeasts, oligosaccharides and herbal formulations.
- The blanket ban on in-feed antibiotic growth promoters would cause havoc in the livestock industries. Mortality and morbidity would run at a very high level and this would be a retrograde step in terms of animal welfare. This exactly reflects the Swedish experience.
- Evidence from the use of vancomycin against *Staphylococcus aureus* strains in the human population indicates that there has been no increase in problems due to agricultural practices.
- Any changes in bacterial populations affecting human health are a direct result of medical practices over the last 25 years; over-prescription of antibiotics to humans is largely to blame.

The argument is also used by the pharmaceutical companies that the current pressure to ban growth promoters is largely a political one coming from the Nordic countries within the EU who wish to create a 'level playing-field' for their own advantage.

Some of these arguments are undoubtedly strong but there are considerable problems with the overall case.

- There is no evidence that, over a period of about 25 years since the mid 1970s, bacterial serotypes have acquired changed resistance to the antibiotics currently in widespread use. The pharmaceutical industry claims that the data exist but this important information has yet to be published.
- The Swedish experience points the way to what ought to happen. That is, by the deployment of best-practice animal production, using appropriate stocking densities and stockmanship, environmental control and nutrition, there is no reason why disease incidence should rise significantly.
- Discounting of the alternatives is largely unjustified. There is a growing database to show that oligosaccharides, for example, can be highly effective alternatives to conventional antibiotic growth promoters; in addition, probiotics are improving all the time.

It is also well understood that even potent antibiotic growth promoters that are consistent in their responses will not work well, if at all, on farms that already have a very high degree of management expertise and input.

Conclusions

Pig and poultry production is now a global market and Europe will have to compete if its farmers are to remain in business. Looking to the future, it is important to recognize the impact of the General Agreement on Tariff and Trade, World Trade Organization (GATT WTO) agreements, which for the first time pulled agriculture into the main stream of GATT in 1995. It must also be borne in mind that the Common Agricultural Policy (CAP) has created a high-cost agricultural system, with the result that European farmers cannot compete in the world market without subsidies and that European consumers have to pay artificially high prices for their food.

The Uruguay Round Agreement (1994) was a major effort to open world agricultural markets, prompting increased trade and dynamic growth. Its effect on Europe will be to reduce its export subsidies, allow other markets access to the EU and reduce its internal support structures. In other words, prices will become closer to those of the world market. In order to compete in this market, Europe must have the same tools as the rest of the world, otherwise its chicken and pork will be imported from the USA or Asia.

It is of vital importance that science is allowed to prevail. Only in this way will European farmers compete on a level playing-field and European citizens be able to eat high-quality home-produced meat, eggs and milk. Local quality assurance schemes depend on there being local producers. If they cannot compete, there will have to be reliance on distant suppliers.

In conclusion, it seems possible that antibiotic growth promoters could be totally withdrawn from use very shortly in the UK. The pharmaceutical companies will inevitably react against this move but it could be that they will be overridden by political expediency. The animal industry will accordingly have to rethink its production technology and practices. In the short term there will be increased disease and this will cause hardship.

The demand for feedstuffs will probably rise and thus there will be firmer demand for feed supplements. In the medium and long term, however, this transition will enable Europe to produce high quality animal products under improved conditions using new technologies of nutrition, environmental control, disease control and management. There will hence be a demand for quality nutritional management and more management consultancy. This latter scenario is what the increasingly powerful consumer really wants.

References

American Society of Animal Science (1986) Public health implications of the use of antibiotics in agriculture. 77th Annual Meeting. *Journal of Animal Science* 62, (Supplement 3), 1–106.

Association of Veterinarians in Industry (1979) *Proceedings of the Conference 'Ten Years after Swann'*. AVI, London.

Council for Agricultural Science and Technology (1981) *Antibiotics in Animal Feeds.* Report No. 88, CAST, Ames, Iowa.

Fiems, L.O., Cottyn, B.G. and Demeyer, D.I. (1991) Animal biotechnology and the quality of meat production. OECD Workshop, Belgium, 7–9 November 1990. In: *Development in Animal and Veterinary Sciences* 25, 261–280.

Food and Drug Administration (1972) *Report of the FDA Task Force on the Use of Antibiotics in Animal Foods.* US Department of Health, Education and Welfare, Washington, DC.

House of Lords Select Committee (1998) *Select Committee on Agriculture and Food Safety, 4th Report.* HMSO, London.

OECD (1995) *Agricultural Outlook 1995–2000.* Organisation for Economic Co-operation and Development, France 81, 77–79.

Swann, M. (1969) Report Joint Committee on the Use of Antibiotics in Animal Husbandry and Veterinary Medicine. Her Majesty's Stationery Office, London.

Uruguay Round (1994) *Global Agreements – Global Benefits.* Office for Official Publications of the European Communities, Brussels, Luxembourg, pp. 5–11, 14–15, 27.

Varley, M.A. and Cole, M.A. (2000) Boosting post-weaning gains. *Pig Farming* 48(9), 2–3.

Verbeke, R. and Viaene, J. (1996) *Environmental Impact of Using Feed Additives. Economic Implications and Legal Environment in Benelux.* Department of Agro-marketing, Faculty of Agricultural Science, University of Ghent, Belgium, 38 pp.

Aerial Pollutants from Weaner Production

13

C.M. Wathes

Silsoe Research Institute, Bio-Engineering Division, Wrest Park, Silsoe, Bedford MK45 4HS, UK

Introduction

At weaning, the piglet may be faced with several changes in its physical environment. There should be no reason for a sub-optimal thermal environment provided that the weaner building has been designed correctly and its control equipment is functioning satisfactorily; the specifications for environmental temperature have been understood for several decades (Bruce, 1981) and the principles and practice of ventilation are well known (Randall and Boon, 1994). However, the newly weaned piglet will have been exposed from birth to aerial pollutants that would not arise in its ancestral habitat and the exposure will continue throughout its life. These pollutants provide an additional physical stressor to the many others that impinge upon the piglet at weaning.

In the modern weaner building, the environmental control system, especially ventilation, now has to satisfy a number of objectives that may have conflicting requirements. On the one hand, a slow ventilation rate is needed in winter to reduce heat loss from the building, though this may lead to a high concentration of aerial pollutants. Conversely, fast ventilation rates in the summer are required to keep the building and piglets cool but the generation rate of aerial pollutants may be higher as a result. Furthermore, intensive pig production contributes to global emissions of important aerial pollutants, such as ammonia and nitrous oxide, that impact adversely upon the countryside. Thus the traditional solution of discharging unwanted aerial pollutants to the atmosphere is no longer tenable given new legislation on pollutant emissions, e.g. Integrated Pollution Prevention and Control (IPPC). There is also very strong evidence (e.g. Donham *et al.*, 1995; Reynolds *et al.*, 1996) for occupational respiratory disease in those who work with pigs; this is believed to arise from chronic exposure over several years to the complex mixture of aerial pollutants in piggeries.

There are, therefore, several reasons why the generation of aerial pollutants in a weaner building should be minimized. This chapter considers some of the mechanisms by which aerial pollutants affect the health and performance of weaner pigs, as well as the importance of weaner production as a source of environmental pollution.

A Natural History of Aerial Pollutants in Weaner Buildings

The air within a weaner building seethes with a dense miasma of bio-aerosols and gases. Both composition and concentration of the miasma vary according to animal husbandry and the design and management of the building. Table 13.1 lists the concentration and emission rates of the common aerial pollutants from short-term measurements made over 24 h in a large survey of 64 pig houses in four countries in northern Europe. Amongst all classes of pigs, the mean mass concentration of dust and endotoxin was highest in weaner buildings with slats. For comparison, the current UK occupational exposure limit for human health is 10 mg m^{-3} for total inhalable dust and 4 mg m^{-3} for the respirable fraction of dust (Health and Safety Executive, 1997), while Donham *et al.* (1995) recommended 2.5 and 0.23 mg m^{-3}, respectively. Both of these recommendations were made on grounds of human health. The composition of the dust was not determined in this survey. Table 13.2 shows that dust can be characterized in a variety of ways, though most workers restrict themselves to mass concentration – presumably because of the tedium and expense involved in identifying and classifying individual dust particles. Painstaking work by Heber *et al.* (1988) showed that, in samples from finishing buildings, feed was the major source of particles over 5 μm; starch particles had a geometric median diameter (GMD) of 12.5 μm and a mass median diameter (MMD) of 21.0 μm and the corresponding values for grain meal were 8.6 μm and 17.9 μm, respectively. For all dust particles, the GMD was 2.6 μm and the MMD was 18.5 μm. The finer respirable particles are difficult to identify microscopically and Donham *et al.* (1986) suggested that faecal material is the predominant source of particles of about 1–2 μm diameter, while hair and skin account for only 1% and 10% of all particles with a diameter between 11 and 16 μm (Honey and McQuitty, 1979). With the exception of these quantitative studies, most authors are content to state merely that the main sources of pig dust are feed, faeces, bedding and skin squames.

Most studies of noxious gases in piggeries have focused upon ammonia, partly because it is toxic but also because of its role in acid rain. However, over 100 gaseous compounds have been identified in the air of livestock buildings (Hartung, 1988); most are simple odorants (which may still give rise to complaint amongst neighbours) but some are greenhouse gases. The concentrations of most of these gases are usually in the range of parts per million or lower with the exception of carbon dioxide, where the concentration can be 5–10 times higher than ambient when the ventilation rate is slow. The mean concentrations of ammonia in Table 13.1

Table 13.1. Mean concentrations and emissions of aerial pollutants in pig housing in northern Europe

	Mean concentration					Mean emission rate		
	Inhalable dust (mg m^{-3})	Respirable dust (mg m^{-3})	Inhalable endotoxin (ng m^{-3})	Respirable endotoxin (ng m^{-3})	Ammonia (ppm)	Inhalable dust emission (mg h^{-1} per pig	Respirable dust emission (mg h^{-1} pig	Ammonia emission (mg h^{-1} per pig)
Sows on litter (*n* = 16)								
England	0.63	0.16	38.0	2.2	5.1	57	23.1	303
Germany	1.64	0.12	566.7	52.4	12.5	301	18.2	1298
Sows on slats (*n* = 32)								
Denmark	3.49	0.46	25.8	4.2	8.7	408	60.6	730
England	0.86	0.09	32.6	0.9	11.0	59	5.7	503
Germany	1.13	0.11	7.8	6.4	10.2	47	5.6	325
Netherlands	1.20	0.13	64.4	2.3	17.8	64	7.4	535
Weaners on slats (*n* = 32)								
Denmark	3.37	0.15	193.5	19.9	5.3	43	1.6	45.8
England	5.05	0.43	41.4	9.8	7.8	17	1.5	26.0
Germany	2.80	0.29	14.4	2.7	4.5	24	2.3	22.0
Netherlands	3.74	0.32	351.3	32.6	4.6	78	7.4	26.6
Finishers on litter (*n* = 16)								
Denmark	1.21	0.10	178.0	21.0	9.1	92	7.1	394
England	1.38	0.15	134.0	9.9	4.3	57	6.5	108
Finishers on slats (*n* = 34)								
Denmark	2.08	0.16	100.0	7.7	14.9	74	7.0	391
England	2.67	0.29	106.0	8.8	12.1	55	8.4	185
Germany	2.31	0.18	99.7	10.4	14.3	78	7.4	308
Netherlands	2.61	0.24	101.2	12.6	18.2	67	4.3	385

Original sources: dust, Takai *et al.* (1998); endotoxin, Seedorf *et al.* (1998); ammonia, Groot Koerkamp *et al.* (1998).
Each of 64 buildings was surveyed over 24 h once in winter and in summer; two extra buildings were included in the German survey in summer.

Table 13.2. Important properties of airborne dust (Wathes *et al.*, 2000).

Physical	Concentration of particles by number and mass Size distribution of particles by number and mass
Chemical	Chemical composition, particularly of toxins and allergens Source materials
Microbiological	Number of viable and non-viable bacteria, viruses and fungi, including fungal propagules Endotoxin content

mask the short-term fluctuations of hourly concentrations in weaner buildings, which ranged between 17.9 and 36.7 ppm in the four countries (Groot Koerkamp *et al.*, 1998).

The final category of aerial pollutants in weaner buildings is microorganisms and their components. The majority of these will be non-pathogenic gram-positive bacteria (Crook *et al.*, 1991) at a concentration of approximately 10^6 colony-forming units (CFU) m^{-3}. Smaller numbers ($\leq 10^5$ CFU m^{-3}) of gram-negative bacteria and fungi will also be found (Seedorf *et al.*, 1998). The majority of airborne microbes will be non-pathogenic: some opportunistic pathogens (such as *Pasteurella multocida*) and primary pathogens (e.g. African swine fever virus and *Bordetella bronchiseptica*) can be isolated from the air in numbers that depend on the shedding rate from the host and their viability whilst airborne. Endotoxins arise from the breakdown of the outer cell wall of gram-negative bacteria and have been implicated in occupational respiratory disease in pig stockmen; typical concentrations range between 14 and 351 ng m^{-3} for the inhalable fraction and 2.7 and 32.6 ng m^{-3} for the respirable fraction (Table 13.1).

The variety of the sources and types of aerial pollutants in weaner buildings poses several problems for abatement. Firstly, clear specifications have yet to be set in terms of animal production and health, though Wathes *et al.* (2000) aim to provide these specifications for airborne dust and ammonia and Donham *et al.* (1995) proposed tolerable limits for aerial pollutants to maintain human health. Secondly, it is not clear whether attempts to control the burden of one pollutant may exacerbate exposure to another. The development of abatement techniques is a topic of active research. Good progress has been made in the use of oil spraying to reduce airborne dust (e.g. Takai *et al.*, 1993), which works by minimizing the resuspension of dust after it has settled within the building. Although its adoption is less advanced, one promising technique to reduce ammonia emissions from pig buildings and waste stores is dietary manipulation to lower the excretion of urea and proteins (Phillips *et al.*, 1998). Environmental control in a weaner building therefore means more than a thermostat operating a ventilation system; in the future, integration of both thermal and air quality criteria will be necessary. One important example of the interaction between these factors concerns the pattern of airflow. Work by Randall (1975) and others established the principles of mechanical ventilation for livestock buildings using air temperature as the control variable. Randall (1975) showed that a stable airflow pattern was critical to the maintenance of an optimum thermal environment and subsequently devised control systems based on sound physical principles. However, the importance of minimizing exposure to aerial pollutants had not been appreciated when the work was done and it is only now being recognised that the two criteria may lead to conflicting requirements for a ventilation system.

Mechanisms by which Aerial Pollutants Affect Weaner Pigs

Avoidance of aerial pollutants – initial studies with ammonia

The ancestors of the domestic pig evolved in a woodland habitat in which pollutant gases were not present at concentrations typically found in modern weaner buildings. There can be no reason *a priori* for the pig to have developed adaptive behaviour when faced with these gaseous pollutants. On the other hand, as the pig roots through the woodland soil, it is likely to be exposed to a heavy burden of inhaled dust particles, which are filtered effectively by the turbinates in the snout. While Woods *et al.* (1993) have shown that the concentration of dust measured at a horse's muzzle is much higher than that at fixed location within a stable, there are no complementary studies of the dust inhaled by the pig during rooting. Furthermore, the nature of the inhaled dust from soil and from a piggery will be quite different.

Ammonia gas is an irritant that, in humans, is detectable at 5–50 ppm, causes irritation of mucous surfaces at 100–500 ppm after 1 h and is rapidly lethal after exposure to 10,000 ppm (Nordstrom and McQuitty, 1976). Although the occupational exposure limit is 35 ppm for a short-term exposure of 15 min or less and 25 ppm over 8 h (Health and Safety Executive, 1997), the initial reaction of most people to such atmospheres is avoidance, followed by habituation if the exposure is prolonged.

A similar avoidance of ammonia has been observed in juvenile pigs (Jones *et al.*, 1996). In a free-choice preference test, pigs made fewer visits of shorter duration to ammoniated atmospheres vs. fresh air (Table 13.3). Overall, 80% of their time was spent in an atmosphere of 10 ppm or lower, indicating a clear preference for fresh air. Although only a small proportion of time was spent in 20 or 40 ppm ammonia, the length of each visit suggested a delayed aversion to ammonia. In subsequent experiments, either single or pairs of juvenile pigs were given a forced choice between either thermal comfort or fresh air (Jones *et al.*, 1999). Heat was provided along with 40 ppm ammonia (HP) in one compartment while the other was unheated and contained fresh air (FA). As the air temperature fell below the animals' lower critical temperature (16.3–21.0°C), the single pigs became increasingly motivated for warmth rather than fresh air. The mean duration of the visits to HP was six times longer than to FA, over an air temperature range from 0 to 15°C (Table 13.4). Paired pigs were also given a choice between HP and FA, which was provided in four compartments so that individuals could make separate choices. In this case, motivation for companionship was stronger than the preference of an individual for an alternative environment. As before, the paired pigs increasingly preferred HP over FA as the air temperature fell.

Table 13.3. Back transformation (from a logit) of mean relative time spent, visit number and visit duration by all pigs to each ammonia concentration ($n = 8$ pigs) (Jones *et al.*, 1996).

	Nominal ammonia concentration (ppm)			
	0	10	20	40
Time spent (%)	53.4	26.9	7.1	5.1
Visit number	46.2	37.1	21.7	17.5
Visit duration (min)	101.4	72.0	39.6	32.1

Table 13.4. Mean and standard error (SE) of the number and average duration of visits made to each option (Jones *et al.*, 1999).

	Single pigs ($n = 8$)				Paired pigs ($n = 8$)			
	Heated 40 ppm ammonia (HP)		Unheated fresh air (FA)		Heated 40 ppm ammonia (HP)		Unheated fresh air (FA)	
Choice option	Mean	SE	Mean	SE	Mean	SE	Mean	SE
Visit number	27.1[a]	0.1	25.9[b]	0.1	21.4[a]	0.1	20.1[b]	0.1
Average visit duration (min)	207.6[a]	19.4	30.5[b]	19.4	264.5[a]	19.4	29.1[b]	19.4

[a,b]Within an experiment, means with different superscripts are significantly different ($P < 0.001$).

These findings demonstrate that juvenile pigs prefer to maintain thermal comfort rather than endure a cold environment of fresh air. The reasons for the delayed aversion are unknown but, clearly, sudden exposure to such high concentrations of ammonia was not sufficiently aversive for the animals to leave immediately. Jones *et al.* (1996) suggested that the animals may have gradually developed a sense of malaise, which eventually drove them to seek fresh air: presumably, the domestic pig has not evolved a set of behavioural and physiological mechanisms that would allow it to make the necessary adaptive responses in the presence of noxious atmospheres of ammonia.

Aerial pollutants and olfaction

Olfaction mediates many aspects of a pig's interaction with its social and physical environment, e.g. sexual attraction, maternal bonding, individual recognition and food location. Jones *et al.* (2000, 2001) and Kristensen *et al.* (2001) addressed the question of whether acute and chronic exposures to atmospheric ammonia affect olfactory perception in the pig by either masking or desensitization of olfactory receptors, respectively. Several tests of olfactory perception were devised and applied

Table 13.5. Effects of acute and chronic exposure to ammonia on olfactory perception in the juvenile pig.

Perceptual test	Ammonia exposure	Results	Source
Location of buried food parcels	Acute – 40 ppm ammonia during test	No effect	Jones *et al.*, 2000
Olfactory acuity for *n*-butanol	Acute – 40 ppm ammonia during test	No effect	Jones *et al.*, 2001
	Chronic – 40 ppm ammonia for ≈ 3.5 weeks prior to test	Reduced acuity in 3 of 6 pigs	Jones *et al.*, 2001
Olfactory cues used in social recognition	Chronic – 40 ppm ammonia for ≈ 2 weeks prior to test	No effect	Kristensen *et al.*, 2001

either after chronic exposure to ammonia at 40 ppm for several weeks or during acute exposure to the same gas (Table 13.5).

The mechanisms by which ammonia exposure could affect olfactory perception include: (i) changes in respiration rate, which may alter the volume of air reaching the olfactory receptors; (ii) preferential occupation of receptors by ammonia at the expense of other odorants; (iii) suppression of the perceived intensity of another odorant; and (iv) chemical reaction of ammonia with other odorants to produce new compounds. However, there was only limited support for the desensitization of olfactory receptors by chronic ammonia exposure and no support for the masking hypothesis. This one abundant gaseous pollutant of weaner and finisher houses does not appear to interfere with olfactory perception.

Aerial pollutants and respiratory disease

The effects of aerial environment in pig productivity were reviewed by de Boer and Morrison (1988). The major conclusions still apply today and were that: (i) the tolerance limits for aerial exposure have not been defined; (ii) potential interactions between aerial pollutants have rarely been examined; (iii) dust plays an important part in the aetiology of disease; (iv) dusts and gases may reduce productivity directly, or indirectly by affecting health; (v) respiratory diseases are of great economic importance worldwide; and (vi) the key features of building design and management to control pollutant exposure are not fully understood.

There is good clinical evidence that poor air quality affects the incidence and severity of common endemic respiratory diseases, such as porcine reproductive and respiratory syndrome, swine influenza and enzootic pneumonia. These diseases are of commercial importance, with no effective vaccine available against many respiratory pathogens. The effects of respiratory disease on pig growth and food conversion ratio (FCR) are substantial. Muirhead and Alexander (1997) state that FCR and the number of days to reach 90 kg are numerically increased by 0.1–0.3 and 4–15 days for *Actinobacillus pleuropneumonia*, 0.1–0.2 and 4–15 days for atrophic rhinitis and 0.05–0.1 and 3–12 days for enzootic pneumonia, respectively, during the period of chronic disease.

Much of the early research on lesions induced by exposure to ammonia and dust used concentrations far in excess of those found in piggeries (see Table 13.1) for short durations in the absence of specific respiratory pathogens (Done, 1991). For example, Drummond *et al.* (1980) reported tracheal and turbinate exudation at 500 ppm ammonia, while Doig and Willoughby (1971) observed tracheal epithelial hyperplasia at 100 ppm ammonia and either 200 mg corn starch m^{-3} or 10 mg corn dust m^{-3}. Conversely, Diekman *et al.* (1993) found no difference in the percentage of lung consolidation and snout grade in gilts exposed to low (4–12 ppm) or moderate (26–45 ppm) ammonia concentrations. The most convincing epidemiological evidence is that of Robertson *et al.* (1990), who found a strong association between commercial concentrations of aerial pollutants and the incidence and severity of atrophic rhinitis.

More recently, Hamilton *et al.* (1996) showed that ammonia exposure of weaned pigs not only raises the severity of turbinate atrophy induced by *P. multocida* but also that the damage is maximal at 10–15 ppm and decreases at concentrations above 25 ppm. This surprising result was explained as the net effect of two mechanisms: (i) enhanced colonization of the nasal cavity by *P. multocida* during ammonia exposure, with ammonia providing a source of nitrogen for the bacteria (Hamilton *et al.*, 1998a); and (ii) separate but additional turbinate atrophy following ammonia exposure alone (Hamilton *et al.*, 1996). Whether these mechanisms also apply to other specific respiratory diseases is not yet known but these results, in the first instance, have important implications for specification of an acceptable concentration of ammonia in weaner houses.

The mechanisms by which dust affects respiratory disease are likely to be different from those for ammonia. Organic dusts will be immunogenic (Rylander, 1986) while inorganic dusts may block mucociliary clearance. In a related study, Hamilton *et al.* (1998b) reported an increase in turbinate atrophy with ovalbumin dust exposure in weaned pigs following *P. multocida* infection. Simultaneous exposure to both ovalbumin dust (20 mg m^{-3} total mass) and ammonia (50 ppm) caused greater turbinate atrophy than exposure to either pollutant alone (Hamilton *et al.*, 1999).

The mechanisms by which aerial pollutants are involved in the aetiology of porcine respiratory disease are complex (Wathes, 1998) and require consideration of specific pathogens, commensal respiratory microflora and host-specific factors as well as the nature of the pollutants themselves. Simply put, the question is whether the incidence and severity of respiratory disease in weaner pigs are greater when combined with chronic exposure to aerial pollutants. This is being addressed in a large experiment by Wathes *et al.* (2000).

Environmental Impact of Aerial Pollutant Emissions from Weaner Production

Table 13.6 lists the common aerial pollutants emitted from pig buildings and the reasons for concern over their impact upon the environment. In the UK, pig production is not responsible for the bulk of gaseous emissions that are emitted from livestock housing and manure stores and during manure spreading: pigs account for about 14, 3 and 3% of the total emissions of ammonia, methane and nitrous oxide, respectively (Phillips and Pain, 1998). The dominant sources of these gases are cattle, followed by poultry.

Interest is now being shown in livestock production as a source of particulate aerosols because of the association that has been demonstrated between fine dust (PM10 – particulate matter less than 10 μm in size) and human respiratory disease

Table 13.6. Common aerial pollutants emitted from intensive pig housing (Phillips and Pain, 1998).

Type of gas	Mechanism(s) of production	Reasons for concern
Ammonia	Enzymic degradation of urine, or, in the case of poultry, uric acid. Microbial (anaerobic) degradation of faeces	Contributes to acid rain Upsets natural ecosystems by deposition of N Increases need for N fertilizer on farmland, which brings both water pollution and economic penalties Implicated in the aetiology of environmental respiratory diseases of livestock
Methane	Enteric fermentation, especially in ruminants. Microbial (anaerobic) degradation of excreta	Greenhouse gas
Nitrous oxide	Incomplete microbial denitrification or nitrification of mixed bedding and excreta	Greenhouse gas Harms ozone layer
Carbon dioxide	Animals' metabolism. Microbial action on excreta	Asphyxiant, if allowed to accumulate Greenhouse gas, though this source is mostly non-fossil in origin
Hydrogen sulphide	Microbial (anaerobic) degradation of faeces	Toxic
Odour (can contain traces of well over 100 gases)	Microbial degradation, especially anaerobic	Nuisance

in urban areas. An inventory of PM10 emissions from livestock production in the UK has not been published, but this source – of which pigs comprise a significant part – is probably about 10% of total emissions. Equally there is uncertainty whether dusts from pig production offer a health hazard over and above that provided by urban PM10.

Traditionally, the solution to aerial pollution within a pig building was to discharge the pollutants to the atmosphere via the ventilation exhaust air. This policy is no longer acceptable, given the above concerns. Indeed, awareness of the strength of agricultural sources of greenhouse and other gases has prompted European legislation to reduce emissions from intensively housed livestock. The European Union Directive on Integrated Pollution Prevention and Control (Anon., 1996) has been implemented and embraces pig farms comprising at least 750 sows or 2000 growing pigs of over 30 kg liveweight. It requires these pig farmers to limit the emissions of gases by the best available technology not entailing excessive cost (BATNEEC). The EU Directive itself does not specify limits or operating procedures: each Member State has been asked to draw up its own system. Forthcoming EU legislation includes the Acidification Strategy, which aims to protect sensitive ecosystems in Europe by reducing atmospheric deposition to less than the 'critical loads'. In turn, 'national ceilings' will be placed on the emission of ammonia and other gases. Once these ceilings have been fixed, each Member State will need to apportion the reductions between different sources.

Conclusions

Concern over aerial pollutants from weaner production has arisen on three grounds: pig health and performance, stockman health and environmental pollution. While the evidence for the first concern is tentative, that for the others is much stronger and implies that action must be taken to minimize emissions. This task is hampered by the lack of quantitative emission targets and, in the case of dust, identification of the noxious or toxic component(s). Nevertheless, some abatement techniques show promise and the challenge will be to integrate them into the overall system of environmental control in the weaner house.

References

Anon. (1996) Council Directive 96/61/EC of 24th September 1996, concerning integrated pollution prevention and control. *Official Journal of the European Communities*, No. L257 (10 October 1996), 26–40.

de Boer, S. and Morrison, W.D. (1988) *The Effects of the Quality of the Environment in Livestock Buildings on the Productivity of Swine and Safety of Humans.* University of Guelph, Ontario, 121 pp.

Bruce, J.M. (1981) Ventilation and temperature control criteria for pigs. In: Clark, J.A. (ed.) *Environmental Aspects of Housing for Animal Production.* Butterworths, London, pp. 197–216.

Crook, B., Robertson, J.F., Travers, S.A., Botheroyd, E.M., Lacey, J. and Topping, M.D.

(1991) Airborne dust, ammonia, microorganisms and antigens in pig confinement houses and the respiratory health of exposed farm workers. *American Industrial Hygiene Association Journal* 52, 271–279.

Diekman, M.A., Scheidt, A.B., Sutton, A.L., Green, M.L., Clapper, J.A., Kelly, D.T. and van Alstine, W.G. (1993) Growth and reproductive performance, during exposure to ammonia, of gilts afflicted with pneumonia. *American Journal of Veterinary Research* 54, 2128–2131.

Doig, P.A. and Willoughby, R.A. (1971) Response of swine to atmospheric ammonia and dust. *Journal of the American Society Veterinary Medical Association* 159, 1353–1361.

Done, S.H. (1991) Environmental factors affecting the severity of pneumonia in pigs. *Veterinary Record* 128, 582–586.

Donham, K.J., Scallon, L.J., Popendorf, W., Treuhaft, M.W. and Roberts, R.C. (1986) Characterisation of dusts collected from swine confinement buildings. *American Industrial Hygiene Association Journal* 47, 404–410.

Donham, K., Reynolds, S., Whitten, P., Merchant, J., Burmeister, L. and Popendorf, W. (1995) Respiratory dysfunction in swine production facility workers: dose-response relationships of environmental exposures and pulmonary function. *American Journal of Industrial Medicine* 27, 405–418.

Drummond, J.G., Curtis, S.E., Simon, J. and Norton, H.W. (1980) Effects of aerial ammonia on growth and health of young pigs. *Journal of Animal Science* 50, 1085–1091.

Groot Koerkamp, P.W.G., Metz, J.H.M., Uenk, G.H., Phillips, V.R., Holden, M.R., Sneath, R.W., Short, J.L., White, R.P., Hartung, J., Seedorf, J., Schröder, M., Linkert, K.H., Pedersen, S., Takai, H., Johnsen, J.O. and Wathes, C.M. (1998) Concentrations and emissions of ammonia in livestock buildings in Northern Europe. *Journal of Agricultural Engineering Research* 70, 79–95.

Hamilton, T.D.C., Roe, J.M. and Webster, A.J.F. (1996) The synergistic role of gaseous ammonia in the aetiology of *P. multocida*-induced atrophic rhinitis in swine. *American Journal of Clinical Microbiology* 34, 2185–2190.

Hamilton, T.D.C., Roe, J.M., Hayes, C.M. and Webster, A.J.F. (1998a) Effects of ammonia inhalation and acetic acid pretreatment on the colonisation kinetics of toxigenic *Pasteurella multocida* within the upper respiratory tract of swine. *Journal of Clinical Microbiology* 36, 1260–1265.

Hamilton, T.D.C., Roe, J.M., Hayes, C.M. and Webster, A.J.F. (1998b) Effect of ovalbumin aerosol exposure on colonisation of the porcine upper airway by *Pasteurella multocida* and effect of colonisation on subsequent immune function. *Clinical and Diagnostic Laboratory Immunology* 5, 494–498.

Hamilton, T.D.C., Roe, J.M., Hayes, C.M., Jones, P., Pearson, G.R. and Webster, A.J.F. (1999) Contributory and exacerbating roles of gaseous ammonia and organic dust in the etiology of atrophic rhinitis. *Clinical and Diagnostic Laboratory Immunology* 6, 199–203.

Hartung, J. (1988) Tentative calculations of gaseous emissions from pig houses by way of the exhaust air. In: Nielsen, V.C., Voorburg, J.H. and Le Hermite, P. (eds) *Volatile Emissions from Livestock Farming and Sewage Operations.* Elsevier Applied Science, London, pp. 54–58.

Health and Safety Executive (1997) *Occupational Exposure Limits.* EH40/97. Health and Safety Executive, London.

Heber, A.J., Stroik, M., Faubion, J.M. and Willard, L.H. (1988) Size distribution and identification of aerial dust particles in swine finishing buildings. *Transactions of the American Society of Agricultural Engineers* 31, 882–887.

Honey, L.F. and McQuitty, J.B. (1979) Some physical factors affecting dust concentrations in a pig facility. *Canadian Agricultural Engineering* 21, 9–14.

Jones, J.B., Burgess, L.R., Webster, A.J.F. and Wathes, C.M. (1996) Behavioural responses of pigs to atmospheric ammonia in a chronic choice test. *Animal Science* 63, 437–445.

Jones, J.B., Webster, A.J.F. and Wathes, C.M. (1999) Trade off between ammonia exposure and thermal comfort in pigs and the influence of social contact. *Animal Science* 68, 387–398.

Jones, J.B., Carmichael, N.L., Wathes, C.M., White, R.P. and Jones, R. (2000) The effects of acute simultaneous exposure to ammonia on the detection of buried odourised food by pigs. *Applied Animal Behaviour Science* 65, 305–319.

Jones, J.B., Wathes, C.M., Persaud, K.C., White, R.P. and Jones, R.B. (2001) Acute and chronic exposure to ammonia and olfactory acuity for *n*-butanol in the pig. *Applied Animal Behaviour Science* 71, 13–28.

Kristensen, H.H., Jones, R.B., Schofield, C.P., White, R.P. and Wathes, C.M. (2001) The use of olfactory and other cues for social recognition by juvenile pigs. *Applied Animal Behaviour Science* 72, 321–333.

Muirhead, M.R. and Alexander, T.J. (eds) (1997) *Managing Pig Health and Treatment of Disease.* 5M Enterprise, Sheffield, UK.

Nordstrom, G.A. and McQuitty, J.B. (1976) Manure gases in the animal environment – a literature review (with particular reference to cattle housing). *University of Alberta Research Bulletin* 76-1, 80 pp.

Phillips, V.R. and Pain, B.F. (1998) Gaseous emissions from the different stages of European livestock farming. In: Matsunaka, T. (ed.) *Proceedings of the International Workshop on Environmentally-Friendly Management of Farm Animal Wastes,* Sapporo, Japan, 25–29 November 1997, pp. 67–72.

Phillips, V.R., Cowell, D.A., Sneath, R.W., Cumby, T.R., Williams, A.G., Demmers, T.G.M. and Sandars, D. (1998) A review of ways to abate ammonia emissions from livestock buildings and waste stores. Unpublished report. Silsoe Research Institute, Bedford, 80 pp.

Randall, J.M. (1975) Prediction of airflow patterns in livestock buildings. *Journal of Agricultural Engineering Research* 20, 199–215.

Randall, J.M. and Boon, C.R. (1994) Ventilation control and systems. In: Wathes, C.M. and Charles, D.R. (eds) *Livestock Housing.* CAB International, Wallingford, UK, pp. 149–182.

Reynolds, S., Donham, K., Whitten, P., Merchant, J., Murmeister, L. and Popendorf, W. (1996) Longitudinal evaluation of dose-response relationships for environmental exposures and pulmonary function in swine production workers. *American Journal of Industrial Medicine* 29, 33–40.

Robertson, J.F., Wilson, D. and Smith, W.J. (1990) Atrophic rhinitis: the influence of the aerial environment. *Animal Production* 50, 173–182.

Rylander, R. (1986) Lung diseases caused by organic dusts in the farm environment. *American Journal of Industrial Medicine* 10, 221–227.

Seedorf, J., Hartung, J., Schröder, M., Linkert, K.H., Phillips, V.R., Holden, M.R., Sneath, R.W., Short, J.L., White, R.P., Pedersen, S., Takai, H., Johnsen, J.O., Metz, J.H.M., Groot Koerkamp, P.W.G., Uenk, G.H. and Wathes, C.M. (1998) Concentrations and emissions of airborne endotoxins and micro-organisms in livestock buildings in Northern Europe. *Journal of Agricultural Engineering Research* 70, 97–109.

Takai, H., Moller, F., Iversen, M., Jorsal, S.E. and Bille-Hansen, V. (1993) Dust control in swine buildings by spraying of rapeseed oil. In: *Proceedings of Fourth International*

Livestock Environment Symposium. American Society of Agricultural Engineers, Warwick, UK, pp. 726–733.

Takai, H., Pedersen, S., Johnsen, J.O., Metz, J.H.M., Groot Koerkamp, P.W.G., Uenk, G.H., Phillips, V.R, Holden, M.R., Sneath, R.W., Short, J.L., White, R.P., Hartung, J., Seedorf, J., Schröder, M., Linkert, K.H. and Wathes, C.M. (1998) Concentrations and emissions of airborne dust in livestock buildings in Northern Europe. *Journal of Agricultural Engineering Research* 70, 59–77.

Wathes, C.M. (1998) Environmental control in pig housing. In: Done, S., Thomson, J. and Varley M. (eds) *15th International Pig Veterinary Society Congress,* Birmingham, UK, Vol. I. Nottingham University Press, pp. 257–265.

Wathes, C.M., Demmers. T.G.M., Richards, P.A., Teer, N., Taylor, L.L. and Goodman, J. (2000) *A Facility for Controlled Exposure of Pigs to Airborne Dusts and Gases.* Paper No. 004111, presented at ASAE Annual International Meeting, July 9–12. ASAE, St Joseph, Mississippi, 23pp.

Woods, P.S.A., Robinson, N.E., Swanson, M.C., Reed, C.E., Broadstone, R.V. and Derksen, F.J. (1993) Airborne dust and aeroallergen concentration in a horse stable under two different management systems. *Equine Veterinary Journal* 25, 208–213

Behaviour of the Young Weaner Pig

14

S. Held and M. Mendl

Centre for Behavioural Biology, Department of Veterinary Clinical Science, University of Bristol, Langford House, Langford BS40 5DU, UK

Introduction

Weaning describes the transition in young mammals from total nutritional and social dependence on the mother to total independence from her. This is characteristically a gradual process involving a series of transitional stages (Galef, 1981). Independence from the mother even with respect to any one single resource (let alone all resources) is rarely achieved abruptly. Rat pups, for example, move from total dependence on maternal milk to a mixed diet of milk and solid food, to a diet of only solid food, which has been provided by adults, to feeding trips with adults, to independent feeding trips (Galef, 1981). The weaning process is also commonly staggered, as the young may begin to show adult-like behaviours to meet different needs at different ages. Rat pups start to be able to control their body temperature independently from their mother before they start to take in solid feed (Galef, 1981). The age at which total independence is finally achieved under natural conditions depends on a complex interplay between the potentially conflicting interests of the mother and her offspring (Trivers, 1974; Bateson, 1994). The optimal time to finish the weaning process from the mother's point of view would be when she has invested enough in her current young to maximize their chances of surviving to breeding age while she still retains energy levels that are high enough to allow her to breed successfully again. The young usually have an interest in maintaining their high demands on the mother for longer than is optimal for her. The balance between various factors such as the nutritional state of the mother, the probability that she will breed again, the nutritional state of the young and the amount of care still provided by the mother after weaning is thought to determine the end of the weaning process (Bateson, 1994; Babbit and Packard, 1990).

These principles apply as much to pigs as to other mammals living under natural conditions. They may also be relevant when weaning practice and ages are determined for piglets under commercial conditions (Fraser *et al.*, 1995a) but, in contrast to gradual natural weaning at a pace determined by the sow and her litter, commercial weaning tends to be abrupt and at an age determined by the producer. Piglets are commonly removed from the mother when they would still be suckling, maintaining a strong social attachment to her, and relying on her for 'social security' and protection under natural conditions. They are thus weaned when they are still behaviourally highly reliant on the sow.

The story of the behaviour of the commercial weaner is the story of how the piglet adapts to the discrepancy between natural and commercial weaning. The objective of this chapter is to highlight some of the ways in which conventional commercial weaning practices depart from the natural weaning process. Attention will also be paid to how this affects several aspects of the behaviour of young weaners, and how some of these effects can be ameliorated by changes in management practice before, during and after weaning.

Behaviour of the 'Semi-natural' Weaner

Information on the behaviour of piglets born and weaned in near-to-natural conditions comes from studies of mixed-sex groups of domestic pigs of commercial breeds that were released from intensive conditions into large outdoor enclosures. These enclosures provide natural habitats such as woodland, bog, small streams and grassy, rocky and shrub areas. Handling and management are kept to a minimum. Natural resources in the enclosures tend to be supplemented with commercial feed and some basic shelter such that the conditions are commonly referred to as 'semi-natural'. Two long-term release studies have yielded most of the information currently available on the behaviour of domestic pigs in semi-natural conditions. One was conducted by Wood-Gush and co-workers in Scotland (e.g. Newberry and Wood-Gush, 1985, 1986; Stolba and Wood-Gush, 1989), the other in Sweden by Jensen, Wood-Gush and co-workers (e.g. Jensen, 1986, 1988; Jensen and Rećen, 1989; Petersen *et al.*, 1989; Stangel and Jensen, 1991). These studies showed that the natural weaning process in domestic pigs is gradual, prolonged and flexible within and between litters.

Under semi-natural conditions, sows give birth in nests away from their family group. The process that ultimately leads to nutritional and social independence and integration of piglets into the sow's family group starts in the first fortnight after birth. After the first week, the number of sucklings initiated by the sow begins to fall and continues to decline gradually, as does the overall number of sucklings (Jensen, 1988; Jensen and Rećen, 1989). Jensen (1988), for example, found that the average percentage of all observed sucklings initiated by the sow fell from about 60% in week 1 to about 15% in week 10 (based on six gilts and their litters). The sow also starts to terminate more and more sucklings herself by walking away, or preventing the piglets from attaching to the udder by lying on it (Newberry and

Wood-Gush, 1986; Jensen and Rečen, 1989), and she spends increasing amounts of time away from the nest to forage and feed (Jensen and Redbo, 1987; Stangel and Jensen, 1991). One study found that the time the mother spent away from the nest increased from about 10% on day 1 to about 40% on day 10 *post partum* (based on ten sows and their litters) (Stangel and Jensen, 1991).

During the first few days of their lives, the piglets stay in or close to the nest and to each other. Towards the end of their first week they begin to follow their mother on short trips (Newberry and Wood-Gush, 1986; Jensen and Redbo, 1987; Petersen *et al.*, 1989; Stolba and Wood-Gush, 1989). These early forays are mainly spent exploring the nest surroundings (Stolba and Wood-Gush, 1989). When out of the nest, littermates stay close to each other. If they start to follow their mother away from the nest, they soon leave her to return to their nest (Jensen and Redbo, 1987). During this period (week 1 after birth), the piglets also tend to have their first social (nose-to-nose) contact with non-littermates when other piglets or group-members visit the nest site (Jensen and Redbo, 1987; Petersen *et al.*, 1989). From about the second week onwards, the piglets start to follow the sow all the way to the feeding sites (Newberry and Wood-Gush, 1986; Jensen and Redbo, 1987; Petersen *et al.*, 1989). As during earlier and shorter trips away from the nest, they move around in their litter group, closer to each other on average than to their mother or the nest (Newberry and Wood-Gush, 1986).

The sow and her litter leave the nest for good to move closer to the group during the second week after birth (Jensen, 1986; Jensen and Redbo, 1987; Petersen *et al.*, 1989). From then on, the distance to the group gradually decreases, and mixing of the piglets with non-littermates (particularly with litters of similar ages) increases steadily until the sow and her piglets are fully integrated into the group at about 7–8 weeks (Newberry and Wood-Gush, 1986; Petersen *et al.*, 1989; Jensen, 1995). The frequency of social interactions between the piglets and non-littermates rises sharply when they first join the group, then levels off around 4 weeks after birth, before declining to a low, stable rate when integration is achieved (Newberry and Wood-Gush, 1986; Petersen *et al.*, 1989). Mother and littermates remain the most frequent 'nearest neighbours' until about the sixth to eighth week, when the time spent with the mother as the closest neighbour declines sharply (Petersen *et al.*, 1989; Jensen, 1995). Aggression between the new piglets and non-littermates is rare. Even during the early phase of social integration, aggressive interactions and aggressive play behaviours together remain below 20% of all social contacts between the piglets and other group-members, and below 18% over the whole weaning period (Petersen *et al.*, 1989). Nibbling, sucking, chewing or belly-nosing between piglets and others is rare or absent (e.g. Petersen, 1994).

For the piglets, the transition from isolation in the nest to social integration into the group is accompanied by gradual nutritional changes. The frequency of sucklings decreases steadily from the first week onwards (Newberry and Wood-Gush, 1985; Jensen, 1988), with the steepest fall during the first 4 weeks (Jensen, 1988). Newberry and Wood-Gush (1985) also found that the frequency of nursings with milk let-down decreased after 6 weeks. Feeding on solids is well established when the piglets are 4 weeks old and increases considerably between the sixth and tenth weeks after birth (Newberry and Wood-Gush, 1985; Jensen, 1995). Petersen *et al.* (1989), for example, found that piglets started to graze when they were between 24 and 36 days old, and to feed on pelleted feed between 28 and 39 days old.

Littermates may vary considerably in their natural weaning pace. For example, the proportion of sucklings in which some littermates are missing increases significantly from week 1 to week 12 (Jensen, 1988; Jensen and Reĉen, 1989), indicating that some piglets start to skip milk meals earlier than others. Stolba and Wood-Gush (1989) observed that suckling intervals at 12 weeks after birth lasted 2 h on average, but some piglets had stopped suckling altogether, which indicates that they were nutritionally independent of the sow. Such individual differences do not seem to be related to the sex or birthweight of the piglets, but rather to the productivity of the teats they occupy early in lactation. Jensen (1995) found that piglets that had occupied highly productive teats during the first 3 days after birth tended to continue to rely on milk as their main source of food for longer, to stay closer to their mothers and to spend less time feeding on solids later on than their littermates. Littermates that had fed from less productive teats showed the opposite behaviour (Jensen, 1995). This points to an individual compensatory trade-off between continued suckling and taking in solids, depending on the amount of milk taken in or teat productivity early in lactation. The weaning process is completed with the whole litter nutritionally independent from the mother, when the piglets are between 9 and 17 weeks old (e.g. Newberry and Wood-Gush, 1985; Jensen and Redbo, 1987; Stolba and Wood-Gush, 1989). Uncompensated seasonal differences in the availability of natural solid feeds may account for some of the variability in the weaning process between litters (e.g. Newberry and Wood-Gush, 1985).

Behaviour of the 'Commercial' Weaner

Commercial weaning age and pre- and post-weaning management of the piglets tend to vary between systems. However, common to most conventional weaning systems is the abrupt removal of the piglets from the sow before natural weaning is achieved, and their subsequent introduction into a new pen containing weaners from other litters. Piglets respond to this drastic change in their nutritional, physical and social environment with a variety of characteristic behaviours. Weaning age as well as pre- and post-weaning practices, such as sow-controlled systems or premixing piglets before weaning, can influence the expression of these behaviours.

Vocalizations

Removal from the sow is accompanied by characteristic high-pitched squeals and lower-pitched grunt calls by the piglets (Weary and Fraser, 1995). The piglets start to call immediately after being separated from their mother (Weary and Fraser, 1997). Separation calls are most frequent in the first couple of days after weaning. For example, Weary and Fraser (1997) found that the call rate fell from an average of 8.2 calls min^{-1} per piglet on weaning day to 1.6 calls on day 4 after weaning (based on averages of three piglets from each of 22 litters). They also found that piglets weaned at 3 weeks vocalized more during the first week after separation (average 3.6 calls min^{-1}) than piglets weaned at 4 or 5 weeks (2.9 and 2.3 calls min^{-1}) (Weary and Fraser, 1997). Piglets weaned at different ages under commercial conditions vary not only in their call rate, but also in call pitch. Early-weaned piglets call at higher frequencies than piglets that are weaned when they are older (Weary and Fraser, 1997; Weary *et al.*, 1999).

Such differences in call rate and pitch could be caused by developmental changes in vocalization behaviour as piglets grow older, rather than by the effects of early weaning (Weary *et al.*, 1999). However, a series of studies has shown that separation calls are an indicator of the piglets' need for the sow. In comparisons of piglets of the same age, but of high and low bodyweight and growth rate, and piglets of the same age and weight that had just missed a milk meal or not, groups of piglets were created at the same stage of development, but with different levels of need for the nutritional resource 'sow' (Weary and Fraser, 1995). Piglets that were lighter, had lower growth rates or had just missed a milk meal called more frequently and at a higher pitch upon separation from their mother than their well-fed, less 'needy' counterparts (Weary and Fraser, 1995). Since all piglets called for their mothers (though at different rates), nutrition was not the only resource that they required from their mother (Weary *et al.*, 1999). A further study, which compared the vocalizations of piglets separated from their mothers and isolated in rooms at 14°C or 30°C, showed that higher call rate and pitch also signal a greater thermal need (Weary *et al.*, 1997).

For separation calls to have evolved as signals of need for the sow, the sow must receive the signal as such and behave appropriately. Weary *et al.* (1996) showed that separation calls do indeed elicit an appropriate behavioural response from the sow. Sows called and also orientated and moved towards the source of the separation calls. This would lead to a reunion with the calling piglet if the sows were allowed access. They reacted quicker and stronger the more frequent and high pitched were the piglets' calls (Weary *et al.*, 1996).

Feeding behaviour

Young commercial weaners often show a characteristic growth check in the first week after weaning that is thought to be associated with abrupt and 'premature' weaning (e.g. Pajor *et al.*, 1991; see also Chapter 8). As growth after weaning is largely limited by voluntary feed intake and by feed quality (Pluske *et al.*, 1995), some of the factors that influence the feed intake behaviour of newly weaned piglets will be discussed here (for reviews see also Pluske *et al.*, 1995; Chapter 8).

Housing conditions and 'environmental enrichment'

Several studies have investigated the effects of housing conditions in general and of adding objects or foraging substrates to weaner pens on the feed intake and growth of newly weaned and mixed pigs. Such manipulations are often referred to as 'environmental enrichment' because it appears that they enrich the environment, rather than because there is firm evidence that they are beneficial for the animals (Newberry, 1995). The findings are inconsistent, which is likely to be a reflection of differences in weaning practices and in conditions that are compared. In one study, Ekkel *et al.* (1995) compared the feed intake and growth in piglets that were raised in 'specific stress-free' (SSF) or control conditions. Under SSF conditions, piglets remained in the same pen from birth to finish. At weaning at 26 days, the sow was removed from the pen and the young weaners stayed in their litter groups. Space allowance was 0.55–0.7 m^2 per animal. Control conditions reflected common practice: piglets were mixed at weaning and housed on slatted floors at 0.3 m^2 per animal up to the end of the nursery stage at 61 days old. After weaning, cortisol levels and skin lesions were significantly higher in control weaners, and daily feed intake over the whole nursery period up to day 60 was also increased. This was reflected in significantly lower weight gains in the control weaners over the same period (Ekkel *et al.*, 1995). A similar comparison by Beattie *et al.* (1995) did not find such differences in feed intake and growth rate: unmixed weaners that had been reared from birth in control pens spent the same amount of time feeding over the whole farrow-to-finish period as weaners in enriched pens. Control litters stayed in standard farrowing crates from birth until weaning on day 42. 'Enriched' litters were moved to a straw pen with their mothers 3 days after birth. At weaning, control piglets were moved to flat-deck cages with metal floors. 'Enriched' piglets were moved to a large, functionally divided pen comprising five different areas: a peat area, a straw area, a sleep area, a feed area and a drink area. Performance measures were compared for three different stages: lactation (0–6 weeks), weaner (7–13 weeks) and finishing (14–20 weeks). While there were differences in aggressive behaviours, skin lesion scores and stress parameters (see also below), no significant differences were found between the two rearing conditions in feed intake or daily weight gain at any stage in the rearing process.

The effect of housing conditions and enrichment may also depend on genotype. Hill *et al.* (1998) compared various behavioural and performance measures in PIC C-15 and PIC EXP-94 weaners kept in five different enrichment conditions. The amount of time that weaners of both lines spent feeding and their growth rates remained unaffected by enrichment conditions during the nursery stage (4–8 weeks old). Differences emerged only later in the finishing stage, when average daily weight gains in EXP-94 weaners were higher when toys were provided than in the other enrichment or in the control conditions. It was suggested that a greater sensitivity to the housing environment in the EXP-94 pigs may underlie the observed breed difference.

Weaning age

Generally, all weaners tend to spend less time feeding on the day after weaning than on subsequent days, regardless of weaning age. However, this effect is stronger the younger the piglets are at weaning. Worobec *et al.* (1999) found that piglets weaned at 1 week old spent virtually no time feeding in the first 2 days after weaning (0.6 ± 0.5% of time). This increased to 6.1 ± 0.5% at 6 weeks after weaning (when they were 7 weeks old). Piglets weaned at 2 weeks old spent 2.7 ± 1.0% of their time feeding immediately after weaning, which was significantly more than the feeding time straight after weaning in weaners 1 week old. Piglets that were 4 weeks old at weaning spent even more time feeding on the first 2 days after weaning (4.6 ± 0.4%). At 8 weeks, percentage feeding times in the piglets that had been weaned at 4 weeks were comparable to percentage times of the 1-week weaners (6.3 ± 0.5%; Worobec *et al.*, 1999). The initial delay in feeding in the early-weaned piglets was reflected in significantly lower weight gains in the first 7 weeks of life. Gonyou *et al.* (1998) showed similar effects of weaning age on times spent feeding after weaning. Weaners aged 2 weeks spent less than 1% of their time feeding on the day just before weaning and on the day after, while piglets weaned when they were 4 weeks old ate for about 4% of their time on both days. Creep feed had been provided to all weaners from 1 week before weaning. Growth rates and relative feed intake over the first week after weaning were correspondingly lower in the earlier weaned piglets. There was also a pronounced growth check 3 days after weaning. Six days after weaning, feeding time in the 2-week weaners had increased to 10%, which was reflected in higher relative feed intake and growth rates than immmediately after weaning (Gonyou *et al.*, 1998). Thus, efficient feed intake seems to develop gradually with age (and associated gut maturity) and experience, with the younger piglets taking longer to commence feeding on solids (see also Appleby *et al.*, 1991; Fraser *et al.*, 1995b).

Weaning management and within-litter differences

When supplementary creep feed is provided during lactation, piglets typically start to eat it when they are 2–3 weeks old (e.g. Pajor *et al.*, 1991). This common practice is thought to decrease the post-weaning growth check by facilitating a more gradual change in diet. However, there is large within- and between-litter variation in the amount of feed consumed before weaning, and the effect of providing creep feed on post-weaning feeding behaviour and growth rates is not clear (Pluske *et al.*, 1995).

Weaning management and individual characteristics of the piglets are thought to influence creep feed intake and post-weaning feeding behaviour and performance. For example, in 'sow-controlled' systems, feed intake by the weaners tends to be greater in the first two days after weaning than in control weaners (e.g. Pajor *et al.*, 1991). In 'sow-controlled' systems, the sows can leave the nest area containing the piglets for a 'piglet-free' area by stepping over a piglet-proof barrier. This allows sows to regulate suckling rates as they might under natural conditions, which is denied them in conventional farrowing crates where piglets have continuous access to their mothers. Studies have shown that sows in sow-controlled systems down-regulate suckling rates from early on in lactation and consequently lose less weight over the course of one lactation and may return to oestrus sooner after weaning than sows in conventional farrowing systems (e.g. Bøe, 1991; Pajor, 1998, cited in Pajor *et al.*, 1999). Pajor *et al.* (1999) also found that feed intake after weaning increased significantly more in weaners originating from sow-controlled pens than in weaners from conventional control farrowing crates during the first 2 weeks following weaning at 35 days old (655 ± 31 g day^{-1} per pig vs. 499 ± 29 g day^{-1} per pig). As expected, differences in feed intake were reflected in growth rates, with control piglets losing more weight in the first 2 days after weaning (–217 ± 33 g day^{-1} per pig vs. –89 ± 59 g day^{-1} per pig) and gaining less on average over the first 2 weeks after weaning (371 ± 18 g day^{-1} per pig vs. 471 ± 20 g day^{-1} per pig) (Pajor *et al.*, 1999).

Allowing piglets from adjacent farrowing pens to mix before weaning also affects feed consumption and growth rates after weaning. Weary *et al.* (1999), for example, let the piglets of three litters mingle when they were approximately 11 days old. Feed consumption and weight gains after weaning of ten such pre-mixed litters and ten previously unmixed control litters were measured. All piglets were weaned at 28 days old and placed into weaner pens containing either littermates or non-littermates. For piglets that had been pre-mixed during lactation, non-littermates consisted of piglets from the other two litters with whom they had mingled during pre-mixing. In the control piglets, non-littermates were unfamiliar. In both the pre-mixed and control groups, feed consumption and weight gain increased in the 2 weeks after weaning. However, piglets in pre-mixed groups ate more in the 2 days immediately after weaning and progressively more during the following 2 weeks than control litters. This, again, was reflected in tendency towards higher weight gains over the same period in pre-mixed litters. A similar effect was found in a study by Wattanakul *et al.* (1997), in which premixed piglets also consumed more food after weaning than control piglets.

Common to many studies on the effect of creep feed consumption on post-weaning feeding behaviour and growth are large differences between littermates (e.g. Algers *et al.*, 1990; Pajor *et al.*, 1991; Appleby *et al.*, 1992). In some studies, piglets that had had the largest weight gains early on in lactation (because of the high productivity of the teats they occupied) spent less time feeding on solid feed before and after weaning than piglets that had been less thrifty during lactation (e.g. Algers *et al.*, 1990; Appleby *et al.*, 1992). Algers *et al.* (1990) recorded the feeding and social behaviour and relative weight gains of littermates from ten litters from birth to 9 weeks old. Creep feed was provided from when the piglets were 1 week old. All piglets were weaned at 6 weeks by removing the sow. Seven days after weaning, piglets were mixed into groups comprising 50% unfamiliar individuals and 50% littermates. Large individual differences emerged in pre-weaning, creep feed consumption, post-weaning feed consumption and weight gain, and behaviour. Although times spent feeding varied from weaning onwards, differences in post-weaning weight gains between littermates did not emerge until mixing took place 1 week after weaning. After mixing, piglets that had occupied highly productive teats early on in lactation put on least weight. They were also more submissive in social encounters at and after mixing. It was suggested that piglets that had occupied more productive teats during lactation had met most of their nutritional requirements pre-weaning with milk, while littermates on less productive teats ate more creep feed pre-weaning to compensate for their nutritional (milk) shortfall. These latter piglets were therefore more experienced in competing at the feeders before weaning and thus competed more effectively after weaning and after mixing. Piglets with higher weight gains due to better teats pre-weaning were less able to compete for feed after weaning, as indicated by their submissive and less aggressive behaviour; as a result, they spent less time feeding and put on less weight (Algers *et al.*, 1990). Similar compensatory relationships between pre- and post-weaning feeding behaviour and performance were observed by Jensen (1995) in piglets weaned under natural conditions as described above.

However, the level of creep feed intake and of subsequent post-weaning feeding seems to depend not only on individual piglet requirements for compensatory creep feeding, but also on their developmental maturity. Pajor *et al.* (1991) found that piglets with high birthweights, i.e. piglets that were more fully developed from birth, tended to eat more creep feed and have higher growth rates pre-weaning than littermates with lower birthweights. Creep feed was provided from 10 days after birth onwards, and all piglets were weaned at 4 weeks old by removing the sow from the farrowing pens. In the post-weaning period (days 28–42), variation in weight gains between individual piglets was high (range of 907–5136 g), with all but two piglets losing weight immediately after weaning. Birthweight (which was correlated to creep feed intake), rather than creep feed intake alone, predicted post-weaning weight gains (Pajor *et al.*, 1991).

Both developmental maturity and the need for compensatory feeding thus seem to determine individual levels of creep feed intake and post-weaning feeding. It has been suggested that developmental maturity may be the more important factor very early in lactation, with lighter and less mature piglets less able to digest creep feed than their heavier littermates (Pajor *et al.*, 1991). With time, all piglets reach a level of developmental maturity, at which even the smaller and less thrifty piglets can digest creep feed and hence compensate for a shortfall in milk by feeding on creep feed (Pajor *et al.*, 1991; Appleby *et al.*, 1992); or, as Fraser *et al.* (1995b) put it rather more elegantly: 'Creep feeding … seems to follow a variation of Marx's famous dictum: they start to feed according to their ability, and continue according to need.'

Social behaviour and aggression following weaning

Commercial weaning often involves abrupt and dramatic changes to the social environment of the weaner pig. The mother is removed and piglets may be separated from some or all of their littermates and mixed with unfamiliar animals to form new groups. This contrasts with the gradual integration of non-littermates typically observed from days 10 to 12 onwards of life under semi-natural conditions (e.g. Jensen and Redbo, 1987). Piglets respond strongly to separation from other family members, as shown by the studies of vocalization described earlier. Abrupt removal of familiar individuals, especially the mother, appears to cause distress, but the abrupt introduction to other individuals that often accompanies commercial weaning usually leads to high levels of aggressive behaviour, with consequences for welfare and production. This section therefore examines the effects of this process on social behaviour during the period following weaning. In the studies cited below, weaning age and the time elapsing between weaning and mixing both vary but, as a rule, the emphasis will be on those studies where weaning takes place between 2 and 5 weeks of age, and when mixing takes place at the same time as or, at the latest, within a week of weaning.

Meese and Ewbank (1973) described the time course of events following post-weaning mixing. Aggressive incidents peaked 1–2 h following mixing, during which time the animal that would become dominant was responsible for the majority of the aggression. At around 3 h there was a subsidence in activity as the whole group rested; a second, smaller peak of aggression followed at 4–5 h, during which the next most dominant animal appeared to be the most aggressive. Interspersed with aggression were bouts of investigatory behaviour directed to the facial, belly and anogenital regions. Vigorous aggression was rarely observed at 24 h and a stable social hierarchy seemed to have been established by 48 h.

This pattern of behaviour, especially the intense aggression occurring during the first few hours after mixing, is commonly observed (e.g. Friend *et al.*, 1983; Rushen, 1987; Björk, 1989). Once individuals have established their relative social status, it appears that aggression is no longer required to sort out most disputes. Instead, these appear to be settled through 'knowledge' of relative social status. Access to resources such as food troughs, lying areas and drinkers is decided by subtle interactions, which often involve active avoidance of one individual by another of lower rank. The group becomes more settled and interactions mainly comprise nosing behaviour, mild knocks and bites, threat and avoidance behaviour, and lying in close proximity to others.

Establishment of a stable social hierarchy thus brings to an end the aggression and fighting that characterize the period during which individuals are uncertain of their relative social status (e.g. Rushen, 1988). The faster that this can be achieved, the lower the levels of aggression are likely to be. Uncertainty of relative status is obviously greatest when pigs are from different litters, and levels of post-weaning aggression are much higher in groups containing pigs from more than one litter than in groups that have not experienced mixing (Friend *et al.*, 1983). However, aggression does occur in unmixed groups, suggesting that being separated from the mother and some littermates and being moved to a new pen might induce a small amount of social upheaval (e.g. Wood-Gush and Csermely, 1981; Friend *et al.*, 1983; Algers *et al.*, 1990). Competition for food following withdrawal of maternal milk may be an important cause. Aggression related to food competition is much less vigorous and less prolonged than the vigorous fighting seen between unfamiliar individuals.

Health, welfare and production consequences of mixing at weaning

Aggression, and even the mere presence of a dominant or aggressive individual, has been shown to induce physiological stress responses in many species, including pigs (e.g. Mendl *et al.*, 1992; Hessing *et al.*, 1994; Marchant *et al.*, 1995; Otten *et al.*, 1997; Tuchscherer *et al.*, 1998; de Jong *et al.*, 1999; Mendl, 2000). Because mixing induces aggression, there will probably be elevated stress responses and their sequelae (which could include temporary growth checks and changes in immune function).

Ekkel *et al.* (1995) found that aggression during the 1 h following weaning was much lower in unmixed compared with mixed groups, as were salivary cortisol levels on the day of weaning. During the 5 weeks following weaning, lower levels of ear lesions and coughing, higher growth rates and stronger cellular immune reactions to intradermal PHA injection were observed in the unmixed groups. The presence or absence of mixing was probably responsible for at least some of these findings, especially those to do with aggression and related lesions, but the environmental features that differed between the two treatments may also have had some effects.

Other studies have shown different effects of post-weaning mixing on health and production. For example, Friend *et al.* (1983) found no differences in growth between mixed and unmixed piglets during the 4 weeks following weaning at 28 days. However, Graves *et al.* (1978) observed higher aggression and slower growth rates in mixed piglets who were fed limited amounts.

Because of the potentially detrimental effects of aggression at mixing, it is important to understand the factors that affect the levels of aggression observed. The following sections examine some of these factors and how this knowledge may be used to devise ways of minimizing aggression and hence improving health, welfare and production in newly weaned and mixed pigs.

Influences on the occurrence of aggression at post-weaning mixing: individual characteristics

One set of influences on the amount of aggression observed when pigs are mixed is the actual characteristics of the individual animals.

WEIGHT. A number of studies have provided evidence that the degree of variation in bodyweight between mixed pigs can have an effect on the duration of fights. Rushen (1987) studied pigs weaned at 4 weeks and paired at 5 weeks and showed that when pair members were similar in weight, they fought for longer than when they differed in weight (see also Rushen, 1988). Jensen and Yngvesson (1998) failed to find such an effect. Rushen (1987) also showed that groups of pigs containing individuals of similar weight fought for longer in total than groups containing pigs of a wider range of weights. Moore *et al.* (1994) and Andersen *et al.* (2000) observed similar effects in 10-week-old pigs.

It seems likely that, as fights between newly mixed pigs progress, large asymmetries in weight facilitate assessment of the relative competitive abilities of two competing individuals and shorten the time taken by the losing contestant to decide to stop fighting or to submit. The result is shorter (and presumably less damaging) aggressive encounters and the more rapid establishment of a stable hierarchy.

AGGRESSIVENESS. There has been interest in the idea that the aggressiveness of pigs in competitive interactions with unfamiliar individuals may be an individual characteristic, with some cross-time and cross-situation stability. Most of the work in this area has focused on pigs of around 5–11 weeks of age. A number of tests have been developed to measure aggressiveness (e.g. Hessing *et al.*, 1993; Jensen, 1994; Forkman *et al.*, 1995; Jensen *et al.*, 1995a; Erhard and Mendl, 1997) and some cross-time consistency in this characteristic has been demonstrated. For example, Erhard and Mendl (1997) demonstrated consistency in latency to attack an unfamiliar intruder in pigs tested at 7 and 11 weeks. It has also been possible to predict levels of aggression in newly mixed groups on the basis of the behaviour in a preceding aggression test (e.g. Erhard *et al.*, 1997).

Particularly relevant here is the study by Hessing *et al.* (1993) in which pigs rated as being highly aggressive at 1–2 weeks of age were more likely to initiate fights during the 30 min following mixing at 10 weeks of age. The initial aggressiveness test involved four or more animals and it is difficult to be certain how much particular individuals influenced the behaviour of others (Jensen *et al.*, 1995b). However, if the apparent predictive value of the test is validated, this suggests that pre-weaned pigs may already show signs of stable individual variation in their aggressive behaviour that may influence how they respond to post-weaning mixing. Further work is required to examine whether stable individual differences really are already present during the pre-weaning stage, and what predictive value they have.

Influences on the occurrence of aggression at post-weaning mixing: management practices

A number of studies have examined whether the timing of post-weaning mixing and the way in which it is carried out can affect the levels of aggression observed. Some of these studies have specifically investigated the effectiveness of management methods that are intended to decrease aggression (e.g. 'environmental enrichment' and use of psychoactive drugs).

WEANING AND MIXING AGE. There is some evidence that the aggression shown by young pigs changes as they grow older. Jensen (1994) found that pigs paired with an unfamiliar individual at 1 week of age adopted more offensive fighting positions and settled fights quicker than those paired at 5 or 9 weeks of age; it was suggested that young pigs might be less motivated to fight, lack energy to sustain a long fight, or were better able to assess relative chances of winning and losing than older pigs. Pitts *et al.* (2000) found that pigs paired with unfamiliar individuals at 26 days of age spent more time fighting in total, had a longer first bout of fighting and incurred more injuries than those paired at 5 days; it was suggested that younger pigs may be more tolerant of strangers, but the possibilities mentioned by Jensen (1994), could not be ruled out nor could the suggestion that younger pigs had more room to manoeuvre and display submissive behaviour than the bigger, older pigs who were tested in an arena of the same size.

Whatever the underlying explanation, these findings raise the possibility that earlier weaning and mixing of piglets may result in lower levels of aggression and injury than later mixing. Worobec *et al.* (1999) investigated this in a study in which litters were weaned and mixed at 7, 14 and 28 days. No clear effect was found of weaning age on the levels of aggression observed immediately post-weaning, and at the same ages in pigs weaned at different times. Further work has examined how mixing piglets at various ages before weaning affects aggression, and this is discussed later.

ENVIRONMENTAL ENRICHMENT. Several studies have examined whether adding objects or foraging substrates to pens of newly weaned and mixed pigs can alter levels of aggression and other forms of behaviour. Schaefer *et al.* (1990) weaned pigs into new groups at 28 days and provided them with either a sugar–mineral block in a suspended metal basket, or a stiff rubber belt dangling from a freely pivoting metal bar, or no 'enrichment' object. During the second week following weaning, they observed lower frequencies and durations of various aggressive interactions in the pens containing the objects than in the control pens. They found a similar result when comparing the aggressive behaviour of pigs weaned and mixed at 6 weeks into pens containing either a tyre and chain or no object, and then observed at 11–12 weeks. In neither study were there reports on what happened during the immediate post-mixing period, when fighting is most intense.

Beattie *et al.* (2000) reared pigs from birth until 21 weeks in either 'barren' or 'enriched' pens. The former had fully slatted floors and a smaller space allowance per pig than the latter, which contained straw and were divided into different activity areas. Levels of aggressive and 'harmful social behaviour' were lower in the 'enriched' pens, but the analysis covered all 21 weeks (10 min observation per focal animal per week) and so it was not possible to determine whether enrichment helped to minimise aggression at the time of weaning and mixing. Beattie *et al.* (1995, 1996) reported similar effects of enrichment on unmixed piglets but, in contrast, Fraser *et al.* (1991) found no effect of the presence or absence of straw on aggressive behaviour within groups of 7-week-old unmixed piglets.

In general, the effectiveness of various forms of 'environmental enrichment' in decreasing aggression at mixing is not clear, though there is some evidence that it may decrease levels of milder agonistic interactions in settled groups, perhaps by diverting the pigs' attention from each other. From a theoretical perspective, it seems unlikely that environmental enrichment of the sort considered would have much impact on the aggression that occurs at mixing and which seems to subside when a stable hierarchy is formed. It is not clear why the presence of straw or other objects should speed up hierarchy formation.

MODIFICATION OF PEN STRUCTURE. Allowing the performance of avoidance or submission behaviour helps to bring agonistic encounters to a close (e.g. Rushen and Pajor, 1987; McGlone, 1990). Therefore, pen structures that facilitate these behaviours could help to minimize aggression following mixing of newly weaned pigs. McGlone and Curtis (1985) demonstrated that providing newly weaned and mixed piglets with 'pop-holes' in which they could place their heads and necks during agonistic interactions appeared to help to terminate such interactions, and to decrease aggression during the first 30 min following mixing. Waran and Broom (1993) observed a 40% reduction in aggressive behaviour during the first week following weaning and mixing into flatdeck or strawed pens if the pens contained a barrier around which piglets could retreat to escape aggressors. Barrier use was also correlated with weight gain during this period. In the 4 weeks following mixing, those piglets that received most aggression used the barrier most often. During this time, pigs with a barrier in their pen gained more weight than those in pens containing no barriers. Thus, there is some evidence that facilitating avoidance behaviour through changes to pen structure is a useful way of minimizing aggression following post-weaning mixing.

USE OF PSYCHOACTIVE DRUGS, ODOUR MASKERS AND NATURAL ODOURS. A number of studies have examined the effects of various psychoactive drugs or odours on aggression at post-weaning mixing. Pluske and Williams (1996) found that piglets given amperozide (a drug acting on the limbic system with minimal sedative effects; Barnett *et al.*, 1993) were generally less active and aggressive during the 2 h following weaning than controls. Injury scores also tended to be lower after 4 h and 24 h, but no associated effects on food intake or weight gain were observed. Björk (1989) reported that amperozide decreased aggression and related injuries, and increased average daily weight gain without affecting feed intake, following post-weaning mixing.

Petherick and Blackshaw (1987) noted that apparent aggression-reducing effects of azaperone ('stresnil', a tranquillizer) were often short-lived and acted to postpone rather than to eliminate aggression (see also Björk, 1989; Luescher *et al.*, 1990). Similarly, Friend *et al.* (1983) found that administration of a commercial odour masker resulted in only a 3–5 min delay in the onset of fighting between newly mixed pigs. In general, it is difficult to see how the administration of psychoactive drugs or odour maskers could help to speed up formation of the dominance hierarchy following mixing and hence decrease aggression in this way. However, it is possible that by inducing newly mixed pigs to lie in the same pen close to each other, or by masking their own smell, the pigs increase the similarity of their odours and hence decrease the extent to which they treat each other as unfamiliar. This possibility needs be investigated in a controlled way. In any case, there are ethical questions concerning the use of drugs to minimize aggression resulting from a common management practice, especially if the drugs have other effects; for example, amperozide appears to increase vomiting and cortisol levels (Barnett, 1993).

A different approach is to take advantage of natural odours produced by pigs. Work by McGlone and colleagues (McGlone, 1990) demonstrated that aggression between two unfamiliar pigs could be decreased by painting them with urine collected from fighting pigs. On the other hand, urine from pigs handled (mildly stressed) by humans acted to increase aggression, whereas urine from ACTH-treated pigs (simulating a more intense stressor by stimulating release of large amounts of cortisol from the adrenal gland) increased submissive behaviour (McGlone, 1985; McGlone *et al.*, 1987). It thus appears that a number of pheromonal cues can influence the agonistic behaviour of pigs. For example, the 'submission' pheromone may be cortisol-dependent and released towards the end of a fight by the losing animal. The adult boar pheromone, androstenone, was also found to reduce fighting in young prepubertal pigs, perhaps by acting as a signal of the presence of a dominant adult (McGlone, 1990).

Further knowledge of these pheromones and their role in chemocommunication is required. It is possible that they could influence the levels of aggression observed in newly mixed weaned pigs, but again it is not clear how they might speed up the formation of dominance hierarchies without providing spurious and confusing information to the animals (for example, all animals appear to be signalling submission). Androstenone may have some beneficial effects if it inhibits aggression by signalling the presence of a dominant animal (McGlone, 1990) but such effects may be only short-lived.

Alternative approaches to minimizing aggression between newly weaned and mixed pigs

As is clear from the above, various methods for minimizing post-weaning aggression have limited effectiveness. This is partly because they do not take into account how unfamiliar piglets integrate in the natural environment, or the fact that the establishment of a stable dominance hierarchy is the main event that terminates aggression. Three methods that take account of these points are considered below.

MIXING PIGLETS PRIOR TO WEANING. Mixing piglets prior to weaning is practised in some piggeries (Pluske and Williams, 1996) and is relatively easy to do by, for example, removing the wooden divisions between adjacent farrowing crates. It can be undertaken at ages at which, under natural conditions, piglets would normally be integrating with unfamiliar littermates with low levels of aggression (e.g. days 10–12 onwards; Jensen and Redbo, 1987). A few studies have investigated the effects of this early mixing on behaviour and growth.

Weary *et al.* (1999) mixed piglets from three adjacent litters at 11 days of age. Low levels of aggression were observed. Nursings decreased relative to control non-mixed litters, perhaps due to competition and disturbance from cross-suckling. At weaning (day 28), the pre-mixed litters were placed either into groups of littermates, or into groups containing individuals from all three litters of the pre-mixed group. Similarly, controls either remained in litter groups, or were mixed into new groups containing individuals from three litters (see also above). Head-to-head knocking and eating behaviour were higher in the controls mixed into new groups than in the other three group types. As mentioned earlier, piglets mixed prior to weaning consumed more food following weaning than control animals and showed a tendency towards greater weight gains. The main effects of mixing the piglets prior to weaning thus seemed to be decreased aggression and increased food intake following post-weaning grouping.

Wattanakul *et al.* (1997) conducted a similar experiment in which piglets were mixed at day 11. Unlike Weary *et al.* (1999), they noticed an increase in skin damage in these mixed piglets at days 14 and 21 in comparison with control piglets left in their litter groups. It was assumed that this was due to some aggressive encounters between the newly mixed litters. Cross-suckling was also observed in the mixed litters, though only an average of one piglet per litter performed this behaviour. At day 28 piglets were weaned and placed with piglets from other litters that they either had already met (pre-weaning mixing group) or had not (control).

Aggression during the 2 h following mixing and skin damage a week later were significantly higher in the control group. No differences in post-mixing weight gain were observed. As with Weary *et al.* (1999), the main effect of pre-weaning mixing seemed to be to decrease fighting following regrouping at weaning, but there was also some evidence of aggression between pigs mixed prior to weaning.

Pluske and Williams (1996) also found that allowing piglets to mix at day 10 of life led to a decrease in aggression-related injuries during the 4 h following weaning and regrouping at 29 days. No effects on growth rate or food conversion ratio were observed during the following 2 weeks.

Allowing unfamiliar piglets to mix with each other at around 10 days of life (the time they would usually do this under natural conditions) seems to result in less aggression than mixing them following weaning. Although the different findings concerning the levels of aggression occurring after pre-weaning mixing (Wattanakul *et al.*, 1997; Weary *et al.*, 1999) deserve further investigation, the fact that stress due to mixing can at least be separated in time from the other challenges of weaning should encourage the use of this simple technique.

PRE-EXPOSING ANIMALS PRIOR TO MIXING. During the period of most intense aggression, some individuals may actively retreat from others or submit without fighting (Meese and Ewbank, 1973; Algers *et al.*, 1990) and not all unfamiliar dyads seem to need to fight in order to establish their relative status (Mendl and Erhard, 1997). It is possible, therefore, that pigs are able to assess their relative social status in some way (Mendl and Erhard, 1997). One possibility is that they use olfactory cues. These seem to be important in signalling submission (McGlone, 1990) and may also play an important role in individual recognition (Meese and Ewbank, 1973; Meese and Baldwin, 1975).

If pigs are able to assess their relative competitive abilities, then allowing them to do this before placing them in direct contact with each other could speed up the establishment of a dominance hierarchy and minimize the amount of aggression shown (Mendl and Erhard, 1997). Fraser (1974) found that 10-week-old pigs allowed limited physical contact through barred partitions showed lower levels of biting and chasing when they were placed together than those allowed visual access but no physical contact. Kennedy and Broom (1994) observed a similar effect in gilts. Jensen and Yngvesson (1998) showed that 7–8-week-old pigs that had been weaned a week or so earlier finished pair-wise fights quicker when the pair members had been pre-exposed to each other for 24 h before mixing than when they were not. However, Rushen (1988), working with newly weaned pigs of around 3–4 weeks age, found that pre-exposing them to each other for 72 h before mixing them together resulted in only a tendency for them to fight for a shorter period of time.

Overall, these findings indicate that pre-exposing unfamiliar pigs to each other before mixing them together does not eliminate fighting but may decrease its duration and intensity. The effectiveness of this may depend on the age of the animals, the length of pre-exposure and other factors. In addition, it is unclear whether the precise mechanism underlying such an effect is due to a form of assessment or simple habituation to and increasing familiarity with the other animal.

MIXING ON THE BASIS OF INDIVIDUAL CHARACTERISTICS. As described earlier, weight asymmetries between individuals may decrease the duration of fights that occur following mixing at weaning and hence may speed up the formation of a stable hierarchy. Thus, establishing groups containing pigs of varying weights may result in less intense aggression than the usual procedure of filling pens with pigs of similar weight. However, subsequently the heavier animals could monopolize resources if they were not provided in a way that prevented this (e.g. adequate trough space, lying space). Furthermore, management requirements for pens containing animals of similar weight mean that many farmers will be reluctant to mix pigs of quite different weights deliberately.

An alternative possibility that might alleviate this problem, while also minimizing levels of aggression, is to mix animals according to their 'aggressiveness'. For example, Mendl and Erhard (1997) showed that the number of dyads fighting during 2 h after mixing in 11-week-old pigs was lower in groups containing pigs of different aggressiveness (categorized as fast and slow attackers in an attack latency test) than in groups containing pigs of similar aggressiveness. Levels of skin lesions tended to be highest in the groups comprising only fast attackers (Erhard *et al.*, 1997). Aggressiveness and weight were not clearly correlated in this study (Erhard and Mendl, 1997).

A simple and practical way of assessing aggressiveness would be needed if this approach was shown to be effective for weaner pigs as well as older pigs. One possibility is that a simple test of a piglet's reaction to being restrained on its back for a short period of time may be an indicator of its aggressiveness (Hessing *et al.*, 1993). However, a similar tonic immobility test did not appear to have the same predictive value (Mendl *et al.*, 1998). More work is required to evaluate the potential of this approach.

Belly-nosing and tail-biting

In addition to competitive aggressive behaviour occurring as social hierarchies are established, other forms of non-agonistic social behaviour may cause damage to newly weaned piglets. Examples of this are belly-nosing and tail-biting.

Damaging tail-biting is usually first observed several weeks after weaning has taken place and becomes more common in growing and fattening pigs (C. Moinard, unpublished data). The causes of tail-biting are complex and multifactorial.

Environmental factors have received most attention and these include stocking density, air quality, temperature changes, diet and method of food delivery, provision of foraging substrates and feeder design (e.g. van Putten, 1969; Penny *et al.*, 1981; Fraser, 1987; Fraser *et al.*, 1991; Morrow and Walker, 1994). Although tail-chewing may be observed in young piglets, even prior to weaning, the link between this behaviour and the outbreak of damaging tail-biting has not been examined in great detail. Furthermore, a recent study found no clear relationship between the occurrence of tail-biting and weaning age (Moinard *et al.*, 2000). Tail-biting thus seems to be more of a problem in older growing animals than in the newly weaned pig.

In contrast, belly-nosing is commonly observed in newly weaned piglets and its appearance seems to be directly related to the weaning process (see below). Belly-nosing may result in damage to the belly and nipples of other pigs, and individuals showing high levels of belly-nosing may grow most slowly during the weeks following weaning (Fraser, 1978). Therefore, an understanding of the causes of this behaviour and how to minimize its occurrence is important.

Description and motivational significance

Belly-nosing involves rhythmic rooting movements directed at the belly or flanks of a conspecific (Fraser, 1978). Terms such as 'rooting of pen-mates' or 'harmful nosing of pen-mates' are also used to describe similar behaviour. The timing of weaning affects the level of belly-nosing that occurs. Weary *et al.* (1999) found that piglets weaned at 2 weeks exhibited more belly-nosing during the 4–12 days following weaning than those weaned at 4 weeks. Metz and Gonyou (1990) observed a similar effect, as did Fraser (1978) in pigs weaned at 3 and 6 weeks. Gonyou *et al.* (1998) noted that higher levels of chewing and nosing of other pigs occurred in pigs weaned at 12 days as opposed to 21 days.

The higher levels of belly-nosing observed in pigs weaned at early ages could indicate that belly-nosing is a non-specific behavioural response to the stress of premature separation from the mother. However, the similarity of the behaviour to the sucking and massaging movements usually directed at the udder, and the occurrence of more belly-nosing in early-weaned piglets who are still most dependent on milk, support the idea that belly-nosing is also a form of redirected suckling behaviour.

Worobec *et al.* (1999) observed elevated levels of belly-nosing in early-weaned piglets not only at the time of weaning, but also several weeks later. Piglets weaned at 7 days showed consistently higher levels of belly-nosing at 28 and 42 days than those weaned at 14 or 28 days. These results suggest that, although the motivation to suck may underlie the initial expression of belly-nosing, the behaviour itself may become a habit that eventually becomes emancipated from its original causal factors and is displayed at high levels even when the motivation to suck has probably diminished.

Minimizing the occurrence of belly-nosing

If belly-nosing indicates that piglets are still highly motivated to suck, an obvious solution is to satisfy this motivation. The best way is simply to delay weaning, as indicated by the above studies. Weaning at 6–12 weeks, as takes place under natural conditions, is probably most effective at minimizing belly-nosing (e.g. Petersen *et al.*, 1994). If this is not possible, an alternative approach is to try to redirect belly-nosing from the bodies of conspecifics by increasing oral activity directed at other substrates.

One possibility is to provide a good rooting substrate (e.g. straw). Beattie *et al.* (1996) showed a lower level of 'harmful nosing of pen-mates' in pigs weaned at 6 weeks, kept in their litter groups, and housed in pens with access to a peat area, a straw area and shredded paper in comparison with those weaned into pens lacking rooting substrates, even if these latter provided a high space allowance per pig. This finding was particularly noticeable during the period following weaning (Beattie *et al.*, 1995). Fraser *et al.* (1991) also found a lower level of 'rooting pen-mates' in 7-week-old pigs housed in strawed as opposed to non-strawed pens (see also Beattie *et al.*, 2000). Mckinnon *et al.* (1989) found that piglets weaned into fully slatted pens showed higher levels of belly-nosing than those weaned into pens with solid floors with or without straw. Thus, even a simple solid floor might be a more suitable rooting substrate than slats.

Unlike aggression, belly-nosing may thus be effectively reduced by a form of 'environmental enrichment' that redirects oral behaviour away from conspecifics. Interestingly, alterations to pen design that are able to reduce aggression by facilitating submission have no effect on belly-nosing (e.g. addition of barriers; Waran and Broom, 1993). This emphasizes that attempts to minimize behaviour patterns that lead to health, welfare or production problems must be guided by an understanding of the causes of the behaviour. Different types of solution will be more or less applicable to different behaviours.

Conclusions

The objective of this chapter has been to review some of the ways in which piglets respond behaviourally to weaning under commercial conditions and how their responses are affected by different weaning practices.

It emerges that research in this area has largely focused on the weaner's attempts to cope with the challenge of abrupt nutritional independence and social integration at mixing. Less is known about how well piglets cope with loss of maternal attachment *per se* (Bowlby, 1969). Studies on primates and rodents have shown that the mammalian mother provides more to her offspring during lactation than warmth and milk. She also provides a secure base from which the offspring can start to explore the world beyond 'mother', while still being provided with some 'social security' and protection. Premature separation and isolation from the mother and even a shortfall in maternal care from a present mother can have long-term effects on the behavioural development of humans and other mammals. In some primate species and rats, for example, the young of mothers that show low levels of maternal care are more fearful of unfamiliar stimuli and less able to cope in challenging situations (e.g. Fairbanks, 1996; Liu *et al.*, 1997). In pigs, the long-term behavioural effects of maternal 'detachment' may be less pronounced, since piglets are precocial rather than altricial developers. However, as commercial piglets continue to be weaned at ages when they are still behaviourally dependent on the sow, it seems likely that they, too, will show some negative long-term effects of early maternal separation and lack of maternal care. This is a relevant and important topic for future research.

Finally, this chapter has shown repeatedly how pre-weaning experiences affect the ability to cope with the weaning challenge and post-weaning conditions. It has also emerged that weaning practices that mimic certain aspects of the natural weaning process can ameliorate some of the behavioural problems in young commercial weaners. In the light of this, one might speculate how considerably less restrictive systems such as outdoor farrowing affect piglet post-weaning behaviour. Early indications are that piglets reared outdoors up to weaning fight less and spend more time feeding at weaning (and mixing) into straw yards than piglets reared indoors (Cox and Cooper, 2001). Similarly, de Jonge *et al.* (1996) found that female piglets reared under conventional indoor conditions up to weaning were more aggressive to each other when weaned into littermate pairs than piglets from enriched farrowing pens. These differences carried over into the later post-weaning period and puberty, with the subordinates in pairs originating from conventional conditions (only) showing symptoms of chronic social stress (de Jonge *et al.*, 1996). De Jonge *et al.* (1996) suggested as one possible explanation that rearing piglets in relatively barren conditions negatively affects the development of their social skills, leading to increased aggression levels. It would be useful, for application to intensive indoor conditions, to examine further the developmental mechanisms that underlie these post-weaning differences in social and feeding behaviour.

References

Algers, B., Jensen, P. and Steinwall, L. (1990) Behaviour and weight changes at weaning and regrouping of pigs in relation to teat quality. *Applied Animal Behaviour Science* 26, 143–155.

Andersen, I.L., Andenaes, H., Bøe, K., Jensen, P. and Bakken, M. (2000) The effect of weight asymmetry and resource distribution on aggression in groups of unacquainted pigs. *Applied Animal Behaviour Science* 68, 107–120.

Appleby, M.C., Pajor, E.A. and Fraser, D. (1991) Effects of management options on creep feeding by piglets. *Animal Production* 53, 361–366.

Appleby, M.C., Pajor, E.A. and Fraser, D. (1992) Individual variation in feeding and growth of piglets: effects of increased access to creep food. *Animal Production* 55, 147–152.

Babbit, K.J. and Packard, J.M (1990) Parent–offspring conflict relative to phase of lactation. *Animal Behaviour* 40, 765–773.

Barnett, J.L. (1993) A drug that reduces aggression in pigs – a welfare dilemma. In: Batterham, E.S. (ed.) *Manipulating Pig Production IV.* Australasian Pig Science Association, Attwood, Victoria, p. 131.

Barnett, J.L., Cronin, G.M., McCallum, T.H. and Newman, E.A. (1993) Effects of 'chemical intervention' techniques on aggression and injuries when grouping unfamiliar adult pigs. *Applied Animal Behaviour Science* 36, 135–148.

Bateson, P. (1994) The dynamics of parent–offspring relationships in mammals. *Trends in Ecology and Evolution* 9, 399–403.

Beattie, V.E., Walker, N. and Sneddon, I.A. (1995) Effects of environmental enrichment on behaviour and productivity of growing pigs. *Animal Welfare* 4, 207–220.

Beattie, V.E., Walker, N. and Sneddon, I.A. (1996) An investigation of the effect of environmental enrichment and space allowance on the behaviour and production of growing pigs. *Applied Animal Behaviour Science* 48, 151–158.

Beattie, V.E., O'Connell, N.E., Kilpatrick, D.J. and Moss, B.W. (2000) Influence of environmental enrichment on welfare-related behavioural and physiological parameters in growing pigs. *Animal Science* 70, 443–450.

Björk, A.K.K. (1989) Is social stress in pigs a detrimental factor to health and growth that can be avoided by amperozide treatment? *Applied Animal Behaviour Science* 23, 39–47.

Bøe, K. (1991) The process of weaning in pigs: when the sow decides. *Applied Animal Behaviour Science* 30, 47–59.

Bowlby, J. (1969) *Attachment and Loss*, Vol. 1, *Attachment.* Hogarth Press, London, 428 pp.

Cox, L.N. and Cooper, J.J. (2001) Observations on the post- and pre-weaning behaviour of piglets in commercial indoor and outdoor environments. *Animal Science* 72, 75–86.

Ekkel, E.D., van Doorn, C.E.A., Hessing, M.J.C. and Tielen, M.J.M (1995) The specific-stress-free-housing system has positive effects on productivity, health, and welfare of pigs. *Journal of Animal Science* 73, 1544–1551.

Erhard, H.W. and Mendl, M. (1997) Measuring aggressiveness in growing pigs in a resident–intruder situation. *Applied Animal Behaviour Science* 54, 123–136.

Erhard, H.W., Mendl, M. and Ashley, D.D. (1997) Individual aggressiveness can be measured and used to reduce aggression after mixing. *Applied Animal Behaviour Science* 54, 137–151.

Fairbanks, L.M. (1996) Individual differences in maternal style. *Advances in the Study of Animal Behaviour* 25, 579–611.

Forkman, B., Furuhaug, I.L. and Jensen, P. (1995) Personality, coping patterns, and aggression in piglets. *Applied Animal Behaviour Science* 45, 31–42.

Fraser, D. (1974) The behaviour of growing pigs during experimental social encounters. *Journal of Agricultural Science, Cambridge* 82, 147–163.

Fraser, D. (1978) Observations on the behavioural development of suckling and early-weaned piglets during the first six weeks after birth. *Animal Behaviour* 26, 22–30.

Fraser, D. (1987) Mineral deficient diets and the pig's attraction to blood: implications for tail-biting. *Canadian Journal of Animal Science* 67, 909–918.

Fraser, D., Phillips, P.A., Thompson, B.K. and Tennessen, T. (1991) Effect of straw on the behaviour of growing pigs. *Applied Animal Behaviour Science* 30, 307–318.

Fraser, D., Kramer, D.L., Pajor, E.A. and Weary, D.M. (1995a) Conflict and cooperation: sociobiological principles and the behaviour of sows. *Applied Animal Behaviour Science* 44, 139–157.

Fraser, D., Phillips, P.A., Thompson, B.K., Pajor, E.A., Weary, D.M. and Braithwaite, L.A. (1995b) Behavioural aspects of piglet survival. In: Varley, M.A. (ed.) *The Neonatal Pig: Development and Survival.* CAB International, Wallingford, UK, pp. 287–312.

Friend, Y.H., Knabe, D.A. and Tanksley, T.D. Jr (1983) Behavior and performance of pigs grouped by three different methods at weaning. *Journal of Animal Science* 57, 1406–1411.

Galef, B.G. (1981) The ecology of weaning: parasitism and the achievement of independence by altricial mammals. In: Gubernick, D.J. and Klopfer, P.H. (eds) *Parental Care in Mammals.* Plenum Press, New York, pp. 211–241.

Gonyou, H.W., Beltranena, E., Whittington, D.L. and Patience, J.F. (1998) The behaviour of pigs weaned at 12 and 21 days of age from weaning to market. *Canadian Journal of Animal Science* 78, 517–523.

Graves, H.B., Graves, K.L. and Sherritt, G.W. (1978) Social behavior and growth of pigs following mixing during the growing–finishing period. *Applied Animal Ethology* 4, 169–180.

Hessing, M.J.C., Hagelsø, A.M., van Beek, J.A.M., Wipekema, P.R., Schouten, W.G.P. and Krukow, R. (1993) Individual behavioural characteristics in pigs. *Applied Animal Behaviour Science* 37, 285–295.

Hessing, M.J.C., Scheepens, C.J.M., Schouten, W.G.P., Tielen, M.J.M. and Wiepkema, P.R. (1994) Social rank and disease susceptibility in pigs. *Veterinary Immunology and Immunopathology* 43, 373–387.

Hill, J.D., McGlone, J.J., Fullwood, S.D. and Miller, M.F. (1998) Environmental enrichment influences on pig behavior, performance and meat quality. *Applied Animal Behaviour Science* 57, 51–68.

Jensen, P. (1986) Observations on the maternal behaviour of free-ranging domestic pigs. *Applied Animal Behaviour Science* 16, 131–142.

Jensen, P. (1988) Maternal behaviour and mother–young interactions during lactation in free-ranging domestic pigs. *Applied Animal Behaviour Science* 20, 297–308.

Jensen, P. (1994) Fighting between unacquainted pigs – effects of age and of individual reaction pattern. *Applied Animal Behaviour Science* 41, 37–52.

Jensen, P. (1995) The weaning process of free-ranging domestic pigs: within- and between-litter variations. *Ethology* 100, 14–25.

Jensen, P. and Rećen, B. (1989) When to wean – observations from free-ranging domestic pigs. *Applied Animal Behaviour Science* 23, 49–60.

Jensen, P. and Redbo, I. (1987) Behaviour during nest-leaving in free-ranging domestic pigs. *Applied Animal Behaviour Science* 18, 355–362.

Jensen, P. and Yngvesson, J. (1998) Aggression between unacquainted pigs – sequential assessment and effects of familiarity and weight. *Applied Animal Behaviour Science* 58, 49–61.

Jensen, P., Rushen, J. and Forkman, B. (1995a) Behavioural strategies or just individual variation in behaviour? A lack of evidence for active and passive piglets. *Applied Animal Behaviour Science* 43, 135–139.

Jensen, P., Forkman, B., Thodberg, K. and Köster, E. (1995b) Individual variation and consistency in piglet behaviour. *Applied Animal Behaviour Science* 45, 43–52.

de Jong, F.H., Bokkers, E.A.M, Schouten, W.P.G and Helmond, F.A. (1996) Rearing piglets in a poor environment: developmental aspects of social security. *Physiology and Behavior* 60, 389–396.

de Jong, I.C., Lambooij, E., Korte, S.M., Blokhuis, H.J. and Koolhaas, J.M. (1999) Mixing induces long-term hyperthermia in growing pigs. *Animal Science* 69, 601–605.

Kennedy, M.J. and Broom, D.M. (1994) A method of mixing gilts and sows which reduces aggression experienced by gilts. In: *Proceedings of the 28th International Congress of the ISAE.* National Institute of Animal Science, Foulum, p. 5.12.

Liu, D., Tannenbaum, B., Caldji, C., Francis, D., Freedman, A., Sharma, S., Pearson, D., Plotsky, P.M. and Meany, M.J. (1997) Maternal care, hippocampal glucocorticoid receptor gene expression and hypothalamic–pituitary–adrenal responses to stress. *Science* 277, 1659–1662.

Luescher, U.A., Friendship, R.M. and Mckeown, D.B. (1990) Evaluation of methods to reduce fighting among regrouped gilts. *Canadian Journal of Animal Science* 70, 363–370.

Marchant, J.N., Mendl, M.T., Rudd, A.R. and Broom, D.M. (1995) The effect of agonistic interactions on the heart rate of group-housed sows. *Applied Animal Behaviour Science* 46, 49–56.

McGlone, J.J. (1985) Olfactory cues and pig agonistic behavior: evidence for a submissive pheromone. *Physiology and Behavior* 34, 195–198.

McGlone, J.J. (1990) Olfactory signals that modulate pig aggressive and submissive behavior. In: Zayan, R. and Dantzer, R. (eds) *Social Stress in Domestic Animals.* Kluwer Academic Publishers, Dordrecht, The Netherlands, pp. 86–109.

McGlone, J.J. and Curtis, S.E. (1985) Behaviour and performance of weanling pigs in pens equipped with hide areas. *Journal of Animal Science* 60, 20–24.

McGlone, J.J., Curtis, S.E. and Banks, E.M. (1987) Evidence for aggression-modulating pheromones in prepubertal pigs. *Behavioral and Neural Biology* 47, 27–39.

Mckinnon, A.J., Edwards, S.A., Stephens, D.B. and Walters, D.E (1989) Behaviour of groups of weaner pigs in three different housing systems. *British Veterinary Journal* 145, 367–372.

Meese, G.B. and Baldwin, B.A. (1975) The effects of ablation of the olfactory bulb on aggressive behaviour in pigs. *Applied Animal Ethology* 1, 251–262.

Meese, G.B. and Ewbank, R. (1973) The establishment and nature of the dominance hierarchy in the domesticated pig. *Animal Behaviour* 21, 326–334.

Mendl, M. (2000). How do animals cope with social problems? In: Broom, D.M. (ed.) *Coping with Challenge: Welfare in Animals Including Humans.* Academic Press, New York.

Mendl, M. and Erhard, H.W. (1997) Social choices in farm animals: to fight or not to fight? In: Forbes, J.M., Lawrence, T.L.J., Rodway, R.G. and Varley, M.A. (eds) *Animal Choices.* BSAS, Edinburgh, pp. 45–53.

Mendl, M., Zanella, A.J. and Broom, D.M. (1992) Physiological and reproductive correlates of behavioural strategies in female domestic pigs. *Animal Behaviour* 44, 1107–1121.

Mendl, M., Erhard, H.W. and Christiansen, S.B. (1998) No evidence for strong links between personality traits in pigs. In: Veissier, I. and Boissy, A. (eds) *Proceedings of the 32nd Congress of the ISAE.* INRA, Clermont-Ferrand, France, p. 65.

Metz, J.H.M. and Gonyou, H.W. (1990) Effect of age and housing condition on the behavioural and haemolytic reaction of piglets to weaning. *Applied Animal Behaviour Science* 27, 299–309.

Moinard, C., Mendl, M., Nicol, C.J. and Green, L.E. (2000) Investigations into risk factors for tail biting in pigs on commercial farms in England, UK. In: Salman, M.D., Morley, P.S. and Ruch-Gallie, R. (eds) *Proceedings of the 9th Symposium of the International Society for Veterinary Epidemiology and Economics,* (CD-ROM extract), Breckenridge, Colorado.

Moore, A.S., Gonyou, H.W., Stookey, J.M. and McLaren, D.G. (1994) Effect of group composition and pen size on behavior, productivity and immune response of growing pigs. *Applied Animal Behaviour Science* 40, 13–40.

Morrow, A.T.S. and Walker, N. (1994) A note on changes to feeding behaviour of growing pigs by fitting stalls to single-space feeders. *Animal Production* 59, 151–153.

Newberry, R.C. (1995) Environmental enrichment: increasing the biological relevance of captive enviroments. *Applied Animal Behaviour Science* 44, 229–243.

Newberry, R.C. and Wood-Gush, D.G.M. (1985) The suckling behaviour of domestic pigs in a semi-natural environment. *Behaviour* 95, 11–25.

Newberry, R.C. and Wood-Gush, D.G.M. (1986) Social relationships of piglets in a semi-natural environment. *Animal Behaviour* 34, 1311–1318.

Otten, W., Puppe, B., Stabenow, B., Kanitz, E., Schön, P.C., Brüssow, K.P. and Nürnberg, G. (1997) Agonistic interactions and physiological reactions of top- and bottom-ranking pigs confronted with a familiar and an unfamiliar group: preliminary results. *Applied Animal Behaviour Science* 55, 79–90.

Pajor, E.A. (1998) Parent–offspring conflict and its implications for maternal housing systems in domestic pigs. PhD Thesis, McGill University, Quebec, Canada.

Pajor, E.A., Fraser, D. and Kramer, D.L. (1991) Consumption of solid food by suckling pigs: individual variation and relation to weight gain. *Applied Animal Behaviour Science* 32, 139–155.

Pajor, E.A., Weary, D.M., Fraser, D. and Kramer, D.L. (1999) Alternative housing for sows and litters. 1. Effect of sow-controlled housing on responses to weaning. *Applied Animal Behaviour Science* 65, 105–121.

Penny, R.H.C., Walters, J.R. and Tredget, S.J. (1981) Tail-biting in pigs: a sex frequency between boars and gilts. *Veterinary Record* 198, 35.

Petersen, H.V. (1994) The development of feeding and investigatory behaviour in free-ranging domestic pigs during their first 18 weeks of life. *Applied Animal Behaviour Science* 42, 87–98.

Petersen, H.V., Vestergaard, K. and Jensen, P. (1989) Integration of piglets into social groups of free-ranging domestic pigs. *Applied Animal Behaviour Science* 23, 223–236.

Petherick, J.C. and Blackshaw, J.K. (1987) A review of the factors influencing the aggressive and agonistic behaviour of the domestic pig. *Australian Journal of Experimental Agriculture* 27, 605–611.

Pitts, A.D., Weary, D.M., Pajor, E.A. and Fraser, D. (2000) Mixing at young ages reduces fighting in unacquainted domestic pigs. *Applied Animal Behaviour Science* 68, 191–197.

Pluske, J.R. and Williams, I.H. (1996) Reducing stress in piglets as a means of increasing production after weaning: administration of amperozide or co-mingling of piglets during lactation? *Animal Science* 62, 121–130.

Pluske, J.R., Williams, I.H. and Aherne, F.X. (1995) Nutrition of the neonatal pig. In: Varley, M.A. (ed.) *The Neonatal Pig: Development and Survival.* CAB International, Wallingford, UK, pp. 187–235.

van Putten, G. (1969) An investigation into tail-biting among fattening pigs. *British Veterinary Journal* 125, 511–517.

Rushen, J. (1987) A difference in weight reduces fighting when unacquainted newly weaned pigs first meet. *Canadian Journal of Animal Science* 67, 951–960.

Rushen, J. (1988) Assessment of fighting ability or simple habituation: what causes young pigs (*Sus scrofa*) to stop fighting? *Aggressive Behavior* 14, 155–167.

Rushen, J. and Pajor, E. (1987) Offence and defence in fights between young pigs (*Sus scrofa*). *Aggressive Behavior* 13, 329–346.

Schaefer, A.L., Salomons, M.O., Tong, A.K.W., Sather, A.P. and Lepage, P. (1990) The effect of environment enrichment on aggression in newly weaned pigs. *Applied Animal Behaviour Science* 27, 41–52.

Stangel, G. and Jensen, P. (1991) Behaviour of semi-naturally kept sows and piglets (except suckling) during 10 days postpartum. *Applied Animal Behaviour Science* 23, 223–236.

Stolba, A. and Wood-Gush, D.G.M. (1989) The behaviour of pigs in a semi-natural environment. *Animal Production* 48, 677–683.

Trivers, R.L. (1974) Parent–offspring conflict. *American Zoologist* 14, 249–264.

Tuchscherer, M., Puppe, B., Tuchscherer, A. and Kanitz, E. (1998) Effects of social status after mixing on immune, metabolic, and endocrine responses in pigs. *Physiology and Behavior* 64, 353–360.

Waran, N.K. and Broom, D.M. (1993) The influence of a barrier on the behaviour and growth of early-weaned piglets. *Animal Production* 56, 115–119.

Wattanakul, W., Stewart, A.H., Edwards, S.A. and English, P.R. (1997) Effects of grouping piglets and changing sow location on suckling behaviour and performance. *Applied Animal Behaviour Science* 55, 21–35.

Weary, D.M. and Fraser, D. (1995) Calling by domestic piglets: reliable signals of need? *Animal Behaviour* 50, 1047–1055.

Weary, D.M. and Fraser, D. (1997) Vocal response of piglets to weaning: effect of piglet age. *Applied Animal Behaviour Science* 54, 153–160.

Weary, D.M., Lawson, G.L. and Thompson, B.K. (1996) Stronger sow response to isolation calls associated with greater levels of piglet need. *Animal Behaviour* 52, 1247–1253.

Weary, D.M., Ross, S. and Fraser D.M. (1997) Vocalizations by isolated piglets: a reliable indicator of piglet need directed towards the sow. *Applied Animal Behaviour Science* 53, 249–257.

Weary, D.M., Pajor, E.A., Bonenfant, M., Ross, S.K., Fraser, D. and Kramer, D.L. (1999) Alternative housing for sows and litters 2: effects of a communal piglet area on pre- and post-weaning behaviour and performance. *Applied Animal Behaviour Science* 65, 123–135.

Wood-Gush, D.G.M. and Csermely, D. (1981) A note on the diurnal activity of early-weaned piglets in flat-deck cages at 3 and 6 weeks of age. *Animal Production* 33, 107–110.

Worobec, E.K., Duncan, I.J.H. and Widowski, T.M. (1999) The effects of weaning at 7, 14 and 28 days on piglet behaviour. *Applied Animal Behaviour Science* 62, 173–182.

15

Practical Management and Housing of the Young Weaned Piglet

M. Evans

Pig Production Training Ltd, 3 Trevose Close, Walton, Chesterfield S40 3PT, UK

Introduction

The trend towards 3-week weaning in the early 1970s led towards the development of a variety of types of specialized weaned pig accommodation. Producers quickly recognized that earlier weaning was more stressful to the young pig than conventional 5-week weaning. In particular, the transition from sow milk to solid feed and water proved more of a problem to the younger pig and, as a result, the environment, hygiene and observation of pigs had to be of a high standard to ensure good survival rates and rapid growth.

All-in/all-out (AIAO) procedures have become an essential part of successful nursery management and fully slatted flatdecks/nurseries have evolved which provide a high degree of control over the basic requirements of the newly weaned pig. Since they first appeared in the early 1970s, flatdecks have remained the system of choice for modern pig production industries around the world. Modern nurseries are designed to provide cost-effective accommodation that allows the stockperson to observe readily many thousands of pigs. They aim to meet the basic needs of the pigs in terms of pen space and layout, hygiene and water/feed availability while providing a warm and draught-free but well ventilated environment. Alternative systems aimed at improving the welfare of the pig (e.g. straw-based systems) tend to result in increased cost of production and poorer hygiene.

The changing patterns of global pig diseases have posed major challenges for pig producers and have had a major impact on production system design. Faced with decreased productivity and performance, higher mortality and higher veterinary costs along with decreasing efficacy of antibiotics, the modern pig producer has been forced to move towards new production technologies. Inevitably these new systems solve some of the problems posed by older systems but also provide new management challenges.

Successful outcomes in the nursery department are dependent upon a well-conceived production system, careful planning of pig-flow, the appropriate feeding programme and stockpersons with good stock skills performing the necessary procedures and daily routines correctly and on time.

The Production System

Typical nursery design

Building structure

The building structure is usually highly insulated and designed to minimize heat loss at low outside temperatures.

Heating and ventilation

The nursery is mechanically ventilated all year round. Heaters (gas or electric) provide the high operating temperatures required by newly weaned pigs and thermoneutral temperatures for all nursery pigs at low outside temperatures. Fans draw air in through primary inlets situated at the soffits, gable ends, ridge or exterior of the hallway and into the room through ceiling or wall inlets. The system is fitted with high-temperature or power-cut fail-safe panels and alarms and is controlled by a multi-stage electronic controller.

Flooring

Slatted flooring made of metal or more commonly plastic is used. Slats are mostly self-cleaning and are made of non-porous materials that can dry out quickly following end-of-batch cleaning procedures.

Penning and gating

Penning/gating may be made of vertical or horizontal solid bar panels or solid plastic panels but must be easy to clean. As considerable time is spent moving pigs, it is essential that gates are designed to open, close and be secured easily.

Feeders

A variety of feeder types may be used in nurseries and may be attached to the penning/gating or be positioned in the middle of the pen. An automatic feed system delivers feed to the feeders, adjusted to ensure optimum flow of feed into the trough according to the size of pig being fed while minimizing waste.

Additional feed trays or pan feeders from which feed is more easily accessible may be used to start nursery pigs on feed.

Watering system

Nipple or bowl drinkers are used in nursery pens to provide a freely available source of clean and fresh water. Often a medicator is plumbed into the water line to allow electrolyte, water acidifier or water-soluble medicine to be added to the water supplied to nursery pigs.

Special care equipment

Comfort boards and heat lamps are provided in the special care and treatment pens that are used for small pigs at weaning and for pigs that fall behind after weaning.

Alternative weaned pig accommodation

Verandas, bungalows and kennels

These systems are now less commonly used as first-stage accommodation. They provide a kennelled lying area and usually an open dunging area. Heating cost is low but environment control, pig observation and pen hygiene are poor and this type of accommodation is not a serious alternative for large-scale pig production.

Weaner arcs and straw yards

The expansion of outdoor pig production in the UK has led to an increase in the use of straw-based weaner arcs or yards, where lying and dunging areas are deep-bedded with straw. The straw acts as a buffer against deficiencies in environment and overall management control. Provided segregation of age groups is practised, satisfactory performance can be achieved but cost of production is high (even in the better systems) due to labour and straw costs, and hygiene control becomes difficult.

Recent developments

As producers strive to improve performance and reduce cost of production, new technologies are developed. Some of these survive to become industry standards and others disappear, e.g. specialized housing types for 10–14-day weaning. The following have developed in recent years and time will tell how widely they will be adopted.

Big-pen nurseries

Big-pen systems were initially converted from standard flatdecks by removing pen divisions and utilizing former access passage space. Overall pen area therefore increased, allowing EU producers to maintain output and comply with legal space requirements or increase output to justify expenditure on removal of sow stalls. New big-pen nurseries have since been developed in the EU and in the US as a means of reducing housing cost. Feeders capable of feeding up to 50 pigs have been developed, increasing innovation in types of big-pen systems and group sizes of 50–200 pigs are common. Due to natural variation in weaning weight, systems filled from large supply farms work best. When the group represents a complete week of production, the smallest pigs weaned are seriously disadvantaged and sometimes injured as the group demonstrates 'huddling' behaviour when frightened.

Wean-to-finish (W–F) accommodation

Wean-to-finish accommodation developed out of the stampede into segregated early weaning (SEW). During the build-up of large production systems, pigs were occasionally transferred into finishing at unusually light weights and performed better than those transferred at conventional weights. This gave further credibility to the early adopters of W–F, who started to apply chicken-rearing principles to pig production and the technology accelerated.

There have been some disasters but, in general, cost of production is no worse than for standard nursery/finish technology and has the potential to be better. A major attraction to the system is that it is easier to find contractors willing to operate W–F than conventional nurseries. Nurseries have peaks of activity, i.e. 2 weeks of weaned-pig start-up and 1 week for clean out 6.5 turns per year, whereas W–F systems have the peaks of activity two turns per year.

Planning Pig-flow

Farrow to finish

Traditionally the farrow-to-finish pig unit (one-site production) has been the most cost-effective means of producing pigmeat. Varying degrees of AIAO may be used, from farrowing rooms and first-stage nurseries only, through to all rearing departments on the unit. Those farrow-to-finish units operating AIAO at all stages have worked reasonably well for many years but even they have become increasingly difficult to operate as disease patterns have changed.

Group production

The management of the sow herd in batches of more than 1-week production has been widely used for many years in the EU and the US. It is now less popular in the US, as traditional producers have stopped breeding and moved to being shareholders in larger sow units. In the EU the 3-week group production system is the most common, though 2-week and 2.5-week systems are also used. The system is now applicable in a UK and EU situation as a means of generating larger numbers of weaned pigs, making segregated production a reality. For example, an 800 sow unit farrowing once every 3 weeks will produce a batch of approximately 1100 weaned pigs every 3 weeks. On single site systems this often makes AIAO by building feasible and the 3-week gap between weanings tends to reduce health problems in the rearing herd. It also makes segregated sites more feasible and allows the opportunity to 'stream' the smaller pigs at weaning into separate accommodation.

Segregated production

New production systems have developed around the world to take AIAO to another level by introducing segregation of age groups. SEW technology has been rapidly adopted and aims to use increasing knowledge of piglet immunity, reduce the level of background disease and place location breaks in the production system to prevent recycling disease. The rapid uptake of the technology has resulted in many mistakes being made, but has stimulated a lot of research into practical application of techniques and rapid dissemination of information. Systems in use are set out in Table 15.1

Weaning age

Having experimented with 10–14-day weaning in the 1980s, most EU producers have drifted back to 25–28-day weaning, which gives good sow productivity and ensures that good strong pigs are transferred into nursery accommodation.

The adoption of SEW technology has resulted in pigs again being weaned at under 21 days, while their immunity from maternal antibodies is still high. Some systems have weaned at 10–14 days and certainly smaller units, which wean once a week, have had to accept a spread of 12–18 days. Larger production systems have again recognized the difficulties of forcing weaning age down and, where possible, are now weaning at 18 days to optimize sow productivity and weaned pig performance.

It is likely that the majority of pigs will continue to be weaned in a range of 14–28 days, depending upon legislation, the production system, the disease levels in the sow unit and the quality of nursery facilities available.

Table 15.1. Production systems for segregated early weaning.

Production system	Site	Site use
Two-site	Site 1	Breeding and farrowing
	Site 2	Nursery and finishing
Three-site	Site 1	Breeding and farrowing
	Site 2	Nursery
	Site 3	Finishing
Multi-site	Site 1	Breeding and farrowing
	Sites 2	Nurseries for 1-week pigs
	Sites 3	Finishing for 1-week pigs

Feeding and Nutrition

Considerable research has gone into developing diets for the early-weaned pig and there is now a range of high quality diets that are readily digested by the newly weaned pig. They play an important role in achieving high post-weaning feed intakes and therefore good growth rates while minimizing digestive upsets. This also leads to lower levels of mortality and a reduction in medication costs. This area has been extensively covered in previous chapters.

The biggest challenge facing the pig producer is to ensure that the weaned pig actually eats the feed presented to it.

Reasons for Substandard Results

The technology of weaned pig production has advanced considerably since the mid-1980s, particularly in the US, which pioneered the move to segregated production. Massive restructuring of the US pig industry resulted in major investment in new nursery housing, with excellent environment control, and many countries in the rest of the world are moving towards similar new multi-site production systems. Nutritional knowledge has also continued to advance, allowing the needs of newly weaned pigs to be met even at very young weaning ages.

Despite these major advances in the technology of housing, disease control, nutrition and management, the actual results achieved in practice are inconsistent and range from mediocre to good. In many cases the potential of the available technology is far from realized. There are many reasons for these variable results at commercial level.

Poor uptake of technology

Some countries, including the UK, have not moved so rapidly towards segregated production systems and, as a result, suffer with major problems of disease control. To stand any chance at all of competing in the global pig market, they need to re-invest in new weaned pig technologies.

Poorly designed production systems

In the headlong rush to SEW, many producers have compromised from the ideal multi-site scenario – due to scale of operation, planning constraints or quite simply because they have misunderstood the fundamental concepts involved. Some have mingled pigs from several sow farms into two-site, three-site or multi-site systems or filled nurseries over an extended period (trickle-in). Each level of compromise increases the risk of a disease flare-up and even true multi-site production creates opportunities for acute disease outbreaks. It is possible that a well-designed single-site herd operating good gilt acclimatization, biosecurity and AIAO management can produce pigs more cost effectively than a badly planned SEW system. Multi-site SEW technology is certainly not applicable everywhere (due, for example, to location, space constraints and scale) but it does limit risk to the overall pig operation and allows strategic disease elimination.

Well-designed production system with poorly managed pig-flow

Many problems encountered in the nursery department are a result of management of the pig-flow. Two managers provided with identical facilities are likely to achieve two very different outcomes and the following are areas where things may go wrong.

Farrowing room size

Many units have been developed over time and, as such, do not necessarily have farrowing buildings or rooms that match weekly output or nursery room size. In such systems it often becomes necessary to mix pigs from two age groups to fill nursery rooms. While AIAO has been applied to the nursery rooms, age segregation has not been applied and disease problems affect pig performance.

Extended fostering

Major fostering beyond 24 h after birth, between farrowing rooms, or the use of foster sows for lightweight pigs part-way through lactation or at weaning, can have a major impact on nursery pig performance, again caused by a breakdown of group segregation.

Number of pigs weaned

A major reason for a breakdown in AIAO in the nursery phase is variable weaned pig production. Nurseries are designed for a given number of weaned pigs (e.g. 1000) and yet variation from 700 to 1400 pigs is very typical. In a fixed-time production system, under- or overstocking becomes a problem and results in pigs of different ages being mixed to make best use of building capacity or, alternatively, in lightweight pigs being sold.

This challenge must be faced at the sow farm, where every effort should be made to achieve a consistent output of weaned pigs, through serving and then farrowing the right number of sows each week. Each week's services consist of gilts, weaned sows, hanger-on sows, pregnancy-diagnosis negatives and returns but ultimately it is the gilt services that act as a buffer to ensure that service targets are met. The number of repeats may be unpredictable, particularly in sow farms with unstable health status, and therefore it is often necessary to reduce the expected farrowing rate when calculating target services and effectively to over-serve and, if necessary, cull surplus sows after pregnancy testing. It has become easier to decide what is true surplus since the arrival of real-time ultrasound, where the drop-out rate after pregnancy testing is minimal and very predictable on stable sow farms.

Quality of weaned pigs

Poor throughput planning on the sow farm results in variation in the weight and age of pigs weaned. This situation may be exaggerated by periods of low birth-weight or a breakdown in farrowing house hygiene or lactation feeding routines. Variation increases through the nursery phase and again threatens AIAO principles in a fixed-time production system as efforts are made to maintain capacity and slaughter weight.

Management skills

Successful nursery management relies upon the correct and timely application of a whole series of key individual skills directed at maintaining pig health and improving production efficiency and productivity.

The organization of individual skills becomes particularly relevant in the nursery department, where the nursery stockperson or supervisor may be looking after many thousands of pigs that require differing degrees of attention according to stage of growth. Successful outcomes depend on prioritizing the daily and weekly routines so that the right job is undertaken at the right time.

The following are examples of breakdowns in nursery management routines:

- End-of-batch cleaning procedures are completed behind schedule so that weaned pigs are placed into nurseries that are still wet or barely dry, which leads to a poor pathogen kill and carry-over of disease.
- There is no time to maintain or set up nursery rooms correctly, leading to emergency repairs and injury to weaned pigs.
- Pigs are weaned too late in the day to allow correct loading and sorting of pigs into the nursery or adequate time for correct weaned-pig start-up procedures on day 1.
- The environment is inappropriate for the age of pigs housed.
- Pigs are fed inappropriate feeds for their age.
- The mundane tasks of checking feeder adjustment and water accessibility are poorly carried out.
- Identification and treatment of fall-behind, sick or injured pigs are poor, leading to high mortality and poor performance.

Realizing the Potential of the Weaned Pig

Producers now have the knowledge to design good production systems based on modern high-tech buildings and segregated production technology and to invest considerable amounts in housing, genetics and nutrition. All too often they fail to recognize that the other major input, which is people, warrants major investment.

Many managers and owners expect first-rate performances from their employees without allowing them to develop the necessary skills and understanding required to do the job effectively. Much on-farm training involves telling and demonstrating, without the trainee fully understanding why a job is being done or which parts of it are critical to success. Staff often learn more by trial and error and make mistakes, which are expensive to the business.

To realize the full potential of the weaned pig, it is necessary to have a well-structured training programme with good training materials and trained instructors. This gives consistency of approach and allows trainees to learn practical skills while developing their background knowledge. The benefits of getting it right are improved motivation of staff, a greater commitment to the objectives of the business and reduced staff turnover.

Weaning Systems in Relation to Disease

16

P. Wallgren and L. Melin

Department of Ruminant and Porcine Diseases, National Veterinary Institute, 751 89 Uppsala, Sweden

Introduction

In undomesticated pig populations, piglets are weaned naturally at an approximate age of 16 weeks (Jensen and Recén, 1985). In contrast, piglets are often weaned at between 2 and 6 weeks of age in modern farming systems. It should be remembered that suckling generally is initiated by the piglets themselves from their second week of life (Algers, 1993). Wild sows cope with this by avoiding contact with their offspring during a great part of the day, a privilege denied to the farmed sow. Consequently, weaning at an earlier age is one way of protecting farmed sows from their hungry brood.

It is questionable how mature the farmed piglet is at weaning in terms of growth and immune status. The piglet will abruptly lose access to protective IgA present in the milk at weaning, and enteric immunity among weaners has been discussed in Chapter 10 (see also Bailey *et al.*, 2001).

This chapter will discuss the weaning of domesticated pigs with a special focus on effects in the digestive system. Aspects on rearing systems will also be covered.

Studies in Sweden

Swedish studies were performed with healthy conventional pigs weaned at 5 weeks of age. All animals used were declared free from diseases according to the A-list of Office International des Epizooties, (OIE, Paris, France) and from Aujeszky's disease, porcine epidemic diarrhoea (PED), porcine respiratory and reproductive syndrome (PRRS), salmonella, swine dysentery and transmissible gastro-enteritis (TGE). No growth promoters were added to the standard feed. Post-weaning

(eds M.A. Varley and J. Wiseman)

diarrhoea (PWD) can successfully be prevented by adding high amounts of zinc oxide (ZnO) to the diet (Holm, 1988; Holmgren, 1994). Therefore, some groups were offered feed supplemented with 2500 ppm ZnO, in order to study the protective role of dietary ZnO.

The coliform flora was determined by cultivation on selective media (violet red bile agar or MacConkey agar). The number of viable *Escherichia coli* per gram of faeces was defined as number of coliforms with ability to produce beta-D-glucuronidase and tryptophanase (Report No. 39 20 00, National Veterinary Institute, Uppsala, Sweden). Metabolic fingerprints of the coliform flora were made by the PhenePlate system (PhP-RS plates; Biosys, Stockholm, Sweden) (Möllby *et al.*, 1993; Kühn *et al.*, 1995). The diversity of coliforms within each sample was calculated as diversity indexes (Hunter and Gaston, 1988). The homogeneity between floras of different pigs was calculated as a mean correlation coefficient (Katouli *et al.*, 1992). The presence of rotavirus in the intestine was demonstrated by an ELISA system (de Verdier Klingenberg, 1999).

Experimental exposure to pathogenic strains of *E. coli* (serotype O141: K85, STb, VT2; serotype O147: K89, STb; or serotype O149: K91, K88, Sta, STb and LT) were carried out by spreading 10^6 colony-forming units (CFU) of the challenge strain per square centimetre on the floor of the pen. The design to spread the challenge dose on the floor was chosen in order to simulate a natural exposure to the microorganisms. To simulate intensive weaning, some of the challenged groups were also given ACTH (intramuscular injection, 0.1 ml mg^{-1} per pig; Synachten, Novartis, Basel, Switzerland). The first exposure to *E. coli* took place on the day of weaning as the piglets were transferred to the weaner pen. When the piglets were repeatedly provoked by *E. coli*, the second exposure was given with either the similar strain or other pathogenic strains of *E. coli* 3 days later. To minimize genetic influence during the challenge experiments, piglets were allotted with respect to litter and sex. Thus each experimental group included piglets emanating from identical original litters.

All challenge attempts were carried out during defined conditions at the Animal Department at the National Veterinary Institute in previously emptied and disinfected pens (one pen – i.e. one group – per room). The room temperature was 25°C and the pens were bedded with shavings. The rooms were illuminated for 14 h day^{-1} and the air was circulated at 15 cycles h^{-1} but without inducing draughts. The weaned piglets were offered feed *ad libitum* and had free access to water.

Experimental observations

The number of *E. coli*, as well as the number of coliforms, excreted per gram of faeces declined from around 10^9 at birth to around 10^5 at weaning, and thereafter remained at that level (Fig. 16.1). This was not influenced by supplementation of ZnO in the feed (Jensen-Waern *et al.*, 1998).

The newborn piglets rapidly pick up coliforms from the environment and developed a highly diversified coliform flora (unless expressing diarrhoea). From 1 week of age, the floras were very similar to that of the dam and of the littermates

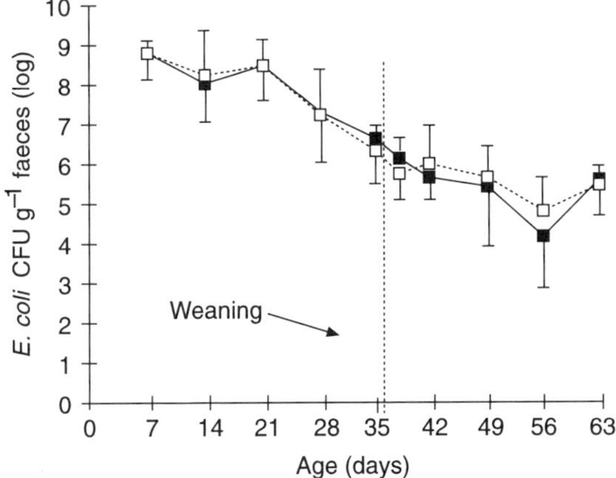

Fig. 16.1. Comparison of the mean number of viable *E. coli* between ten piglets given ZnO from weaning (filled symbols) and nine control animals (open symbols). Both groups remained healthy after weaning.

(Katouli *et al.*, 1995). Piglets maintained a coliform flora with a high diversity within pigs (Katouli *et al.*, 1997) but also with a high homogeneity between pigs until weaning (Melin *et al.*, 1997b).

The enteric flora was severely influenced at weaning among apparently healthy piglets. This was mirrored by a decreased diversity (Fig. 16.2). The homogeneity between the floras of different pigs also declined. The enteric flora recovered within 2–3 weeks, unless the pigs developed post-weaning diarrhoea. The period with disturbed enteric flora could be shortened by several means – for example, by adding high amounts of ZnO to the feed (Fig. 16.2).

Where rotavirus was not present, piglets resisted a challenge with one pathogenic strain of *E. coli* at weaning, i.e. they did not develop PWD (Melin *et al.*, 2000a). However, the enteric flora was disturbed for a longer period among

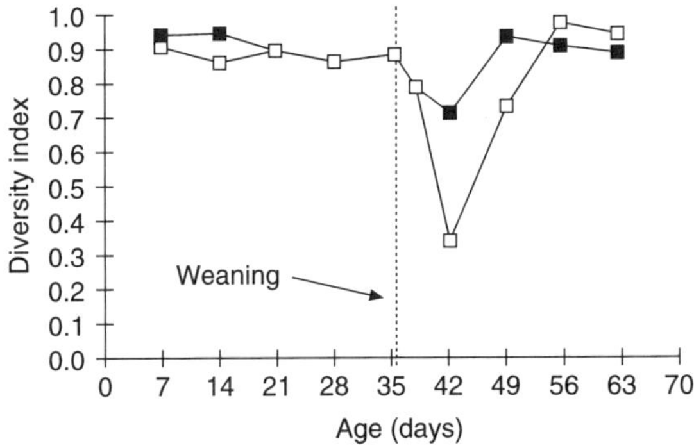

Fig. 16.2. Comparison of the mean diversity of the faecal coliform flora between ten piglets given ZnO from weaning (filled symbols) and nine control animals (open symbols). Both groups remained healthy after weaning.

challenged pigs than among non-challenged control animals (21 vs. 9 days; Fig. 16.3). In contrast, it was found possible to induce PWD by a cascade challenge using three different stains of *E. coli*. (Melin *et al.*, 2000b,c).

In the presence of rotavirus, no control piglet that shed rotavirus developed PWD. Piglets exposed to a sole pathogenic strain of *E. coli* did occasionally develop PWD, but only in association with the shedding of rotavirus. As mentioned above, PWD was recorded among piglets challenged with three pathogenic strains of *E. coli* regardless of rotavirus shedding.

These results support the theory that PWD is a syndrome of multifactorial origin rather than a sole infection. In this context, the documented negative influence of the weaning itself on immune functions may contribute to the development of PWD (Bailey *et al.*, 1992; Hessing *et al.*, 1995; Wattrang *et al.*, 1998; Melin *et al.*, 2000a).

Not all piglets shedding rotavirus and exposed to *E. coli* (one or three serotypes) developed diarrhoea, neither did all piglets exposed to three pathogenic strains of *E. coli* develop PWD. PWD was frequently seen in littermates allocated to different challenge groups, thereby indicating a genetic predisposition to develop PWD. This could be correlated to differentials in immune responses in different litters (Wallgren *et al.*, 1994; Hessing *et al.*, 1995). Treatment with ACTH did not increase the number of pigs affected by PWD but did increase the severity of disease.

Conclusions

Apart from abruptly denying the piglet access to the IgA secreted in sow milk, the weaning of farmed piglets is also associated with a number of other stressors such as altered social environment and feeding system. These stressors certainly contribute to the alterations observed in the enteric flora after weaning. A correlation between a high diversity (in microbial flora) and the stability of a community (gut eco-

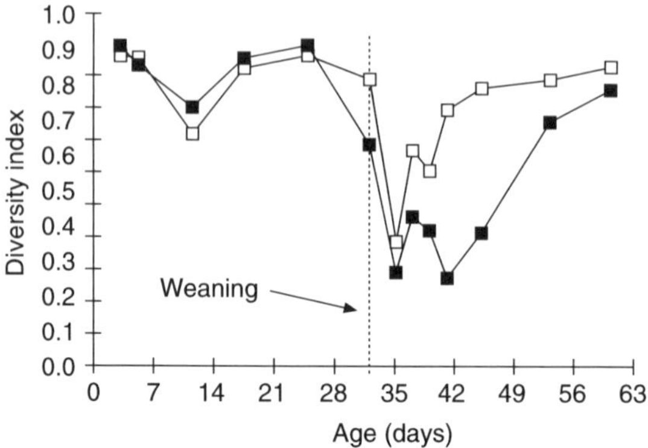

Fig. 16.3. Comparison of the mean diversity of the faecal coliform flora between nine piglets challenged with a pathogenic strain of *E. coli* at weaning (filled symbols) and nine control animals (open symbols). Rotavirus was not detected and both groups remained healthy after weaning.

system) has been suggested (Pielou, 1975): a microbial flora with a high diversity should reflect a high colonization resistance. Indeed, piglets with diarrhoea are known to have an enteric flora with a low diversity (Kühn *et al.*, 1993), and the decreased diversity of the enteric flora induced by the stress at weaning resembles that situation. In addition, the reduced homogeneity between floras of pen-mates after weaning (Melin *et al.*, 1997b) indicates that different clones of bacteria dominate in different pigs. Taken together, the phenomenon of a decreased diversity of the enteric flora within piglets and a decreased homogeneity of floras between individuals kept in the same pen may obviously predispose for the proliferation of pathogenic microorganisms of either internal or external origin.

The hazard of weaning is further exacerbated by the fact that piglets aged 2–6 weeks are quite immature. They generally express a declining maternal (passive) immunity, and are poor responders to specific immunogens (Wallgren *et al.*, 1998). Consequently, they may have problems in combating diseases. The management tactics around weaning should focus on making life easy for the piglets. These measures include preventing the spread of infection, providing the piglets with good thermal comfort, giving them well-designed creep feed and allowing access to that feed for all piglets. Adding antimicrobals (i.e. antibiotics or ZnO) to the feed may hide management errors, but should not be seen as a long-term solution.

The potential of growth promoters (i.e. antibiotics) to compensate for suboptimal management systems during the post-weaning period was well elucidated in Sweden during 1986. A ban on the routine administration of antibiotics in feeds was introduced in October 1985 and implemented on 1 January 1986. The incidence of PWD increased by 100% during 1986 when compared with 1985 (Robertsson and Lundeheim, 1994). Moreover, the age at 25 kg bodyweight was increased by 6 days and the post-weaning mortality increased by 1.2% (Robertsson and Lundeheim, 1994).

The absence of so-called growth promoters obviously demanded more careful rearing strategies on Swedish pig farms. Due to such efforts, productivity today well exceeds that of 1985, despite the absence of routine in-feed medication. During 1997, 40% of the pigs were reared from birth to slaughter in age-segregated systems (R.C. Brendow, Stockholm, 1997, personal communication) and that figure has since increased. Farrowing now takes place in batches in previously emptied and cleaned facilities in 72% of the piglet-producing herds (Löfstedt and Holmgren, 1999). Approximately 65% of Swedish pigs are now allocated to specialized fattening herds at 25 kg liveweight and practically all these units employ AIAO systems (R.C. Brendow, Stockholm, 1997, personal communication).

The Swedish experience elucidates the necessity to control the pathogen load when young piglets are weaned. In herds experiencing problems with PWD, piglets suffering from PWD should be medicated. This may include in-feed medication of batches known to be affected. The Swedish Veterinary Society (SVS) has published guidelines for such in-feed medication (Holmgren *et al.*, 1990; Odensvik *et al.*, 1999) that also include management measures to be undertaken during the treatment period.

Apart from antibiotics, high inclusions of ZnO have been found effective in preventing the development of PWD (Holm, 1988; Holmgren, 1994). Since no inhibitory effect has been demonstrated by parental administration of Zn, the antibacterial effect of ZnO (Jensen, 1987; Melin *et al.*, 1997a) probably has to act locally in the intestine to prevent PWD. Despite the low bio-availability of ZnO, it must be remembered that the piglets offered high amounts of ZnO *per se* will also absorb increased amounts of zinc. Signs of zinc intoxication have been recorded within 4 weeks of oral administration of ZnO (Jensen-Waern *et al.*, 1998) and therefore treatment should never exceed 14 days at an individual level.

If it is necessary to use in-feed medications in order to control PWD, efforts to improve rearing methods should be employed during the treatment period (as also suggested by the SVS guidelines). These measures include reduction of the pathogen load by optimizing the management system and the management routines, as well as optimizing the environment provided for the piglets. All piglets should be provided with good thermal comfort, well-designed creep feed and access to that feed for all piglets. Regrouping of runts within batches of weaned piglets might also be necessary in order to fulfil these demands. In systems that house several weaned litters together, it is relevant to discuss sorting by size and important not to keep too large groups – preferably not more than 60 piglets per group (Löfstedt and Holmgren, 1999).

To conclude, piglets in modern piggeries are weaned at an earlier age than evolution intended. This entails specific demands on production technology around weaning. PWD, although strongly associated with *E. coli*, should be looked upon as a multifactorial syndrome rather than a specific infection. The fact that not all piglets exposed to severely pathogenic strains of *E. coli* at weaning developed PWD highlights the importance of good basic hygiene and of providing the piglets with a good environment at weaning.

References

Algers, B. (1993) Nursing in pigs: communicating needs and distribution resources. *Journal of Animal Science* 71, 2826–2831.

Bailey, M., Clarke, C.J., Wilson, A.D., Williams, N.A. and Stokes, C.R. (1992) Depressed potential for interleukin-2 production following early weaning of piglets. *Veterinary Immunology and Immunopathology* 34, 197–207.

Bailey, M., Haverson, K., Vega-Lopez, M.A., Bland, P.W., Miller, B.G. and Stokes, C.R. (2001) Enteric immunity and gut health. In: *The Weaner Pig. BSAS Occasional Meeting, Nottingham, UK, September 5–7*, p.11.

Hessing, M.J.C., Coenen, G.J., Vaiman, M. and Renard, C. (1995) Individual differences in cell-mediated and humoral immunity in pigs. *Veterinary Immunology and Immunopathology* 45, 97–113.

Holm, A. (1988) *Escherichia coli* induced post weaning diarrhoea in pigs. Dietary supplementation as an antibacterial method? *Dansk Veterinærtidskrift* 71, 1118–1126.

Holmgren, N. (1994) Prophylactic effects of zinc oxide or olaquindox against post weaning diarrhoea in swine. *Svensk Veterinärtidning* 46, 217–222.

Holmgren, N., Franklin, A., Wallgren, P., Bergström, G., Martinsson, K. and Rabe, J. (1990) Guidelines for in feed medication of pigs. *Svensk Veterinärtidning* 42, 407–413.

Hunter, P.R. and Gaston, M.A. (1988) Numerical index of the discriminatory ability of typing systems: an application of Simpson's index of diversity. *Journal of Clinicial Microbiology* 26, 2465–2466.

Jensen, B. (1987) Intestinal microflora, zinc oxide and *E. coli* enteritis in pigs. *Landbonyt.* Aug 5–10.

Jensen, P. and Recén, B. (1985) Maternal behaviour of free-ranging domestic pigs. *Proceedings International Ethological Congress* 4, 5 pp.

Jensen-Waern, M., Melin, L., Lindberg, R., Johannisson, A., Petersson, L. and Wallgren, P. (1998) Dietary zinc oxide in weaned pigs – effects on performance, tissue concentrations, morphology, neutrophil functions and faecal microflora. *Research in Veterinary Science* 64, 225–231.

Katouli, M., Erhart-Bennet, A.S., Kühn, I., Kollberg, B. and Möllby, R. (1992) Metabolic capacity and pathogen properties of the intestinal coliforms in patients with ulcerative colitis. *Microbial Ecology in Health and Disease* 5, 245–255.

Katouli, M., Lund, A., Wallgren, P., Kühn, I., Söderlind, O. and Möllby, R. (1995) Phenotypic characterization of intestinal *Escherichia coli* of pigs during suckling, post-weaning and fattening periods. *Applied Environmental Microbiology* 61, 778–783.

Katouli, M., Lund, A., Wallgren, P., Kühn, I., Söderlind, O. and Möllby, R. (1997) Metabolic fingerprinting and fermentative capacity of the intestinal flora of pigs during pre- and post-weaning periods. *Journal of Applied Bacteriology* 83, 147–154.

Kühn, I., Katouli, M., Lund, A., Wallgren, P. and Möllby, R. (1993) Phenotypic diversity and stability of the intestinal coliform flora in piglets during the first 3 months of age. *Microbial Ecology in Health and Disease* 6, 101–107.

Kühn, I., Katouli, M., Wallgren, P., Söderlind, O. and Möllby, R. (1995) Biochemical fingerprinting as a tool to study the diversity and stability of intestinal microfloras. *Microecology and Therapy* 23, 140–148.

Löfstedt, M. and Holmgren, N. (1999) *Avvänjningsboken* (*The Weaning Book*). SHS Text & Tryck, Hållsta, Sweden, 28 pp.

Melin, L., Holmgren, N., Wallgren, P. and Franklin, A. (1997a) Sensivity to olaquindox and zinc in coliforms isolated from weaned pigs. *Svensk Veterinärtidning* 49, 573–579.

Melin, L., Jensen-Waern, M., Johannisson, A., Ederoth, M., Katouli, M. and Wallgren, P. (1997b) Development of selected faecal microfloras and of phagocytic and killing capacity of neutrophils in young pigs. *Veterinary Microbiology* 54, 287–300.

Melin, L., Katouli, M., Lindberg, Å., Fossum, C. and Wallgren P. (2000a) Weaning of piglets. Effects of an exposure to a pathogenic strain of *Escherichia coli. Journal of Veterinary Medicine B*, 663–675.

Melin, L., Mattsson, S. and Wallgren, P. (2000b) Challenge with pathogenic strains of *E. coli* at weaning. I. Clinical signs and reisolation of the challenge strains. *Proceedings of the International Pig Veterinary Society Congress* 16, 22.

Melin, L., Mattsson, S. and Wallgren, P. (2000c) Challenge with pathogenic strains of *E. coli* at weaning. II. Monitoring of the faecal coliform flora. *Proceedings of the International Pig Veterinary Society Congress* 16, 23.

Möllby, R., Kühn, I. and Katouli, M. (1993) Computerised biochemical fingerprinting – a new tool for typing of bacteria. *Reviews in Medical Microbiology* 4, 231–241.

Odensvik, K., Robertsson, J.Å. and Wallgren, P. (1999) Strategies for in feed medications, including zinc oxide, of pigs with special emphasis on enteric disorders. *Svensk Veterinärtidning* 51, 293–299.

Pielou, E.C. (1975) *Ecological Diversity.* Wiley Interscience, New York.

Robertsson, J.Å. and Lundeheim, N. (1994) Prohibited use of antibiotics as feed additive for growth promotion – effects on piglet health and production parameters. *Proceedings of the International Pig Veterinary Society Congress* 13, 282.

de Verdier Klingenberg, K. (1999) Neonatal calf diarrhoea with special reference to rotavirus infections. Significance, epidemiology and aspects of prevention. Dissertation, Swedish University of Agricultural Sciences, Uppsala, Sweden.

Wallgren, P., Wilén, I.-L. and Fossum, C. (1994) Influence of experimentally induced endogenous production of cortisol on the immune capacity in swine. *Veterinary Immunology and Immunopathology* 42, 301–316.

Wallgren, P., Bölske, G., Gustafsson, S., Mattsson S. and Fossum, C. (1998) Humoral immune responses to *Mycoplasma hyopneumoniae* in sows and offspring following an outbreak of mycoplasmosis. *Veterinary Microbiology* 60, 193–205.

Wattrang, E., Wallgren, P., Lindberg, Å. and Fossum, C. (1998) Signs of infections and reduced immune functions at weaning of conventionally reared and specific pathogen free pigs. *Journal of Veterinary Medicine B* 45, 7–17.

17 Weaning in Practice

J.W.G.M. Swinkels, H.A.M. Spoolder and H.M. Vermeer

Research Institute for Animal Husbandry, PO Box 2176, 8203 AD Lelystad, The Netherlands

Introduction

In the 1980s the Dutch pig industry invested large sums of money in reducing ammonia emissions from pig buildings, to address the environmental concerns raised by public opinion. Research showed that minimizing the emitting surface was not only the most effective but also the cheapest way to achieve a reduction of 50–70% in ammonia emission. In housing systems for the weaned pig, ammonia emissions were reduced from 0.6 kg NH_3 per pig space per year to less than 0.3 kg NH_3 per pig space per year. The emitting surface was minimized with the design of a solid floor area in the pen that covered up to 40% of the total floor area. A larger solid floor area increased the risk of dirtying the floor, which not only gave greater ammonia emission, but also increased the risk of leg problems. A further reduction in emitting surface could be achieved in the pit area, with the design of a sloping channel and a sewer system. In this design the manure is removed weekly from the barn to a closed manure tank.

The outbreak of classical swine fever in 1997 focused public attention on not only the risk of disease outbreaks but also the welfare of the pigs. For almost a year, daily coverage of the epidemic disease in the national news media showed how contaminated pigs were killed at the pig farm and how pigs were housed on overcrowded farms. Responding to public opinion, the 1994 Dutch legislation on pig welfare was revised in 1998 allowing, among other things, more freedom of movement for dry sows and more space for weaned and fattening pigs. The 1998 Dutch legislation became effective immediately when a new facility was renovated or built. All pig farms must comply with the new legislation by 1 January 2008. For weaned piglets the space requirement was increased from 0.3 m^2 to 0.4 m^2 but slat width was maintained at 15 mm. Full slatted floors are allowed when plastic-coated or metal slats are used; otherwise 40% of the floor area has to be solid.

Weaned and fattening pigs may also benefit from a new rule that allows pigs to be mixed only from birth to 1 week after weaning. In current practice, pigs are mixed at weaning (*c.* 4 weeks of age) and mixed again when reallocated/or transported to a fattening barn at *c.* 10 weeks or approximately 25 kg.

The 1998 legislation on pig welfare has an impact on the design of new housing systems for all pig categories. General aspects on housing of weaned pigs were reviewed by Pedersen (1999); therefore this chapter will focus on how to implement the new legislation on pig welfare in low-emission housing systems for weaned pigs.

Farrow-to-finish Production

Implementation of the new rule on mixing could, in the most extreme case, lead to a housing system in which piglets are kept from birth to slaughter (Ekkel *et al.*, 1995): in a small-scale experiment, significant improvements were found in performance, health and welfare when pigs were kept in one housing system, as opposed to being mixed at weaning and mixed and transported at *c.* 10 weeks of age. At the Research Institute for Pig Husbandry (RIPH) in Rosmalen, the farrow-to-finish production system was evaluated using over 40 litters of pigs (Vermeer *et al.*, 1997). Half of the litters were kept in one housing system from birth to slaughter, whereas the other half were reallocated at weaning and again reallocated and mixed at approximately 25 kg. Table 17.1 presents the growth of the pigs for the different stages from birth to slaughter.

Growth was increased by 6% between 8 and 24 kg and 11% between 24 and 40 kg when pigs were not reallocated at weaning and not reallocated or mixed at *c.* 25 kg, respectively. Because growth was similar in the remaining growth stages, the overall improvement in growth was only 3%. Moreover, the improvement in growth was associated with an increase in feed intake; therefore, feed efficiency was similar

Table 17.1. Growth of pigs from birth to slaughter when housed either in the same housing system from birth to slaughter or reallocated twice (at weaning, not mixed; and at approximately 25 kg, mixed) (adapted from Vermeer *et al.*, 1997).

	Birth to slaughter	Reallocated/ mixed	SEM	Significance
Number of litters	20	20		
Number of slaughtered pigs	193	195		
Age at slaughter (days)	175	180		
Bodyweight at slaughter (kg)	105	104.1		
Growth from 1 to 8 kg (g day^{-1})	226	224	7.8	n.s.
Growth from 8 to 24 kg (g day^{-1})	418	393	9.1	$P < 0.10$
Growth from 24 to 40 kg (g day^{-1})	736	655	15.2	$P < 0.01$
Growth from 40 to 105 kg (g day^{-1})	753	755	15.6	n.s.
Growth from 1 to 105 kg (g day^{-1})	590	573	3.8	$P < 0.01$

n.s. = not significant.

between the two groups. In all growth stages morbidity among the pigs that were reallocated and mixed was higher than among the pigs housed in one housing system from birth to slaughter, whereas mortality was similar between the two groups. In an economic evaluation it was found that the improved performance and labour savings (euro 8.20 and 5.00 per pig space, respectively) did not compensate for the increased cost of housing and energy usage (euro 13.20 and 5.45 per pig space, respectively). In practice, economic incentives will favour reallocation of pigs at least once between birth and slaughter. The question is whether reallocation can take place without mixing.

Reallocation and splitting

The 1998 Dutch legislation on pig welfare stipulated that only an intact group of pigs may be mixed from birth to 1 week after weaning. Thereafter, it is only allowed to maintain or split up an intact group of pigs. Mixing unaquainted piglets at weaning does increase aggression but this does not always lead to a decrease in performance (Friend *et al.*, 1983; Pluske and Williams, 1996). Using climatic respiration chambers, Heetkamp *et al.* (1995) found that mixing does have an effect on total heat production and activity-related heat production during the first hour after mixing 8-week-old piglets but no long-term effect on energy partitioning was observed. Regrouping fattening pigs has been shown to affect the duration of aggressive behaviour and long-term performance (Stookey and Gonyou, 1994). It seems that it takes longer to establish a new, stable social structure when previously unaquainted pigs are mixed at an older age.

In a study at the RIPH, forming large groups of piglets at weaning has proved to be beneficial when combined with splitting up the large group at the time of reallocation to the fattening-pig barn (Vermeer and Hoofs, 1994). During the nursery period, piglets reared in a group of 45 or 90 piglets had, respectively, 5% and 8% less growth than piglets reared in a group of ten piglets. On average the depression in weaned pig performance amounted to euro 0.22 per piglet. Performance of the piglets in the groups of ten and 90 was subsequently monitored in the fattening period, where all pigs were fattened in groups of eight (Table 17.2). The eight pigs were either mixed from several pens of ten piglets or split up from a pen of 90 piglets according to their bodyweight.

Fattening pigs reared in a group of 90 had not only higher growth but also a higher lean percentage than pigs reared in a group of ten. Because food intake was similar, food efficiency was numerically improved for the fattening pigs reared in a group of 90 piglets. Altogether the economic performance in the fattening period was improved by euro 1.48 per pig.

Injuries on the skin were subjectively scored on day 4 of the fattening period to assess the level of aggressive behaviour among the fattening pigs. Those reared in small groups and mixed at the start of the fattening period had four times as many skin injuries as pigs reared in large groups and split up at the beginning of the fattening period. Both improved performance and fewer skin injuries suggest that pigs require little time to establish a new stable social structure in a subgroup of former pen-mates.

Table 17.2. Performance of fattening pigs reared either in a small group (ten piglets per pen) or in a large group (90 piglets per pen) (adapted from Vermeer and Hoofs, 1994).

	Small group	Large group	SEM	Significance
Number of pens	31	31		
Number of slaughtered pigs	239	242		
Age at slaughter (days)	182	182		
Bodyweight at slaughter (kg)	110.9	111.6		
Growth (g day^{-1})	760	776	7.3	$P < 0.05$
Food intake (kg day^{-1})	2.24	2.23	0.04	n.s.
Food conversion ratio	2.96	2.87	0.05	n.s.
Lean (%)	54.2	54.8	0.2	$P < 0.05$

n.s. = not significant

Behavioural Needs

Mixing and other aspects associated with weaning and reallocation may impose serious stresses on piglets. Behaviourally, piglets may respond by manipulating pen-mates through head–head/body-knocking, navel-sucking, and flank-, ear- or tail-biting. However, allowing piglets to redirect these behaviours to components of their pen, rather than their pen-mates, may positively affect stress resistance and health. Piglets reared in an environment enriched with a tyre-and-chain toy had fewer ($P < 0.05$) aggressive acts and tended to have a higher (4%) growth than pigs reared in a barren environment (Schaefer *et al.*, 1990). Providing extra space and substrates for manipulation by the piglets also reduced their aggressive behaviour and altered their social behaviour. In a barren environment piglet dominance was demonstrated by aggressive behaviour, whereas in enriched environments the heavy piglets showed dominant behaviour (O'Connell and Beattie, 1999). It should be noted that individual characteristics of piglets influence their aggressive behaviour and that animals differ in its expression (Hessing *et al.*, 1993).

Using mixed and weaned piglets, Blackshaw *et al.* (1997) studied the effect of a fixed or free toy (neck tether covered with hard plastic piping) on growth and aggressive behaviour. The presence of fixed and free toys did reduce aggressive behaviour but growth was not affected. Of the weaned pigs, 75% touched the toy within 5 min of its introduction to the pen. Piglets exhibited more interest in the fixed than in the free toy. In an unpublished RIPH study, a fixed cotton rope and straw from a fixed rack were provided to weaned pigs with intact tails. During the 5-week rearing period, piglets used 19 mm cotton rope or 50 g straw per piglet per week. The cotton rope was mostly consumed, but the straw was removed from the pen trough in the pit area. The presence of the cotton rope and straw equally diminished ($P < 0.05$) tail damage and tended to reduce ($P < 0.10$) aggressive behaviour among the piglets.

Waran and Broom (1993) examined the influence of an opaque barrier on the aggressive behaviour and growth of piglets in a conventional flatdeck pen and a straw pen. There was no difference in aggressive behaviour and growth between the two pen types. In both pen types the presence of the opaque barrier reduced the frequency of aggressive interactions by 40% and improved growth by 15% in the first week after weaning. Supposedly, the barrier improved the environment because it offered an escape route during the period when most aggressive interactions occur.

Enrichment of the rearing pen with toys or a barrier does modify aggressive behaviour but the economic impact of the redirection of aggressive behaviour from pen-mates in the pen is difficult to quantify. To retain its effectiveness in controlling aggressive behaviour, piglets should not become too familiar with the toy (Schaefer *et al.*, 1990). Using a pacifier, or alternating toys that are preferably fixed in the pen, may ensure a stimulating environment for the piglets throughout the rearing period. It has been suggested that a stimulating environment for weaned piglets leads to more vigorous fattening pigs (S.A. Edwards, Aberdeen, 1998, personal communication).

Environment

Extra space allowance for piglets leads to an increase in emitting surface per piglet and subsequently an increase in ammonia emission, because it is expressed as kg NH_3 per piglet space per year. In a study at RIPH, available techniques were used to design a low-emission housing system for a large group of weaned piglets (Fig. 17.1).

The new design was implemented in a compartment that was formerly used to house 70 piglets in small pens (ten piglets per pen) on a surface area of 0.3 m^2 per piglet (Zeeland *et al.*, 1999). In the new design, an alley for checking the piglets was added to the pen. In the middle of the pen a convex solid floor (1.7 m wide) was constructed, with no pit area underneath. On each side of the convex solid floor was a channel 0.45 m deep and 1.1 m wide, including a 0.05 m manure slot adjacent to the wall. The channels were equipped with tri bar metal slats and a sewer system. The feeders and incorporated water nipples were placed on the left side of the pen to stimulate piglets to defecate on the right side of the pen. The channel beneath the feeders was used as a water channel. Before introducing a batch of piglets, the water channel was filled with approximately 10 cm of cleaning water. In both channels plastic-coated boards were placed at an angle of 45° to minimize the emitting surface in the manure channel and the amount of cleaning water in the water channel. The water channel was not emptied during the rearing period, whereas the manure channel was emptied several times. The total surface area was 0.4 m^2, consisting of 0.18 m^2 solid floor, 0.11 m^2 water channel and 0.11 m^2 manure channel.

The ammonia emission was measured continuously in three batches of piglets using B&K type 1302 monitors, according to the measuring protocol of van't Klooster *et al.* (1992), during one winter and two summer periods (Table 17.3).

In all three batches of piglets, the ammonia emission remained below the target level of 0.3 kg NH_3 per piglet space per year that is used for housing systems of small groups of piglets with a total surface area of 0.3 m^2 per piglet. Averaged over the three batches, the ammonia emission was 0.25 kg NH_3 per piglet space per year.

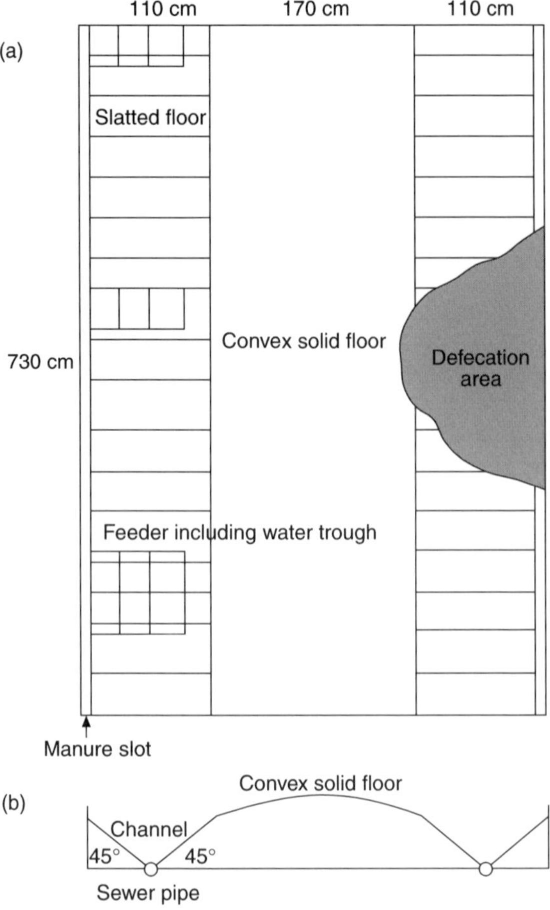

Fig. 17.1. (a) Top view of a housing system for 70 piglets (0.4 m^2 per piglet) and (b) cross-sectional view of the pit area equipped with shallow channels and a sewer system (Zeeland *et al.*, 1999).

The low ammonia emission may be associated with the size of the defecating area in the pen. Observations on pen fouling showed that the piglets defecated in the middle of the pen, primarily on the manure channel (Fig. 17.1). Although not measured in this study, it may be assumed that the combined defecating area of seven pens of ten piglets is larger than that of one pen of 70 piglets.

In an economic evaluation, the investment cost and annual cost of the new design were compared with a conventional design with small groups housed on 0.4 m^2 per piglet. Housing pigs in a large group saved euro 4.09 per piglet space on investment and euro 1.45 per piglet space on annual cost (Zeeland *et al.*, 1999). Extra savings on investment and annual costs may be achieved when farmers succeed in renovating the present rearing facilities for small groups of piglets in facilities for large groups of piglets.

Table 17.3. Ammonia emission measured in a new design of a large group of 70 piglets (see Fig. 17.1) housed on a total surface area of 0.4 m^2 per piglet (Zeeland *et al.*, 1999).

	Batch 1	Batch 2	Batch 3
Date trial commenced (1998)	14 April	11 June	24 July
Date of last measurement (1998)	13 May	13 July	31 August
Length of rearing period (days)	30	33	39
Number of measuring days (days)	30	33	39
Mean temperature in exhaust channel (°C)	25.0	24.3	25.5
Mean outside temperature (°C)	12.6	15.5	16.5
Mean ventilation rate (m^3 h^{-1} per piglet space)	7.3	11.4	12.3
Mean ammonia concentration (mg NH_3 m^{-3})	3.49	2.58	2.95
Ammonia emission (kg NH_3 per piglet space per year)	0.24	0.24	0.27

Dutch environmental standards on ammonia emission can be met in housing systems for large groups of weaned piglets using relatively cheap construction measures. Low-emission housing systems for weaned piglets are fairly barren, to prevent fouling of the pen and to facilitate fast removal of slurry from the pit area to an outside manure storage.

Conclusions

The 1998 Dutch legislation on pig welfare has to be implemented when building new or renovating piglet facilities. All piglet facilities have to be replaced or renovated by 1 January 2008. The increase in space requirement from 0.3 m^2 to 0.4 m^2 per pig and the ban on mixing piglets from 1 week after weaning will have a major impact on future weaning practices. Economic incentives will stimulate the implementation of new designs for large groups of piglets and these will also ensure that ammonia emission is low. The size of the group of piglets will vary depending on the size of the commercial sow farm and the number of pigs per pen in the fattening unit. Recent information on the optimum pen size in fattening units is lacking. It may be, therefore, that piglets can be reallocated as a group to one fattening pen. However, there were more handling problems with large groups (> 20) of fattening pigs than with small groups (≤ 20).

At present the newly designed piglet facilities are fairly barren to avoid fouling of the pen and obstruction of the manure channel. Enrichment of the environment of the weaned pig has been shown to stimulate redirection of aggressive behaviour from pen-mates to toys in the pen. The most effective are pacifiers and other toys that maintain the interest of the piglet during the entire rearing period. The challenge will be to enrich further (e.g. with toys and straw) the new housing systems for large groups of weaned piglets, keeping in mind both public opinion and the international competitive position of the industry.

References

Blackshaw, J.K., Thomas, F.J. and Lee, J.-A. (1997) The effect of a fixed or free toy on the growth rate and aggressive behaviour of weaned pigs and the influence of hierarchy on initial investigation of the toys. *Applied Animal Behaviour Science* 53, 203–212.

Ekkel, E.D., van Doorn, C.E.A., Hessing, M.J.C. and Tielen, M.J.M. (1995) The specific-stress-free housing system has positive effects on productivity, health, and welfare of pigs. *Journal of Animal Science* 73, 1544–1551.

Friend, T.H., Knabe, D.A. and Tanksley, T.D. Jr (1983) Behavior and performance of pigs grouped by three different methods at weaning. *Journal of Animal Science* 57, 1406–1411.

Heetkamp, M.J.W., Schrama, J.W., de Jong, L., Swinkels, J.W.G.M., Schouten, W.G.P. and Bosch, M.W. (1995) Energy metabolism in young pigs as affected by mixing. *Journal of Animal Science* 73, 3562–3569.

Hessing, M.J.C., Hagelso, A.M., van Beek, J.A.M., Wiepkema, P.R., Schouten, W.P.G. and Krukow, R. (1993) Individual behavioural characteristics in pigs. *Applied Animal Behaviour Science* 37, 285–295.

van't Klooster, C.E., Heitlager, B.P. and van Gastel, J.P.B.F (1992) *Measurement systems for emissions of ammonia and other gasses at the Research Institute of Pig Husbandry, Rosmalen.* Research Report P3.92, Research Institute for Pig Husbandry, Rosmalen, The Netherlands, 10 pp.

O'Connell, N.E.O. and Beattie, V.E. (1999) Influence of environmental enrichment on aggressive behaviour and dominance relationships in growing pigs. *Animal Welfare* 8, 269–279.

Pedersen, B.K. (1999) Housing of weaners – meeting their environmental demands. *50th Annual Meeting of the European Association for Animal Production, Session V: Feeding and Management of the Weaned Pig*, Zürich, Switzerland, 14 pp.

Pluske, J.R. and Williams, I.H. (1996) The influence of feeder type and the method of group allocation at weaning on voluntary food intake and growth in piglets. *Animal Science* 62, 115–120.

Schaefer, A.L., Salomons, M.O., Tong, A.K.W., Sather, A.P. and Lepage, P. (1990) The effect of environment enrichment on aggression in newly weaned piglets. *Applied Animal Behaviour Science* 27, 41–52.

Stookey, J.M. and Gonyou, H.W. (1994) The effects of regrouping on behavioral and production parameters in finishing swine. *Journal of Animal Science* 72, 2804–2811.

Vermeer, H.M. and Hoofs, A.I.J. (1994) *The effect of weaner group size on performance and profitability.* Research Report P1.118, Research Institute for Pig Husbandry, Rosmalen, The Netherlands, 28 pp. (In Dutch.)

Vermeer, H.M., Plagge, J.G., Binnendijk, G.P. and Backus, G.B.C. (1997) [*Housing pigs in one pen from birth to slaughter.*] Research Report P1.170, Research Institute for Pig Husbandry, Rosmalen, The Netherlands, 28 pp. (In Dutch.)

Waran, N.K. and Broom, D.M. (1993) The influence of a barrier on the behaviour and growth of early-weaned piglets. *Animal Production* 56, 115–119.

van Zeeland, A.J.A.M., den Brok, G.M., van Asseldonk, M.G.A.M. and Verdoes, N. (1999) [*Ammonia emission of large groups of weaned piglets on a floor area of 0.4 m*^{2} *per piglet.*] Research Report P1.224, Research Institute for Pig Husbandry, Rosmalen, The Netherlands, 28 pp. (In Dutch.)

Index

Page numbers in **bold** refer to illustrations and tables.